Natural and Selected Synthetic Toxins

ACS SYMPOSIUM SERIES **745**

Natural and Selected Synthetic Toxins

Biological Implications

Anthony T. Tu, EDITOR
Colorado State University

William Gaffield, EDITOR
Western Regional Research Center
U.S. Department of Agriculture

American Chemical Society, Washington, DC

Library of Congress Cataloging-in-Publication Data

Natural and selected synthetic toxins : biological implications / [editors] Anthony T. Tu, William Gaffield.

    p.   cm.—(ACS symposium series , ISSN 0097–6156 ; 745)

    Includes bibliographical references and index.

    ISBN 0–8412–3630–5

    1. Toxins—Congresses.

    I. Tu, Anthony T., 1930–   . II. Gaffield, William. III. Series

RA1191 .N38   1999
615′.373—dc21   99–43789
                                                      CIP

The paper used in this publication meets the minimum requirements of American National Standard for Information Sciences—Permanence of Paper for Printed Library Materials, ANSI Z39.48–1984.

PRINTED IN THE UNITED STATES OF AMERICA

Foreword

THE ACS SYMPOSIUM SERIES was first published in 1974 to provide a mechanism for publishing symposia quickly in book form. The purpose of the series is to publish timely, comprehensive books developed from ACS sponsored symposia based on current scientific research. Occasionally, books are developed from symposia sponsored by other organizations when the topic is of keen interest to the chemistry audience.

Before agreeing to publish a book, the proposed table of contents is reviewed for appropriate and comprehensive coverage and for interest to the audience. Some papers may be excluded in order to better focus the book; others may be added to provide comprehensiveness. When appropriate, overview or introductory chapters are added. Drafts of chapters are peer-reviewed prior to final acceptance or rejection, and manuscripts are prepared in camera-ready format.

As a rule, only original research papers and original review papers are included in the volumes. Verbatim reproductions of previously published papers are not accepted.

ACS BOOKS DEPARTMENT

Contents

INDEXES

Preface

The chapters comprising this American Chemical Society (ACS) Series volume are derived from presentations at the Third International Symposium on Natural Toxins, which was convened at the 216th National Meeting of the ACS in Boston, Massachusetts, August 23–27, 1998. The Symposium was sponsored by the Division of Agricultural and Food Chemistry and contained presentations from 26 scientists who represented 11 countries on four continents.

A common characteristic of natural toxins is that minute quantities exert a pronounced deleterious effect on the metabolic and physiologic functions of a living organism. Because natural toxins usually express highly specific and potent effects, they often serve as excellent candidates from which therapeutic drugs may be developed. Natural toxins discussed in the Symposium include compounds derived from marine, fungal, microbial, plant, and animal origins. In addition to the extensive coverage of naturally-occuring toxins, a session on forensic toxicology was incorporated into the Symposium, which covered topics on the toxicological effects of synthetic nerve agents and doping compounds. Although the latter compounds do not occur in Nature, their inclusion in the Symposium resulted from their development as superior alternatives to natural products for purposes of chemical warfare or enhanced athletic performance. Topics discussed in this book of particular timeliness include: (1) nerve agents that were dispersed in Tokyo subway stations in 1995 by terrorists and (2) mycotoxins reported to be incorporated into missiles that could serve as potential sources of economic and political intimidation by renegade countries.

Earlier symposia on natural toxins in this series were convened at American Chemical Society meetings in Washington, D.C. (1990) and Anaheim, California (1995). The editors are very appreciative of support from the Division of Agricultural and Food Chemistry and are pleased to acknowledge the excellent technical and editorial assistance provided by Kelly Dennis and Anne Wilson of ACS Books.

ANTHONY T. TU
Department of Biochmistry and Molecular Biology
Colorado State University
Fort Collins, CO 80525

WILLIAM GAFFIELD
Research Chemist
Western Regional Research Center
U.S. Department of Agriculture
Albany, CA 94710

Chapter 1

Structure–Function Relations of Natural Toxins and Nerve Agents: An Overview

Anthony T. Tu[1] and William Gaffield[2]

[1]Department of Biochemistry and Molecular Biology,
Colorado State University, Fort Collins, CO 80523–1870
[2]Western Regional Research Center, U.S. Department of Agriculture,
Albany, CA 94710

Nature has endowed mankind with sources of an extraordinarily rich array of complex molecular structures and assemblies possessing functions that range from gene regulation and signal transduction to photosynthesis and protein biosynthesis (*1*). The creation of diverse collections of chemical structures followed by critical selection of specific molecules exhibiting a desired biological property allows the evolution of new molecular functions. As the structures and molecular mechanisms involved in the biological processes are elaborated and understood, the insights gleaned, as well as the processes themselves, can be combined with the tools of the physical and biological sciences to create new and novel molecules that incorporate functions found neither in Nature or the laboratory.

Inspection of the Bioactive Natural Products Database shows that the origin of approximately 30,000 biologically active natural products is derived from four primary kingdoms: bacteria (33%), plants (27%), fungi (26%), and animals (13%) (*2*). Recent years have witnessed the isolation of increased numbers of bioactive compounds from fungal and marine organisms; however, the potential for the discovery of new natural products is far from exhausted and compounds of natural origin continue to represent a key source for the discovery of important leads. Furthermore, although combinatorial chemistry and high-throughput screening are the methods of choice for lead-structure optimization, a combinatorial library will not provide the structural diversity inherent in a library generated from natural products.

As an example, natural products possessing physiological properties such as those of constituents of snake, spider, and scorpion venoms, can assist in creating targeted libraries of potential drug candidates (*3*). Venom peptides and proteins that act

1

upon mammalian physiological processes are extremely site-specific in their actions and this specificity is highly valuable in performing as a lead therapeutic or biological probe. Thus, although a protein isolated and purified from snake venom might be therapeutically useful, more likely it will serve as a lead for drug development. Alternatively, a venom protein can function as a biological probe that delineates the molecular surface of its receptor binding site, thus providing key insights that enable the design and development of simpler molecules that mimic the desired receptor site and have the potential to serve as drugs.

The biological implications of natural toxins that are derived from marine, fungal, microbial, plant, and animal sources, along with selected synthetic nerve gases and doping agents, are described in this ACS Symposium Series volume under four subject headings (for earlier natural toxin symposia, see ref. *4,5*). Although the nerve gases and doping agents are not naturally occurring, their development was stimulated by the goal of preparing large quantities of inexpensive substances that have toxic or enhanced performance properties of potency and selectivity greater than those of naturally occurring materials.

Marine, Fungal, and Microbial Toxins

Certain marine species contain deadly toxins; the puffer fish, which is a Japanese culinary delicacy, accumulates the potent neurotoxin tetrodotoxin from dietary dinoflagellates who, in turn, obtain the toxin from marine bacteria whereas corals of the *Palythoa* genus contain palytoxin, a toxin that depolarizes excitable tissue and ruptures blood cells. Extracts from these corals were employed by natives of the Hawaiian Islands for hunting and warfare; however, the relative inaccessibility of marine organisms limited their applicability, resulting in greater use of amphibian and other terrestrial poisons in hunting by early cultures. A coral reef is a treasure chest of novel marine organisms, both toxic and benign. T. Higa (Japan) describes the isolation and characterization of numerous cytotoxic and bioactive compounds from coral reef sponges that are prime candidates for anti-tumor drugs and immunostimulatory compounds. Pardaxin is a unique substance released from a Red Sea flatfish that repels a predatory shark. P.H. Lazarovici (Israel) reports the cellular and molecular mechanism of pardaxin effects on cellular signaling.

Serious contamination of foodstuffs often results from exposure to fungi and their constituent mycotoxins. A notorious example is the fungus *Claviceps purpurea* that produces ergot alkaloids which can contaminate rye and other cereal grains resulting in induction of ergotism in consumers. Other fungal toxins are more insidious, inducing liver or gastrointestinal tumors over a period of years. Prominent among the latter toxins are the aflatoxins which were originally identified in 1960 following the deaths of thousands of turkeys. No direct evidence exists that liver disease in humans can be caused by ingestion of aflatoxins although a high incidence of liver cancer has occurred in portions of Africa where large amounts of aflatoxins are consumed by a population often afflicted with hepatitis. In addition to presenting a small, albeit real, long-term hazard to mankind, mycotoxins offer the potential for inducing fetal abnormalities as demonstrated in laboratory rodents.   R. J. Molyneux (US) describes isolation and characterization of anti-aflatoxigenic constituents of tree nuts, such as pistachios, almonds, and walnuts.   The perception of extreme hazards due to aflatoxin contamination has been exacerbated recently by reports that Iraq has placed the mycotoxin in ballistic warheads and bombs. These missiles are speculated to serve as potential sources of economic and political intimidation through the possible dispersal and contamination of food crops that become either non-consumable or unexportable. Current research on the newest class of mycotoxins, the fumonisins, is reported by W. T. Shier (US). His research suggests that the most abundant fumonisin, FB_1, binds covalently to protein and starch during processing of foods in a manner that may allow the release of a biologically active form of the toxin upon digestion.

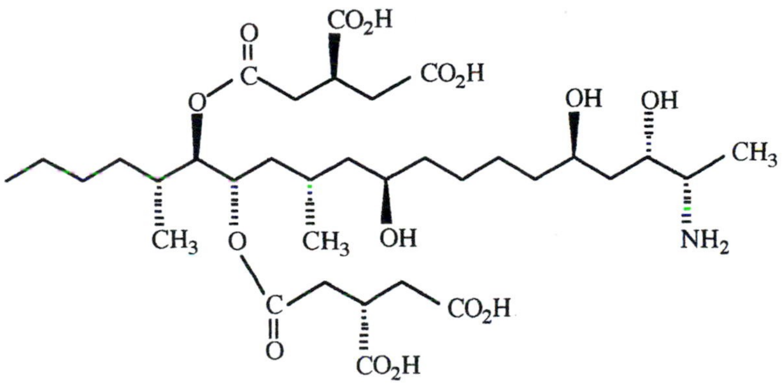

FUMONISIN B_1

The occurrence of pathogenic bacteria in diverse environmental, food processing, and industrial sources has necessitated a sensitive and selective assay for their detection and identification. T. Krishnamurthy (US) has distinguished a number of pathogenic bacteria from their non-pathogenic counterparts by the application of either electrospray ionization or matrix assisted laser desorption ionization mass spectroscopy.

Plant Toxins

Primitive cultures differentiated edible from poisonous plants by a process of trial and error. After appropriate experimentation, they began to exploit the naturally

4

occurring toxins for materials useful in hunting, euthanasia, and murder. From the perspective of the plant, toxins are produced presumably for the purpose of deterring predatory herbivores.

Pyrrolizidine alkaloids (e.g., diester derivatives of retronecine) are potent hepatotoxins and carcinogens isolated from a variety of plants that at low levels can cause genetic damage, cirrhosis, and possibly cancer. The mode of action of these toxic alkaloids occurs via a cytochrome P-450 mediated oxidation which produces a highly active intermediate. Subsequently, the intermediate dehydropyrrolizidine mediates the interstrand cross-linking of DNA by forming a bis-alkylated adduct. R.J. Huxtable (US) reports that human exposure to pyrrolizidine alkaloids in the United States occurs primarily through intentional consumption of herbs thought to provide health benefits. Thus, although comfrey has been shown to contain dangerous levels of toxic pyrrolizidine alkaloids, the herb is still available in some US health-food stores. J. A. Edgar (Australia) states that even though pyrrolizidine alkaloids do not normally occur in plants consumed by humans, they may enter the human food chain from food-producing animals that eat plants containing the alkaloids. For example, humans may be exposed potentially to pyrrolizidine alkaloids upon consuming eggs, milk, or honey.

RETRONECINE

A number of polyhydroxylated indolizidine alkaloids that resemble natural carbohydrates but which incorporate a nitrogen atom in place of an oxygen ring atom, were observed to function as glycosidase inhibitors,i.e., they interfere with the functioning of certain glycosidases that assist in the removal or cleavage of carbohydrate residues from glycoproteins. A related group of nortropane alkaloids, the calystegines, has been identified by R. J. Nash (UK) in various fruits and vegetables. In particular, both the occurrence of calystegines (e.g., calystegine B_2) in potatoes and the likelihood that their effects would be expressed as gastrointestinal distress and neurological disorders suggests that the calystegines might account for certain toxicological aspects normally attributed to steroidal alkaloid glycosides. Several known and new alkaloids of varied structures have been characterized from Australian pasture grasses by S. M. Colegate (Australia) that might be responsible for outbreaks of neurological and sudden death intoxication in range animals.

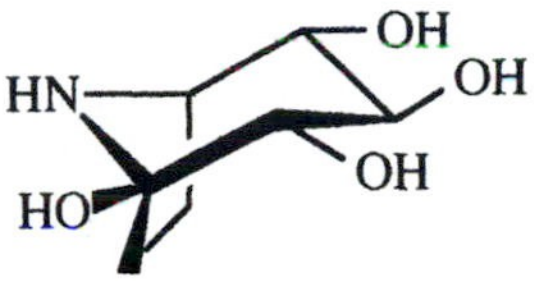

CALYSTEGINE B$_2$

Mankind has long been interested in plant extracts that control fertility and either induce abortion or labor depending upon the period since conception. K. E. Panter (US) describes natural toxins from poisonous range plants that affect reproductive function in livestock. His review focuses on three groups of plants that contain specific classes of toxins affecting different aspects of reproduction; locoweeds from *Astragalus* and *Oxytropis*, labdane resin acids, e.g., the abortifacient isocupressic acid, from Ponderosa pine, and quinolizidine and piperidine alkaloids from *Lupinus* spp. The Greek philosopher Dioscorides recommended a decoction known as 'abortion wine' that contained a *Veratrum* sp. as one of several ingredients. Several Western US plains and mountain Native American tribes employed a decoction prepared from the roots of *Veratrum californicum* as a contraceptive rather than as an abortifacient. In the 1960s, the steroidal alkaloids jervine and its 11-deoxo analog, cyclopamine, were shown to be responsible for induction of cyclopia and related congenital malformations in sheep that had ingested *V. californicum* on a specific day of their gestation. W. Gaffield (US) reports recent molecular biological research that establishes cyclopamine's selective and potent inhibition of Sonic hedgehog signaling. This pathway is responsible for critical aspects of neural tube development that upon disruption will result in cyclopia and other craniofacial defects.

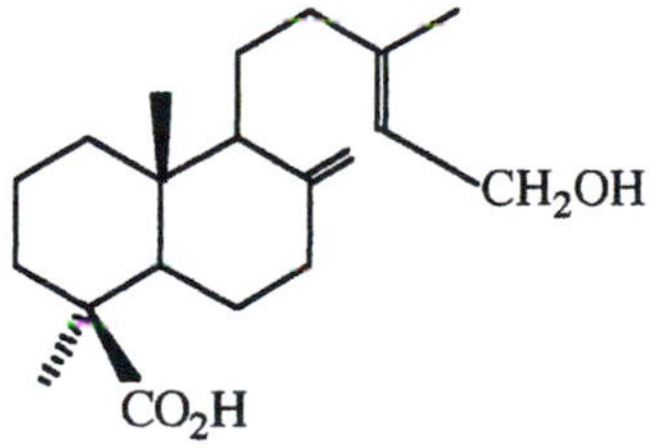

ISOCUPRESSIC ACID

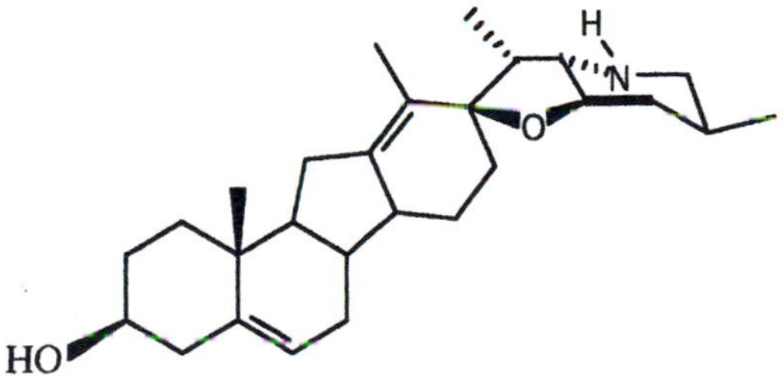

CYCLOPAMINE

Fifteenth century explorers were astonished to observe Cuban natives inserting smouldering leaves from tobacco plants into their nostrils and mouths. Five centuries later, nearly 500,000 Americans die annually from smoking related afflictions and approximately 60% of all hospitalizations are attributed to illnesses related to smoking. K. Stone (US) describes the DNA damaging properties of stable free radicals derived from cigarette tar extracts.

Ricin possesses a sinister history of usage exemplified by its presumed administration to an unsuspecting victim via a needle-tipped umbrella in a highly publicized international espionage incident of the 1970s. R. G. Wiley (US) describes the application of ricin and related toxic plant lectins as tools in molecular neurosurgery for producing highly selective neural lesions.

Sheep and goats in various regions of the world are susceptible to seasonal hepatogenous photosensitization. The photosensitivity is caused by accumulation of a photodynamic compound, phylloerythrin, a porphyrin derived from chlorophyll degradation, whose excretion is prevented by certain types of liver damage. A. Flåøyen (Norway) discusses several plant-derived natural toxins that cause liver damage or dysfunction in livestock resulting in retention of phylloerythrin.

PHYLLOERYTHRIN

A comparison of highly poisonous substances reveals that amphibian and reptile toxins are among Nature's most potent. Snake envenomation is often fatal to a victim; thus, snake venom has been feared since the beginning of human history. However, humans were puzzled as to how such small amounts of venom could exert such powerful effects on biological systems. Toxins derived from poisonous frogs have a fascinating ethnopharmacological history and were used by primitive cultures as arrow poisons. Studies of the mode of action of these poisons have contributed immensely to our understanding of neuropharmacology. Snake venom toxins are effective tools in medical applications as demonstrated by the development of angiotensin-converting enzyme (ACE) inhibitors starting from a component of venom. Other research has employed snake venoms in the development of platelet aggregation and blood clotting inhibitors.

Recently, scientists have been engaged in more sophisticated investigations targeted at delineating the molecular basis of toxic action. The secondary structure of snake venom cardiotoxins as revealed by nuclear magnetic spectroscopy is discussed by T. K. S. Kumar (Taiwan) and the crystal structure of snake venom phospholipase A, that possesses presynaptic toxicity, is described by Z. Lin (China). Illustrating the potential beneficial effects of advanced technology in elaborating the mechanism of snake venom constituents, F. S. Markland (US) reports the application of disintegrin in blocking the adhesion of human breast cancer cells to several extracellular matrix proteins. Thus, disintegrin, isolated from the venom of the Southern Copperhead snake, inhibits *in vivo* progression by a combination of anti-tumor and anti-angiogenic activities. Among all lizards, only the species of *Heloderma* (Gila monster) is venomous. The Gila monster produces a venom approximately as toxic as that of the Western Diamondback rattlesnake. A. T. Tu (US) reports the chemical and functional aspects of Gila monster toxins, with particular emphasis on toxic enzymes.

Nerve Agents and Doping Compounds

The chemical warfare agents, commonly known as nerve gases, are not gases but polar organic liquids at ambient conditions. Most of these nerve gases are esters similar in structure to organophosphorus insecticides that irreversibly react with the enzyme acetylcholinesterase (AChE), inhibiting its control over the nervous system. For example, tabun, a G-type toxin, was discovered in 1936 by a German scientist involved with developing synthetic insecticides. Later before the end of World War II, Germans developed more powerful nerve agents sarin and soman. A single milligram of nerve agent on the skin is sufficient to cause death; fortunately, the G-agents were not used in World War II. In the 1950s, the V-type nerve gases were developed, such as V-X, that are more toxic and persistent. Thousands of tons of these compounds were stockpiled by major countries over the intervening decades which has created current safety and environmental impact concerns with regard to their detoxification and destruction. In 1995, a terrorist group dispersed sarin in a Tokyo subway station resulting in the death of 3 people and the illness of several thousand commuters. One year earlier, the same sect had murdered one of its own members with the deadly nerve gas, V-X. Five chapters of this book describe the implications of these incidents. An overview of the chemical terrorism attacks in Japan is provided by A. T. Tu (US). Additional chapters by Y. Seto (Japan), Y. Ogawa (Japan), T. Okumura (Japan), and H. Tsuchihashi (Japan) describe the biological effects of both sarin and V-X on human victims. Nerve gases not only may be analyzed directly but, in addition, generate unique metabolites in human blood that can provide conclusive evidence of the specific nerve agent administered.

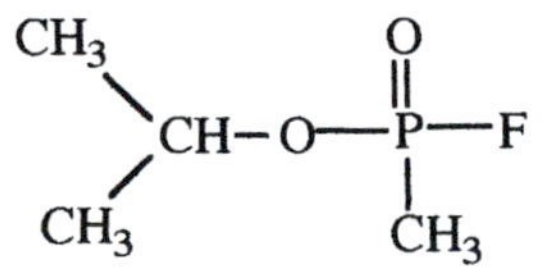

SARIN

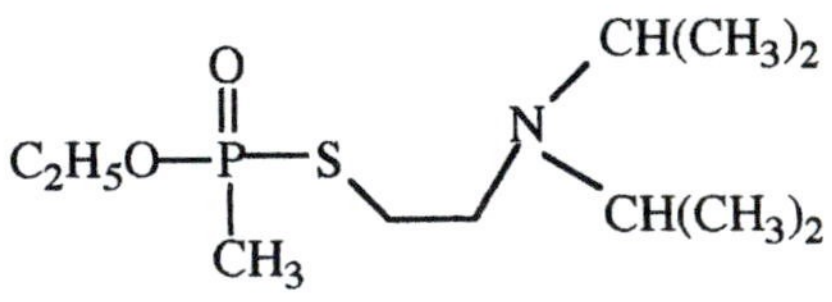

V-X

Mild stimulants such as tea and coffee are primarily of social importance in modern times while more potent stimulants such as cocaine and amphetamines are substances of abuse. In primitive cultures, stimulants were more highly prized for their ability to extend endurance and alleviate hunger than for social or mind-altering properties. R.K. Mueller (Germany) discusses state-of-the-art analytical methodology currently employed to monitor modern athletes, some of whom occasionally ingest controlled substances in order to achieve enhanced athletic performance.

Literature Cited

1. Mann, J. *Murder, Magic, and Medicine*; Oxford University Press: New York, 1994.

2.   Bioactive Natural Product Database, Szenzor Management Consulting Company, Budapest, Hungary, 1996.

3.   Berressem, P. *Chem. Brit.* **1999,** *35*,40.

4.   *Natural Toxins: Toxicology, Chemistry, and Safety*; Keeler, R.F.;Mandava, N.B.; Tu, A.T., Eds.; Alaken, Inc.: Fort Collins CO, 1992

5.   *Natural Toxins 2: Structure, Mechanism of Action, and Detection*; Singh, B.R.; Tu, A.T., Eds.; Plenum Press: New York NY, 1996

MARINE, FUNGAL, AND MICROBIAL TOXINS

Chapter 2

Cytotoxins and Other Bioactive Compounds from Coral Reef Organisms

Tatsuo Higa, Ikuko I. Ohtani, and Junichi Tanaka

Department of Chemistry, Biology, and Marine Science, University of the Ryukyus, Nishihara, Okinawa 903–0213, Japan

Coral reef organisms are good sources of unique molecules having cytotoxicity and other biological activities. Manzamine A, a complex alkaloid which was initially discovered as a cytotoxin from a sponge, has recently been shown to have potent antimalarial activity. Another cytotoxin, misakinolide A, possesses unique action on filamentous actin and is becoming an important reagent in the study of actin cytoskeleton. Based on α-galactosylceramides isolated from an Okinawan sponge, a new type of anticancer agent KRN7000 was developed by Natori and coworkers at Kirin Brewery. KRN7000, an immunostimulatory compound, is highly effective in preventing metastasis of colon cancer to the liver in a mouse model. It has entered human clinical trials. Also discussed are new cytotoxic compounds, cyclic peptides, cupolamides consisting of some uncommon amino acid residues from a sponge and new class of bisindole pigments, iheyamines from an ascidian.

Numerous species of organisms live in tropical waters, particularly on coral reefs. Diversity of species is the most significant feature of the coral reef as a biological community. It is especially rich in the diversity of bottom-dwelling invertebrates such as sponges, soft corals, and tunicates. Except for a few, the majority of these sessile invertebrates have no food value and, hence, have not been considered an economically important resource. However, chemical and pharmaceutical studies in the past three decades have revealed that these organisms unpalatable to man are important sources of bioactive substances which have potential for development into new therapeutic drugs and other useful products.

Because of high diversity and hence of high competition for survival, coral reef organisms, especially lower invertebrates which lack physical means of defense, have evolved a variety of unique toxic substances as their defensive strategy. Many of these compounds have no terrestrial counterparts in chemical structure or biological activities. These facts, together with a high incidence of bioactive compounds, have made tropical coral reef organisms particularly attractive targets of research.

Cytotoxicity screening using animal or human cancer cell lines is often carried out as a first convenient step in the search of lead compounds for new anticancer drugs. When we screened extracts prepared from more than one thousand organisms collected on the coral reefs of Okinawa, the incidence of cytotoxicity observed was greater than 50%. Likewise, incidence of other activities, such as antiviral, is also high with coral reef organisms, indicating that they are good sources of bioactive compounds.

In 1985 we joined the SeaPharm project of the Harbor Branch Oceanographic Institution as an outside collaborator. The main focus of the project was to discover lead compounds for anticancer, antiviral, and antifungal drugs from marine organisms. In this and subsequent collaborations with PharmaMar Research Institution of Spain and Pharmaceutical Research Laboratory of Kirin Brewery, Co., Ltd. of Japan, we discovered a number of cytotoxins and other bioactive compounds from Okinawan marine organisms. Some representative compounds are shown in Figure 1; many of them have novel structures and significant biological activities. Except for iheyamine A, isolation and structure determination of all others in Figure 1 has already been described elsewhere. Here we discuss two compounds, manzamine A and misakinolide A, which were originally reported as cytotoxins and quite recently revealed to have additional interesting biological activities. Also presented are our most recent results in search of new bioactive compounds and the promising anticancer drug KRN7000 developed by Kirin Brewery.

Antimalarial Activity of Manzamine A

Manzamines are complex alkaloidal constituents of sponges. Since our first report of manzamine A (1) from *Haliclona* sp. in 1986 (*1*), about 40 related compounds have been described from more than a dozen species of Pacific sponges (*2*). Many of these compounds are known to be cytotoxic. Manzamine A was highly cytotoxic against P388 mouse leukemia cells with an IC_{50} of 0.07 µg/mL (*1*). However, it showed no *in vivo* efficacy and was eventually dropped from a list of antitumor leads. Since its discovery the unique structure of manzamine A has attracted many synthetic chemists to challenge its total synthesis. The first elegant synthesis has recently been reported by Winkler (*3*).

When Kara of the National University of Singapore screened our marine compounds for antimalarial activity, she found that manzamine A (1) was most active when tested *in vivo* in mice infected with the parasite *Plasmodium berghei*. The antimalarial test for manzamines A and F (2) was conducted in comparison with the existing drugs, artemisinin (3) and chloroquine (4). The results are shown in Table I (Kara, A. U., personal communication). Kara noted that mice treated with a single intraperitoneal injection of manzamine A as low as 50 µM/kg (29.3 mg/kg) of body weight exhibited above 90% inhibition of the asexual erythrocytic stages of *P. berghei* on the first three days after treatment. With a single dose of 100 µM/kg (58.5 mg/kg) of manzamine A two out of five mice survived 60 days of post-treatment. Neither artemisinin nor chloroquine attained this level of efficacy in a parallel test (see Table I). Manzamine A was found to be toxic above 500 µM/kg. On the other hand, manzamine F (2) which has a closely related structure exhibited no effect at a single dose of 100 µM/kg. Scarcity of the sample precluded further testing of manzamine F. Kara's results suggest that manzamine A is a promising new antimalarial agent and that its *in vivo* toxicity is not as high as is implied by its *in vitro* cytotoxicity. We are now working on the structure-activity relationship of manzamines and their derivatives.

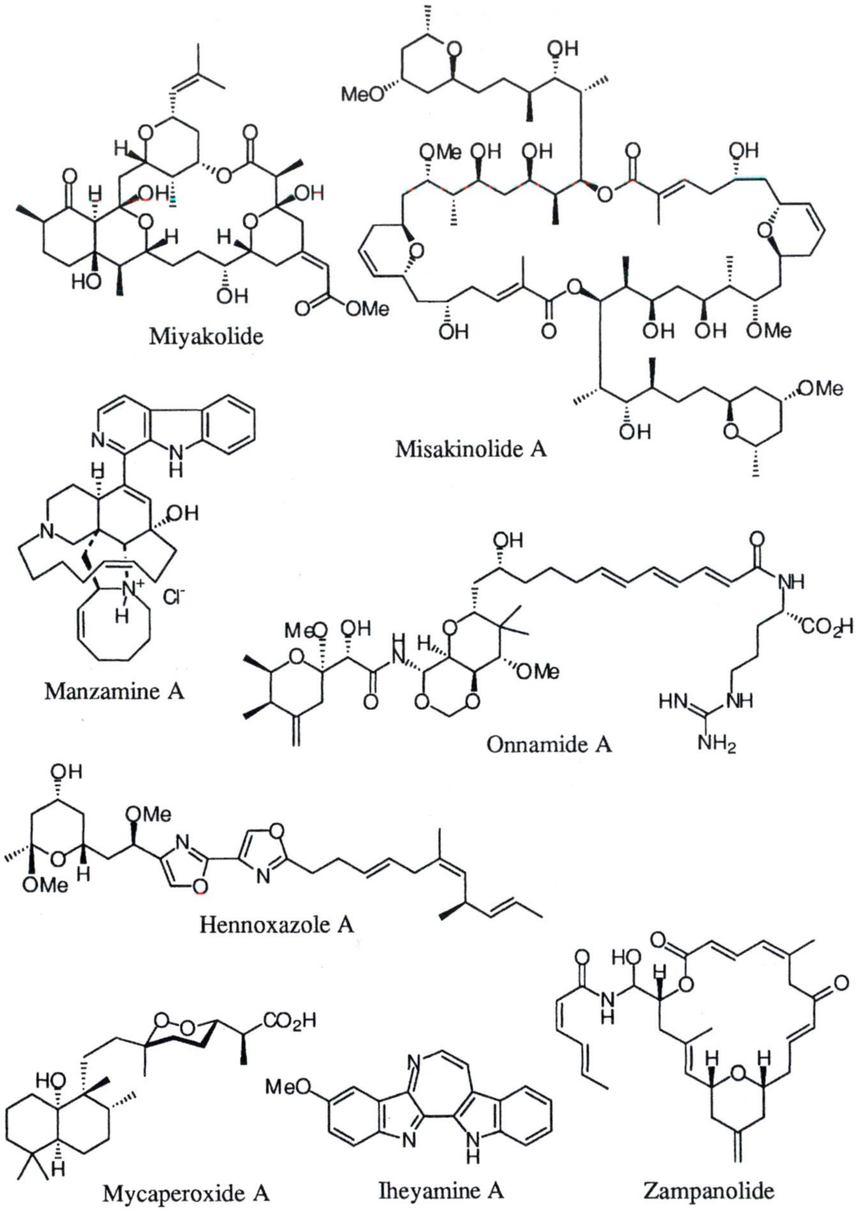

14

Miyakolide

Misakinolide A

Manzamine A

Onnamide A

Hennoxazole A

Mycaperoxide A

Iheyamine A

Zampanolide

Figure 1. Some bioactive compounds from coral reef organisms.

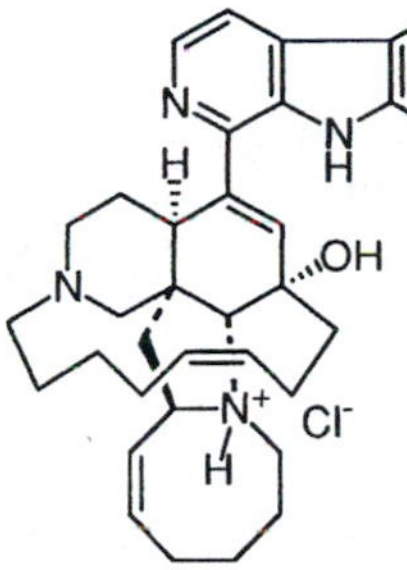

Manzamine A (**1**)

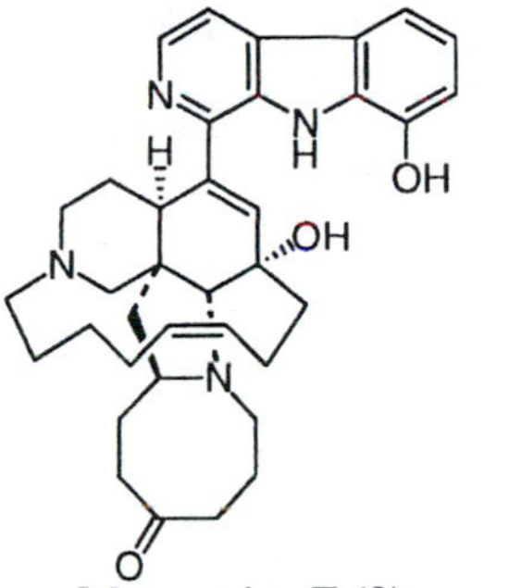

Manzamine F (**2**)

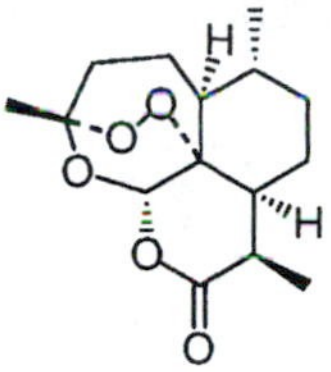

Artemisinin (**3**)

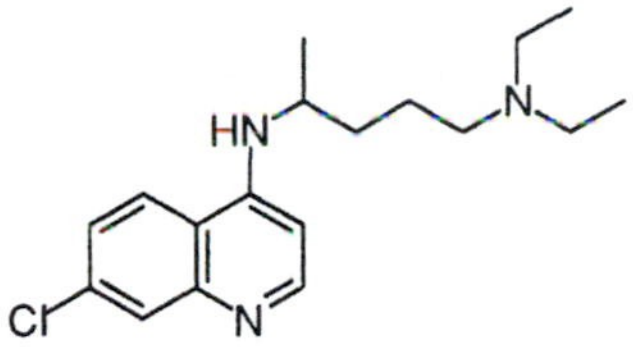

Chloroquine (**4**)

Table I. Antimalarial Activity of Manzamines,
Artemisinin, and Chloroquine[a]

Drug	Dose (μM/kg)	No. of Mice	No. of Surviving Mice on Day							
			0	2	4	6	10	15	25	60
Manzamine A	1000	4	4	1	0					
	500	4	4	2	0					
	100	5	5	5	5	5	5	4	2	2
	50	5	5	5	5	5	5	3	0	
Manzamine F	100	5	5	5	0					
Artemisinin	1000	5	5	5	5	5	5	1	1	1
	500	5	5	5	5	5	1	0		
	100	5	5	5	4	0				
	50	5	5	5	2	0				
Chloroquine	1000	2	0							
	500	2	0							
	100	5	5	5	5	5	0			
	50	5	5	5	5	5	0			
Control		5	5	5	0					

[a]Unpublished data from Dr. A. U. Kara, National University of Singapore. Mice infected with *Plasmodium berghei* were given a single intraperitoneal injection of each drug on Day 2 after infection.

Actin-Capping Activity of Misakinolide A

Misakinolide A (**5**) is also a highly cytotoxic compound isolated from an Okinawan sponge (*4*). The initial 20-membered macrolide structure was revised to a 40-membered dimeric dilactone (*5*). The absolute stereostructure of **5** was determined by chemical degradation and correlation to swinholide A (**6**), another related macrolide, in collaboration with Kitagawa's group at Osaka University (*6*). Misakinolide A exhibited potent cytotoxicity with IC_{50} 4-10 ng/mL against HCT-8 human colon cancer, A549 human lung cancer, and P388 cell lines. It also showed *in vivo* antitumor activity against P388 mouse leukemia with a T/C value of 145% at a dose of 0.5 mg/kg. Furthermore, misakinolide A revealed antifungal activity against *Candida albicans* with an MIC 5 μg/mL. No further steps have been taken to develop this compound as an antitumor lead.

Swinholide A (**6**) was first reported by Kashman (*7*) as a 22-membered macrolide isolated from the Red Sea sponge *Theonella swinhoei*. The structure was later revised by Kitagawa to a 44-membered dimeric dilactone (*8*). The only difference between the two structures of **5** and **6** lies at the presence of two additional C=C double bonds in **6**. In all other points the two compounds are identical. However, their biochemical properties are quite different as noted recently by Spector and coworkers (*9*).

Misakinolide A (5)   Swinholide A (6)

Both misakinolide A and swinholide A bind to two actin subunits with virtually the same affinity. Swinholide A severs actin filaments, while misakinolide A does not sever but caps the barbed end of F-actin. This property of misakinolide A is unique, and no other compounds have ever been known to exhibit such action. Misakinolide A is now considered to be a very important tool in the study of the organization and dynamics of the actin cytoskeleton. As a reagent, according to Spector and other biochemists, misakinolide A is much more interesting than swinholide A. However, it is less well known among chemists than swinholide A. Two total syntheses of swinholide A have already been reported (*10-11*), while no synthesis of misakinolide A has been described.

Anticancer Agent KRN7000

In our collaboration with Kirin Brewery, Natori and coworkers discovered a series of new galactosylceramides exhibiting *in vivo* antitumor activity from the Okinawan sponge *Agelas mauritianus* (*12*). The new compounds named agelasphins (e.g., **7**) are unique as they contain an unprecedented α-galactoside. These α-galactosylceramides showed good *in vivo* antitumor activity with T/C values of 160-190% against B16 murine melanoma and potent immunostimulatory activity in a mixed lymphocyte reaction assay. Importantly, agelasphins showed no *in vitro* cytotoxicity against B16 melanoma cells at 20 μg/mL. Acute toxicity in rats (intravenous injection) was also weak with LD_{50} >10 mg/kg (*13*).

In their extensive studies on the structure-activity relationship the Kirin scientists prepared a number of synthetic analogues of agelasphin and examined their activities. Important features for the manifestation of higher activity are among others: the α-linkage of a pyranose, presence of the hydroxyl groups at C-3 and C-4, but not necessarily at the C-2' position in the ceramide portion, and long chain for both the fatty acid and the long-chain base. They selected a synthetic cerebroside coded KRN7000 (**8**) as a drug candidate (*14*). This compound shows good antitumor activity against several tumor models. It is also effective in preventing metastasis of colon cancer to the liver. When tested in mice inoculated with Colon 26 cells into spleen, all control mice and those treated with the anticancer drug adriamycin died by day 40 because of the metastasis of the cancer to liver, while all those treated with KRN7000 still survived on Day 100 with no sign of liver metastasis (*15-16*).

Agelasphin 9a (**7**)   KRN7000 (**8**)

KRN7000 is not cytotoxic by itself but helps to kill cancer cells by activating immune systems (*17*). It protects cells from radiation damage (*18*), enhances natural killer cell activity (*19*), increases platelet and white blood cell count (*20*), and proliferates bone marrow cells (*20*). Studies on the mechanism of action have been published (*21-23*). Human clinical trials of this drug began in Europe in October, 1998.

Cytotoxic Cyclic Peptides from a Sponge

An extract of the sponge *Theonella cupola* exhibited *in vitro* antiviral activity against human immunodeficiency virus (HIV). We therefore started to isolate the active principle. However, before we could complete isolation of the constituents, our collaborator suspended the anti-HIV assay. Nevertheless, we were able to isolate two nucleosides from an active fraction and several cyclic peptides from an inactive portion of the extract. We do not know, whether these nucleosides are responsible for the anti-HIV activity of the crude extract.

An acetone extract of the sponge was partitioned between ethyl acetate and water. The organic layer showed anti-HIV activity, cytotoxicity against P388, and antibacterial activity against *Bacillus subtilis*. Separation of the layer yielded two nucleosides. The aqueous layer was concentrated to dryness and extracted with methanol. The methanol-soluble material was separated to yield cyclic peptides, cupolamides A-G. The isolation scheme is shown in Figure 2.

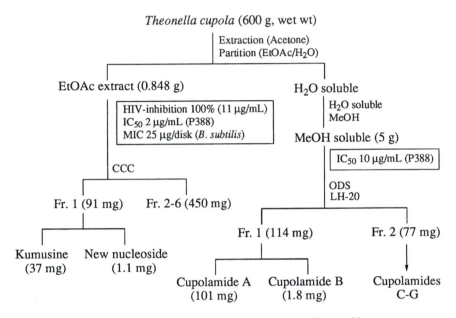

Figure 2. Isolation scheme for nucleosides and cyclic peptides
from *Theonella cupola*.

One of the nucleosides was identified by spectroscopic data as the known kumusine (24) [= trachycladine (25), 9] which was reported by two independent groups. It is an unusual nucleoside having a chlorine atom on the adenine moiety, a methyl group at C-2' and dehydroxylation of the sugar at C-5'. The other nucleoside was shown to be a new related compound which will be published elsewhere. Kumusine has been described to have cytotoxic (24-25) and immunosuppressive (24) activity. Although the two nucleosides have not yet been tested, they might have been responsible for the observed anti-HIV activity of the organic layer.

Of the cyclic peptides, only the structure of cupolamide A (10) will be discussed here. The gross structure of cupolamide A was determined by 2D NMR connectivity studies and was shown to be a cyclic heptapeptide. The stereochemistry was determined by application of Marfey's method on each residue generated by hydrolysis. Thus, cupolamide A (10) was shown to be composed of two L-Val, one D-Leu, one D-Ser, and three uncommon amino acid residues: D-homoarginine, *trans*-4-hydroxy-L-proline, and L-2,4-diaminobutanoic acid. Regarding the position of the sulfate group, there were three possible sites for the ester linkage with the hydroxyl on either serine, proline, or the phenyl ring. It was unambiguously demonstrated to be on the proline residue by comparing chemical shift values of the relevant portions of 10 and those of its diacetate. Cupolamide A was also discovered independently by Scheuer from the same sponge collected in Indonesia (26). It showed cytotoxicity against P388 with an IC_{50} 7.5 µg/mL.

Cupolamide B was a closely related derivative of cupolamide A. Cupolamides

C-G belonged to a different series of cyclic peptides, microsclerodermins (*27*). In fact some of them were identical with those reported recently by Faulkner (*28*). Microsclerodermins are reported to have antifungal activity.

Kumusine (trachycladine, **9**)   Cupolamide A (**10**)

Cytotoxic Pigments from an Ascidian

Like sponges, ascidians are prolific sources of novel bioactive compounds (*29*). Many of them are nitrogenous, *viz.* cyclic peptides and polynuclear aromatic alkaloids based on the pyrido[2,3,4-*k,l*]acridine nucleus.   Recently we found at a coral reef of Iheya, Okinawa, a purple ascidian, *Polycitorella* sp. whose extract exhibited strong cytotoxicity.   Separation of the extract yielded two new pigments named iheyamines A (**11**) and B (**12**) in addition to the known potent cytotoxins, e.g., halichondramide (**13**).   Halichondramide (*30*) and related compounds have been known from sponges and some nudibranchs which prey on sponges.   It is the first time to be isolated from an ascidian.   Both iheyamine A and B showed moderate cytotoxicity with IC_{50} at the level of 1 μg/mL against P388, A549, and HT29 cell lines (Sasaki, T.; Ohtani, I. I.; Tanaka, J.; Higa, T. *Tetrahedron Lett.*, in press).

The structures of the iheyamines were determined by spectroscopic analysis. The iheyamines are a new class of natural purple pigments consisting of a fused nucleus of a bisindole with an azocine moiety.   Existing bisindole pigments from marine sources include Tyrian purple from certain species of snails (*31*), caulerpin from green algae (*32*), and 11-hydroxystaurosporine from an ascidian (*33*).

Conclusion

As demonstrated by the above examples, search of cytotoxins and other bioactive compounds from coral reef organisms has been rewarded by discovery of a variety of novel structures.   Like manzamine A and misakinolide A, which were discovered as cytotoxins and have been abandoned in the antitumor lead selection process, some of these compounds may be found to be useful as other drugs or as research tools. There are a number of marine-derived toxins which are commercially available as biochemical reagents.   Some of them have played indispensable roles in the studies of physiology, cell biology, and related disciplines.   Moreover, many novel marine compounds have been targets of total syntheses.   In case of the manzamines at least twelve groups of synthetic chemists around the world have engaged or are currently

20

working on the synthesis (2). The search of cytotoxins from coral reef organisms has a broad impact beyond mere discovery for anticancer lead compounds.

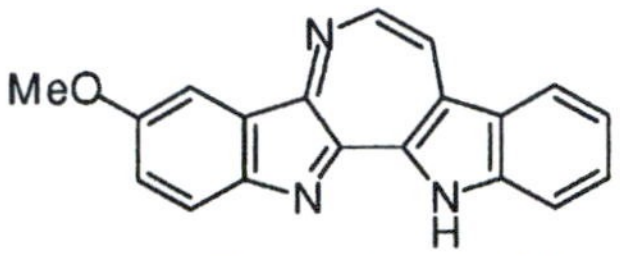

Iheyamine A (11) 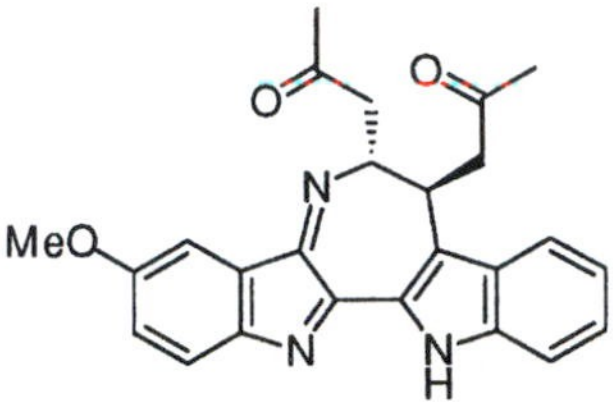   Iheyamine B (12)

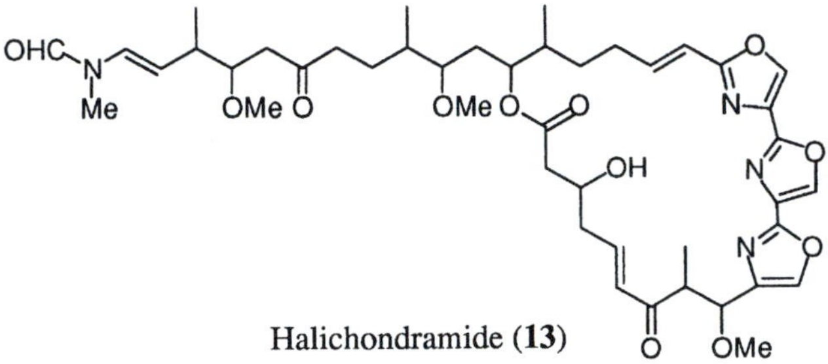

Halichondramide (13)

Acknowledgment

We thank Dr. A. Ursula Kara, National University of Singapore for providing us with antimalarial test results for manzamine A prior to publication and Dr. Takenori Natori, Pharmaceutical Research Laboratory, Kirin Brewery, Co. Ltd. for sending us all publications and other information on KRN7000 and related compounds. We are also indebted to Professor Paul J. Scheuer for informing us of the work of his group on cupolamide and publishing it jointly. In our laboratory the peptide work was carried out by Lea Bonnington and the isolation of iheyamines by Tsutomu Sasaki. PharmaMar of Spain conducted cytotoxicity assays.

References

1. Sakai, R.; Higa, T.; Jefford, C. W.; Bernardinelli, G. *J. Am. Chem. Soc.* **1986**, *108*, 6404-6405.
2. Magnier, E.; Langlois, Y. *Tetrahedron* **1998**, *54*, 6201-6258.
3. Winkler, J. D.; Axten, J. M. *J. Am. Chem. Soc.* **1998**, *120*, 6425-6426.
4. Sakai, R.; Higa, T.; Kashman, Y. *Chem. Lett.* **1986**, 1499-1502.
5. Kato, Y.; Fusetani, N.; Matsunaga, S.; Hashimoto, K.; Sakai, R.; Higa, T.; Kashman, Y. *Tetrahedron Lett.* **1987**, *28*, 6225-6228.
6. Tanaka, J.; Higa, T.; Kobayashi, M.; Kitagawa, I. *Chem. Pharm. Bull.* **1990**, *38*, 2967-2970.
7. Carmely, S.; Kashman, Y. *Tetrahedron Lett.* **1985**, *26*, 511-514.
8. Kobayashi, M.; Tanaka, J.; Katori, T.; Matsuura, M.; Yamashita, M.; Kitagawa, I. *Chem. Pharm. Bull.* **1990**, *38*, 2409-2418.

9. Terry, D. B.; Spector, I.; Higa, T.; Bubb, M. R. *J. Biol. Chem.* **1997**, *272*, 7841-7845.
10. Paterson, I.; Yeung, K. S.; Ward, R. A.; Smith, J. D.; Cumming, J. G.; Lamboley, S. *Tetrahedron* **1995**, *51*, 9467-9486.
11. Nicolaou, K. C.; Patron, A. P.; Ajito, K.; Richter, P. K.; Khatuya, H.; Bertinato, P.; Miller, R. A.; Tomaszewski, M. J. *Angew. Chem. Int. Ed.* **1996**, *35*, 847-868.
12. Natori, T.; Koezuka, Y.; Higa, T. *Tetrahedron Lett.* **1993**, *34*, 5591-5592.
13. Natori, T.; Morita, M.; Akimoto, K.; Koezuka, Y. *Tetrahedron* **1994**, *50*, 2771-2784.
14. Morita, M.; Motoki, K.; Akimoto, K.; Natori, T.; Sakai, T.; Sawa, E.; Yamaji, K.; Koezuka, Y.; Kobayashi, E.; Fukushima, H. *J. Med. Chem.* **1995**, *38*, 2176-2187.
15. Nakagawa, R.; Motoki, K.; Ueno, H.; Iijima, R.; Nakamura, H.; Kobayashi, E.; Shimosaka, A.; Koezuka, Y. *Cancer Research* **1998**, *58*, 1202-1207.
16. Natori, T.; Akimoto, K.; Motoki, K.; Koezuka, Y.; Higa, T. *Folia Pharmacol. Jpn.* **1997**, *110*, 63-68.
17. Kobayashi, E.; Motoki, K.; Uchida, T.; Fukushima, H.; Koezuka, Y. *Oncology Research* **1995**, *7*, 529-534.
18. Motoki, K.; Kobayashi, E.; Morita, M.; Uchida, T.; Akimoto, K.; Fukushima, H.; Koezuka, Y. *Bioorg. Med. Chem. Lett.* **1995**, *5*, 2413-2416.
19. Kobayashi, E.; Motoki, K.; Yamaguchi, Y.; Uchida, T.; Fukushima, H.; Koezuka, Y. *Bioorg. Med. Chem.* **1996**, *4*, 615-619.
20. Motoki, K.; Morita, M.; Kobayashi, E.; Uchida, T.; Fukushima, H,; Koezuka, Y. *Biol. Pharm. Bull.* **1996**, *19*, 952-955.
21. Yamaguchi Y.; Motoki, K.; Ueno, H.; Maeda, K.; Kobayashi, E.; Inoue, H.; Fukushima, H.; Koezuka, Y. *Oncology Research* **1996**, *8*, 399-407.
22. Kawano, T.; Cui, J.; Koezuka, Y.; Toura, I.; Kaneko, Y.; Motoki, K.; Ueno, H.; Nakagawa, R.; Sato, H.; Kondo, E.; Koseki, H.; Taniguchi, M. *Science* **1997**, *278*, 1626-1629.
23. Kawano, T.; Cui, J.; Koezuka, Y.; Toura, I.; Kaneko, Y.; Sata, H.; Kondo, E.; Harada, M.; Koseki, H.; Nakayama, T.; Tanaka, Y.; Taniguchi, M. *Proc. Natl. Acad. Sci. USA* **1998**, *95*, 5690-5693.
24. Ichiba, T.; Nakao, Y.; Scheuer, P. J.; Sata, N. U.; Kelly-Borges, M. *Tetrahedron Lett.* **1995**, *36*, 3977-3980.
25. Searle, P. A. Molinski, T. F. *J. Org. Chem.* **1995**, *60*, 4296-4298.
26. Bonnington, L. S.; Tanaka, J.; Higa, T.; Kimura, J.; Yoshimura, Y.; Nakao, Y.; Yoshida, W. Y.; Scheuer, P. J. *J. Org. Chem.* **1997**, *62*, 7765-7767.
27. Bewley, C. A.; Debitus, C.; Faulkner, D. J. *J. Am. Chem. Soc.* **1994**, *116*, 7631-7636.
28. Schmidt, E. W.; Faulkner, D. J. *Tetrahedron* **1998**, *54*, 3043-3056.
29. Davidson, B. D. *Chem. Rev.* **1993**, *83*, 1771-1791.
30. Kernan, M. R.; Molinski, T. F; Faulkner, D. J. *J. Org. Chem.* **1988**, *53*, 5014-5020.
31. Christophersen, C.; Wätjen, F.; Buchardt, O.; Anthoni, U. *Tetrahedron Lett.* **1997**, *38*, 1777-1748.
32. Maiti, B. C.; Thomson, R. H. *In* Marine Natural Products Chemistry.; Faulkner, D. J.; Fenical, W. H. Eds.; Plenum Press: New York, 1977; pp 159-163.
33. Kinnel, R. B.; Scheuer, P. J. *J. Org. Chem.* **1992**, *57*, 6327-6329.

Chapter 3

Cellular Signaling in PC12 Affected by Pardaxin

S. Abu-Raya[1], E. Bloch-Shilderman[1], H. Jiang[2], K. Adermann[3], E. M. Schaefer[4], E. Goldin[5], E. Yavin[6], and P. Lazarovici[1,7]

[1]Department of Pharmacology, School of Pharmacy,
Faculty of Medicine of the Hebrew University, Jerusalem, Israel
[2]Section of Growth Factors, NICHD and [5]NINDS,
National Institutes of Health, Bethesda, MD 20892
[3]Niedersaechsisches Institut fuer Peptide-Forschung,
Feodor-Lynen-Strasse 31, D–30625, Hannover, Germany
[4]Promega Company, 2800 Words Hollow Road, Madison, WI 53711–5399
[6]Department of Neurobiology, The Weizmann Institute
of Science, Rehovot 76100, Israel

Pardaxins, a family of polypeptide, excitatory neurotoxins, isolated from the gland's secretion of *Pardachirus* fish are used as pharmacological tools to investigate calcium-dependent and calcium-independent signaling leading to neurotransmitter release. In PC12 cells, a neural model to study exocytosis, pardaxin forms voltage-dependent pores which are involved in pardaxin-induced increase in intracellular calcium and calcium-dependent dopamine release. Pardaxin-induced calcium-independent, dopamine release from PC12 cells is attributed to the stimulation of the arachidonic acid cascade and the increased release of the arachidonic acid metabolites produced by the lipoxygenase pathways. Pardaxin rapidly stimulates MAPK phosphorylation activity, which is proposed to be involved in pardaxin-induced arachidonic acid and dopamine release. It seems likely that pardaxin delayed stimulation of stress-kinases JNK and p38 will play a prominent role in triggering cell death. Collectively, these results demonstrate that pardaxin selectively modulates neuronal signaling to achieve massive exocytosis and neurotoxicity.

The ordered growth, and functioning of multicellular organisms requires the transfer of information from one cell to another and from one part of the cell to another. This information comes from a variety of sources both within and outside the cell. Hormones, growth factors and neurotransmitters interact with specific receptors on the plasma membrane or in the cytoplasm. Inside the cell, information is relayed and transduced by intracellular messengers including calcium, cyclic nucleotides, diacylglycerol, inositol polyphosphates, etc. (*1*).
Finally, balanced interactions between these different signaling pathways ensure that

[7]Corresponding author.

the required biological response occurs. It is not surprising, therefore, that many toxins interfere with the chemical communication in the body.

A simple way to achieve intercellular communication in the nervous system involves release of neurotransmitter which signals through receptor molecules that exist on the same or neighbouring cell. This principle lies at the heart of the signaling process utilized by the synapse, the basic unit on which an integral nervous system is built (2). At synapses, the elementary signaling event involves a depolarization dependent calcium influx into the presynaptic terminal that triggers the release of the neurotransmitter to be sensed by receptors in the postsynaptic cells (3).

Neurotoxins are commonly defined as chemicals which interfere with synaptic activity. They can be classified into ionic channel toxins which modify ionic conductance, presynaptic toxins which affect neurotransmitter release, and post-synaptic toxins which block the neurotransmitter receptors (4). For the last two decades we investigate the mechanism of action of pardaxin, a presynaptic neurotoxin isolated from the Red Sea flatfish Pardachilus marmoratus (5). Elucidation of pardaxin's mode of action is essential to its effective use as a pharmacological tool and is also expected to provide new understanding into the molecular steps involved in the mechanism of neurotransmitter release.

Isolation, Characterization and Synthesis of a Novel Pardaxin Isoform

Sole fish of the genus Pardachirus produce an exocrine secretion from epithelial glands located at the base of the dorsal and anal fins (Figure 1a). This secretion contains toxic components that play an important role in the defense of predatory fish. Beside aminoglycosteroids (6), peptides, know as pardaxins (5), represent the major components of the secretion (7-9). Pardaxins have been shown to form voltage-dependent ion channels (10-12) at low concentrations in a variety of artificial membranes and cytolysis at high concentrations (5, 13). The membranal activity of pardaxins such as binding, insertion and rearrangement for pore formation, is dependent on the α-helix content as well as the intramolecular interaction between the amino- and carboxy-terminal domains (12,14,15). Thus, this peptide family is structurally and functionally related to other membranally active peptides, e.g. cecropins (16), magainins (17) and melittin (18). To date, pardaxins have been isolated and characterized from the western Pacific Peacock sole, P. pavoninus (P1, P2 and P3) (7), and the Red Sea Moses sole, P. marmoratus (P4) (5,9) (Figure 1b). All know pardaxins contain one single peptide chain composed of 33 amino acids and show a high tendency of aggregation in aqueous solution (5,9). Their sequences are very homologous, differing only at positions 5, 14, or 31 (Figure 1b). The membranal effects of pardaxins are rather sensitive to structural variations such as the net charge, attached side chains, the stereochemistry of α-carbons and chain truncation (19,13,20).

Recently we reported the isolation of a novel pardaxin isoform (P5) from P. marmoratus (49). The primary structure of the native peptide has been determined by amino acid sequencing, endoproteinase cleavage and mass spectrometric techniques. The exchange of Gly31 of Pardaxin P4 by Asp (Figure 1b) is the major difference between P4 and P5. The sequence of the novel pardaxin P5 differs by one

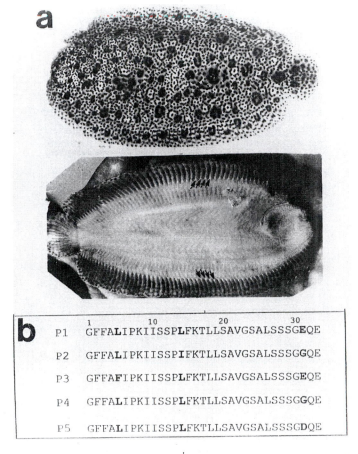

	1 10 20 30
P1	GFFALIPKIISSPLFKTLLSAVGSALSSSGEQE
P2	GFFALIPKIISSPIFKTLLSAVGSALSSSGGQE
P3	GFFAFIPKIISSPLFKTLLSAVGSALSSSGEQE
P4	GFFALIPKIISSPLFKTLLSAVGSALSSSGGQE
P5	GFFALIPKIISSPLFKTLLSAVGSALSSSGDQE

Figure 1a. Lateral views of the *Pardachirus marmoratus* fish; Arrows indicate the location of the toxic glands, at the basis of the fins.

Figure 1b. Amino acid alignment of pardaxins. P1, P2 and P3 are from *P. pavoninus*; P4 and P5 are from *P. marmoratus*. Variable residues are bold-faced.

point mutation from the pardaxin P1 isolated from *P. pavoninus* (Figure 1b). The change Asp31→Glu31 would require a mutation of the third base of the Asp31 codon. In addition, a mutation C→U, the first base of the Leu5 codon of P1, would lead to the sequence of P3, another pardaxin found in *P. pavonius*. In contrast, no single point mutation of either P1, P3 or P5 would lead to P2. The pardaxin P2 is genetically closest to P5, differing in position 31 (Asp→Gly:C→A exchange of the first base of the Leu14 codon). We suggest that the pardaxin isoform P4 may provide the mutation gap between P5 and P2. It would be one point mutation away from P5 (Asp31→Gly31) and another point mutation away from P2 (Leu14→Ile14). However, the investigation of this evolutionary structure-function relationship of paradaxins requires further evaluation using other *Pardachirus* species.

PC12 Pheochromocytoma Cells: A Neuronal Model to Study Secretion

Over the past two decades, the clonal PC12 cell line has become a widely used preparation for various model studies on neurons and adrenal gland chromaffin cells (*26*). Regulated exocytosis of catecholamines has been largely investigated in bovine adrenal chromaffin cells (*27*) and in rat PC12 cells (*28*). The secretory vesicles of these cells, the chromaffin granules, similar to synaptic vesicles store dopamine, norepinephrine, ATP and various proteins. The essential role of calcium in catecholamine secretion from chromaffin cells has been well established (*27*). Cytosolic calcium ($[Ca]_i$) can be increased due to membrane depolarization, opening of receptor-operated channels or release of calcium from intracellular stores (*27*). The concentration of ($[Ca]_i$) is strictly regulated, and it is thought that the increase in calcium concentration within the microdomain of the active exocytotic zone allows vesicles to fuse and release their catecholamines content.

In contrast to calcium-dependent neurotransmitter release, the signal transduction pathways of calcium-independent neurotransmitter release (*29-31*) have received scant attention.

Characterization of Pardaxin-induced Dopamine Release: The Role of Calcium

The ability of Pardaxin versus to other secretagogues to induce dopamine release from PC12 cells in the presence or absence of extracellular calcium ($[Ca]_0$) is presented in Figure 2a. By subtracting the basal release, Pardaxin (6 μM) stimulated dopamine release by $26 \pm 2\%$ and $20 \pm 1\%$ of total content in the presence or absence of $[Ca]_0$, respectively. In calcium-containing medium, carbachol (10 μM) bradykinin (1 μM) and KCl (50 mM) stimulated dopamine release by $9.5 \pm 2\%$, $13 \pm 3\%$ and $15 \pm 2.5\%$ of total content, respectively. In the absence of $[Ca]_0$ these compounds did not induce dopamine release. The dose and time course dependence of Pardaxin-induced dopamine release were determined in calcium-containing medium. As shown in Figure 2b, Pardaxin caused dopamine release in a concentration and time dependent manner. The most pronounced effect was obtained at 10 μM. At this concentration, Pardaxin stimulated dopamine released by $37 \pm 3\%$, $51 \pm 2.5\%$ and $58 \pm 2.5\%$ of total content after 15, 30 and 60 min, respectively. These data indicate Pardaxin, in contrast to KCl, carbachol and bradykinin, induced dopamine release in

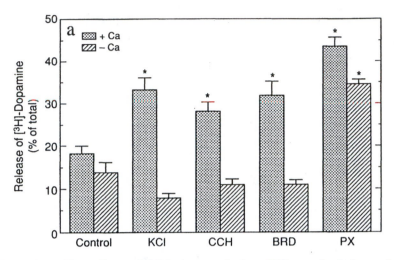

Figure 2a. The effect of $[Ca]_o$ on pardaxin-, KCl-, carbachol-, and bradykinin-induced dopamine release from PC12 cells. PC12 cells (5×10^5 cells/well) preloaded with ^{3}H-dopamine were incubated for 20 min at 37°C in the presence (+ Ca) or absence (- Ca + 1 mM EGTA) of $[Ca]_o$ with KCl (50 mM), carbachol (CCH, 10 µM), bradykinin (BRD, 1 µM), Pardaxin (PX, 6 µM) or left untreated (control).

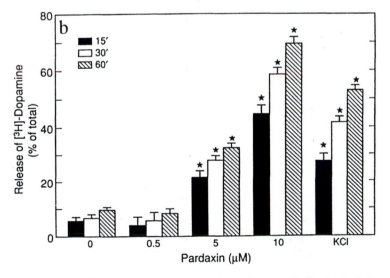

Figure 2b. Time-course and dose-dependence of Pardaxin-induced dopamine release from PC12 cells. PC12 cells (5×10^5 cells/well) preloaded with ^{3}H-dopamine were incubated at 37°C with 50 mM KCl or various concentrations of Pardaxin for the indicated times in the presence of $[Ca]_o$.

the absence of [Ca]$_o$ (Figure 2a), similarly to Pardaxin-stimulated calcium-independent catecholamine release from bovine adrenal medullary chromaffin cells (31).

To clarify the effect of Pardaxin on intracellular calcium, acetoxymethyl ester of fura 2 (Fura 2-AM) loaded PC12 cells were treated with Pardaxin in the presence or absence of extracellular calcium [Ca]$_0$ (Figure 3a). Removal of extracellular calcium reduced the baseline level of intracellular calcium [Ca]$_i$ (Figure 3a,b), as previously described (32). After the addition of Pardaxin in calcium-containing medium, there was a gradual and sustained increase in cytosolic calcium (Figure 3a, +Ca). In the absence of extracellular calcium, Pardaxin did not augment [Ca]$_i$ (Figure 3A,-Ca). Since Pardaxin did not increase [Ca]$_i$ in calcium-free medium we can conclude that Pardaxin did not release calcium from intracellular stores. In order to examine the possibility that the intracellular calcium stores were depleted under these conditions (absence of extracellular calcium and presence of 1mM ethyleneglycol-bis-(β-amino-ethyl ether)N,N'-tetra acetic acid (EGTA)), experiments with thapsigargin and KCl were performed in the presence or absence of extracellular calcium (Figure 3b). Thapsigargin, an inhibitor of the endoplasmic reticulum pump, is a compound known to release calcium from intracellular stores (33). KCl causes an influx of calcium through depolarization-induced activation of voltage-sensitive calcium channels (34). As shown in Figure 3b treatment of the cells in calcium containing buffer with 30 mM KCl, resulted in a rapid increase in [Ca]$_i$, which plateaued after 1 min and then gradually declined. In calcium-free medium KCl did not lead to an elevation in [Ca]$_i$. Stimulation of PC12 cells with thapsigargin (1 μM) caused a sustained rise in [Ca]$_i$, the magnitude of which is similar to KCl-induced signal, while the onset is slightly delayed (Figure 3b). In calcium-free medium, thapsigargin yielded a transient elevation of [Ca]$_i$ peaked after 1 min and then rapidly returned to basal level, confirming that calcium redistribution from intracellular stores is maintained. The influx of calcium elicited by Pardaxin in the presence of extracellular calcium (Figure 3a, +Ca) should enhance dopamine release. To test this possibility, we determined the effect of 1,2-bis(2-aminophenoxy) ethane N,N,N',N'-tetra-acetic acid (BAPTA-AM), a membrane permeant chelator of cytosolic calcium. In the presence of [Ca]$_0$, BAPTA (10 μM) inhibited by 63% KCl-induced dopamine release and by 40% Pardaxin-induced dopamine release. However, Pardaxin-induced dopamine release in the absence of [Ca]$_0$ was not affected by BAPTA treatment (Figure 3c). These data suggest that calcium influx is partially involved in Pardaxin-induced dopamine release and that mobilization of calcium from intracellular stores is not the mechanism by which Pardaxin stimulate dopamine release (Abu-Raya, S. et al. *J. Pharm. Exp. Therap.*, in press).

Pardaxin Stimulation of the Arachidonic Acid Cascade in PC12 Cells

As the triggering of neurotransmitter release induced by Pardaxin could be due to phospholipases stimulation, we explored this possibility by treating PC12 cultures labeled with arachidonic acid ([³H]AA) with several concentrations of Pardaxin, in the presence or absence of extracellular calcium (35). In the presence of extracellular calcium, 1 μM and 10 μM Pardaxin stimulated AA release 2.3 and 10-fold, versus

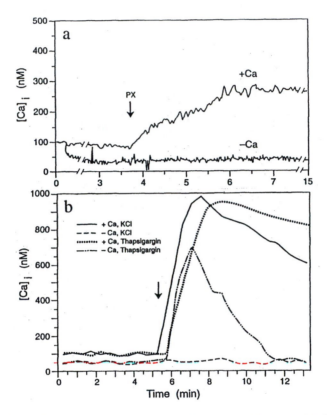

Figure 3a,b. The effect of Pardaxin, thapsigargin, and KCl on [Ca]$_i$ in the presence or absence of [Ca]$_o$. PC12 cells were loaded for 45 min with 5 μM Fura 2-AM and the fluorescence experiments were carried out at room temperature using a concentration of about 10^6 cells/ml in the presence of 2 mM CaCl$_2$ (+Ca), or calcium-free buffer supplemented with 1 mM EGTA (-Ca). Pardaxin, (PX, 3a), thapsigargin and KCl (3b) were added at the concentrations 6 μM, and 30 mM, respectively. Arrow indicates the time addition of the compound.

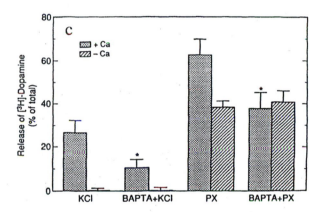

Figure 3c. The effect of BAPTA-AM on Pardaxin and KCl-induced dopamine release from PC12 cells in the presence or absence of [Ca]$_o$. PC12 cells (5×10^5 cells/well), preloaded with ^{3}H-dopamine, were incubated for 1hr at 37°C with BAPTA-AM (10 μM). the cells were then treated with KCl (50 mM) or Pardaxin (PX, 6 μM) for an additional 20 min.

the control, respectively (Figure 4c). In the absence of extracellular calcium, 1 μM and 10 μM Pardaxin stimulated AA release 1.6 and 7.1-fold, versus the control, respectively (Figure 4c). Stimulation of AA release by pardaxin (5 μM) was detected after 5 min of incubation, whereas maximal stimulation was measured after 30 min of incubation. The same time course of AA release was observed in the presence or absence of extracellular calcium. The effect of Pardaxin on prostaglandin E_2 (PGE_2) (an arachidonic acid metabolite produced by the cyclooxygenase pathway) release in the presence or absence of extracellular calcium is presented in Figure 4d. Treatment of the cultured with 5 μM and 15 μM Pardaxin for 30 min in calcium-containing medium, stimulated PGE_2 release by 5- and 13-fold over that of the control, respectively). In the absence of extracellular calcium, PGE_2 release in the presence of Pardaxin (5-15 μM) was between 50-70% of that obtained in the presence of extracellular calcium (Figure 4d). Similar to native Pardaxin, synthetic Pardaxin (5 μM) stimulated PGE_2 release about 3- fold over that of the control. To verify the selective effect of Pardaxin on PGE_2 release, three Pardaxin structural analogues, without helical structure (20) were tested and proved to be ineffective in stimulating PGE_2 release. (Abu-Raya, S. et al., *J. Pharm. Exp. Therap.*, in press.)

To verify that AA release induced by Pardaxin is mediated mainly by phospholipase A_2 (PLA_2) activation, the effect of Pardaxin on ^{32}P-phospholipid content of PC12 cells was examined. As shown in Figure 4a,b the level of the major phospholipid species phosphatidylethanolamine (PE), phosphatidylinositol (PI), and phosphatidylserine (PS), was not changed after Pardaxin treatment. As Pardaxin did not affect PI level this data indicate that phospholipase C activation is not involved in Pardaxin-induced arachidonic acid cascade. In addition no formation of phosphatidic acid (PA) was observed. Since Pardaxin did not lead to PA generation, it appears that phospholipase D (PLD) is also not activated and/or involved in Pardaxin-induced AA release. However, a reduction by about 30% in phosphatidylcholine (PC) level was observed after Pardaxin treatment which was accompanied by about 40% increase in lysophosphatidylcholine (LPC) formation (Figures 4a,b) indicating PLA_2 activation. As Pardaxin affect PC level, but not PS or PE, these data suggest that PC is one of the major phospholipid sources utilized by Pardaxin for AA mobilization in PC12 cells.

Characterization of Pardaxin-induced Dopamine Release: The Role of Arachidonic Acid and Eicosanoids

Since AA and eicosanoids may act as intracellular second messengers and affect synaptic transmission (35), we examined the effect of inhibitors for the AA cascade in PC12 cells on Pardaxin-induced dopamine release. In PC12 cells treated with indomethacin (a cyclooxygenase inhibitor) there was an increase by about 10-20% in dopamine release in response to 1 μM and 5 μM Pardaxin in the presence, or absence of $[Ca]_0$. At 10 μM Pardaxin, the indomethacin effect was not significant. These results suggest that the cyclooxygenase pathway is not involved in Pardaxin-induced dopamine release.

To determine the involvement of the lipoxygenase pathway in Pardaxin-induced dopamine release, several lipoxygenase inhibitors were tested. As

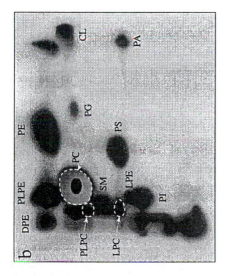

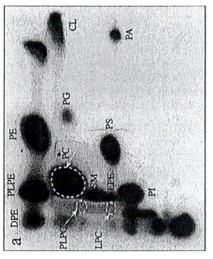

Figure 4a,b. Thin layer chromatography analysis (TLC) of [^{32}P] orthophosphate incorporated into the major phospholipids in PC12 cells before (a) or after (b) Pardaxin treatment. PC12 cells (2 x 10^6 cells/dish) were labelled with [^{32}P] orthophosphate, washed and treated with 5 μM Pardaxin for 15 min at 37°C. The phospholipids were extracted with n-propanol and separated by TLC. The plates were exposed for autoradiography and scanned. PE - phosphatidylethanolamine; PS - phosphatidylserine; PC - phosphatidylcholine; CL - cholesterol; PI - phosphatidylinositol; LPC - Lysophosphatidylcholine; PLPC - Phospholysophosphatidylcholine; PA - phosphatidic acid.

c

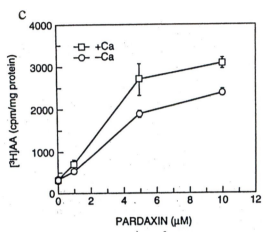

Figure 4c. Dose-dependent stimulation of [³H]AA released from PC12 cells by Pardaxin in the presence (+Ca) or absence (-Ca) of extracellular calcium (+ 1 mM EGTA).

d

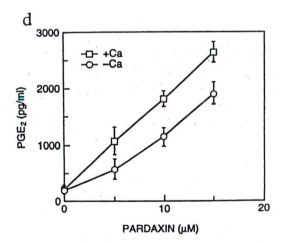

Figure 4d. Dose-dependent stimulation of PGE$_2$ release from PC12 cells by Pardaxin in the presence (+Ca) or absence (-Ca) of extracellular calcium(+1 mM EGTA).

shown in Table 1, esculetin (20 µM), 2-(12-hydroxydodeca-5,10-diynyl) -3,5,6 -trimethyl-1,4-benzoquinone (AA861) (10 µM) and nordihydroguaiaretic acid (NDGA) (100 nM) inhibited Pardaxin-induced dopamine release by about 50% in calcium-containing medium. A parallel inhibition of Pardaxin-induced 5-hydroxyeicosatetranoic acid (5-HETE) release by these compounds was also observed (Table 1). At 1 µM and 5 µM, NDGA further inhibited (by about 85%) Pardaxin-induced dopamine release, which was accompanied by a marked reduction (by about 90-95%) in 5-HETE release (Table 1). In the absence of extracellular calcium, NDGA (1 µM) and AA861 (10 µM) inhibited Pardaxin-induced dopamine release by about 90% (data not shown).

To further confirm the involvement of arachidonic acid cascade in Pardaxin-induced dopamine release we tested the effect of two PLA_2 inhibitors. 4-bromophenacyl bromide (BPB) (30 µM) and Mepacrine (50 µM) inhibited by 47% and 60% respectively the Pardaxin-induced dopamine release in the presence of $[Ca]_0$, and by 73% and 87%, in the absence of $[Ca]_0$, respectively. In parallel cultures, BPB and mepacrine inhibited by 75-90% Pardaxin-induced AA release independently of $[Ca]_0$ (Table 2). These results suggest the involvement of PLA_2 activation in neurotransmitter release by Pardaxin.

In other studies with PC12 cells, it was suggested that PLA_2 is involved in exocytosis (36,37). Therefore, it is very tempting to assume that AA or derived eicosanoid mediates Pardaxin-induced dopamine release. The lipoxygenase inhibitors NDGA, esculetin (38) and AA861 (39) strongly inhibited both Pardaxin-induced dopamine release and 5-HETE release (Table 1). These results indicate that the lipoxygenase products may be involved in Pardaxin-induced dopamine release. The involvement of the lipoxygenase pathway in hormone and neurotransmitter release has been reported in several studies (38,40,41). However, in these and other investigations the involvement of the lipoxygenase pathway was examined under physiological conditions (in calcium-containing medium). In the present experiments, we presented evidence that the lipoxygenase pathway may be involved in dopamine release also in the absence of extracellular calcium.

While the role of calcium in catecholamine release has been widely investigated, the intracellular signaling pathways whereby eicosanoids activate release in the absence of an increase in $[Ca]_i$ are unknown. It would be interesting to know whether AA metabolites act as intracellular second messengers or in an autocrine fashion. The possibility that the generation of lipid products such as eicosanoids directly regulate the synaptic vesicles fusion process, independent of calcium, also merits careful examination. Pardaxin can provide a new pharmacological tool for clarifying these issues.

Detection of JNK, p38 and MAPK Enzyme Activation Using Anti-Dual-Phosphopeptide Antibodies

Signaling in eukaryotic cells typically originates with the introduction of a particular stimulus into the extracellular space where it can activate, either directly or indirectly, a specific receptor. These stimuli can be quite diverse in nature, comprising several distinct classes of biological molecules including hormones,

Table I.

The effect of lipoxygenase inhibitors on Pardaxin-induced release of 5-HETE and [³H]-dopamine

Inhibitor	5-HETE release[a]	[³H] dopamine release[b]
	% of Pardaxin alone	
Esculetin (20 μM)	48 ± 7*	51 ± 6*
AA861 (10 μM)	42 ± 2*	52 ± 5*
NDGA (100 nM)	43 ± 6*	54 ± 3*
NDGA (1 μM)	10 ± 4**	17 ± 2**
NDGA (5 μM)	5 ± 2**	15 ± 3**

[a]   PC12 cells were preincubated with the lipoxygenase inhibitor in calcium-containing medium at the indicated concentration. The cells were further incubated at 37°C with Pardaxin (5 μM) for 15 min. The values presented are mean ± S.E.M.*p<0.01

[b] Cells prelabelled with [³H]-dopamine were incubated with the lipoxygenase inhibitor in calcium-containing medium at the indicated concentration. Then the cells were further incubated at 37°C with pardaxin (5 μM) for 15 min. *p<0.01

Table II.

The effect of PLA₂ inhibitors on Pardaxin-induced release of AA and dopamine in the presence or absence of extracellular calcium

Compound	[³H]-AA release (% of control)		[³H]-dopamine release (% of control)	
	+Ca	-Ca	+Ca	-Ca
PX (control	100	100	100	100
PX+Mepacrine	10 ± 4*	11 ± 2*	40 ± 7*	13 ± 2*
PX+BPB	26 ± 6*	25 ± 6*	53 ± 9*	27 ± 3*

PC12 cultures were labeled with [³H]-dopamine or [³H]-AA. After washing, the cultures were treated for 30 min with mepacrine (50 μM), 4-bromophenacyl bromide (BPB, 30 μM) or left untreated, in the presence of calcium (+Ca), or in the absence of calcium and 1 mM EGTA (-Ca). The cultures were then incubated with 5 μM Pardaxin (PX) for 15 min at 37°C. The values presented are mean ± S.E.M.*p<0.01

growth factors, cytokines, toxins, osmotic stimuli as well as ultraviolet light. Differences in the nature and consequences of signaling initiated by such diverse agents and ligands results from the differential modulation of multiple signaling elements in distinct combinations. Indeed, all known effectors of signaling cascades can be regulated by more than one stimulus. Moreover, overlapping pathways can be activated by different stimuli, further contributing to the regulation of signaling specificity.

One group of enzymes under intense investigation as part of the effort to understand signaling cascades that are activated by a number of extracellular stimuli is the extracellular signal-regulated protein kinase (ERK), also referred to as the mitogen-activated protein kinase (MAPK), superfamily of enzymes. This large group of protein kinases contains over a dozen members that participate in many eukaryotic regulatory pathways (42). These enzymes comprise at least three parallel, yet interwoven, signal transduction cascades that are differentially regulated in response to mitogens, growth factors, cytokines and various forms of stress (Figure 5a). Each cascade consists of a minimum of three enzymes activated in series: a MAP kinase/ERK kinase or MEKK (a MEK activator), a MAP kinase/ERK kinase or MEK (a MAP kinase activator), and a MAP kinase/ERK homologue (43).

The first and best studied enzymes of the ERK/MAPK superfamily reside in the classical mitogen-activated (MAPK) subfamily pathway (Figure 5a). This cascade is composed of MEKK isoforms (Raf-1, B-Raf, A-Raf), MEK1 or MEK2, and ERK1 or ERK2 (43). The JNK and p38 enzymes comprise two additional subfamilies, whose members are potently activated by a variety of stress-related stimuli. The JNK (c-Jun N-terminal protein kinase) enzymes, also referred to as stress-activated protein kinases (SAPKs), are activated by a variety of stress-related stimuli (e.g., heat shock, osmotic imbalance, endotoxin, UV and protein synthesis inhibits) and cytokines (44). The p38 kinase, also referred to as High Osmolarity Glycerol response kinase (HOG), is activated by a variety of stress-related stimuli, as well as cytokines and insulin (45). p38 and the p38-related kinases (p38α, p38β, p38γ, also referred to as p38, p38-2 and ERK6, respectively, and p38δ) represent a rapidly expanding group of enzymes that are being targeted for potential treatment of a variety of disease states (42). The MAP kinases are unusual in that MEK-mediated phosphorylation of both a tyrosine and a threonine residue, in the Thr-X-Tyr motif (Thr-Glu-Tyr for MAPK, Thr-Pro-Tyr for JNK, and Thr-Gly-Tyr for p38) within the "phosphorylation cleft" of each enzyme is required to stimulate their activities (46).

The need to study coordinated signaling of the ERK/MAPK superfamily cascades in response to a variety of extracellular stimuli has become increasingly critical to our understanding of both fundamental cell biology and disease, as well as in the pursuit of new drugs and toxins with increased therapeutic value. Recent advances in synthesis of phosphopeptides and the ability to produce highly selective and high-affinity polyclonal antibodies to phosphorylated enzymes is a significant breakthrough in the study of complex signal transduction cascades. The Anti-ACTIVE series of rabbit polyclonal antibodies that preferentially recognize the dually phosphorylated active form of the MAPK, JNK and p38 enzymes are highly effective in detecting enzyme activation in Western blotting and immunocytochemistry using PC12 cells. In PC12 cells we have studied coordinated

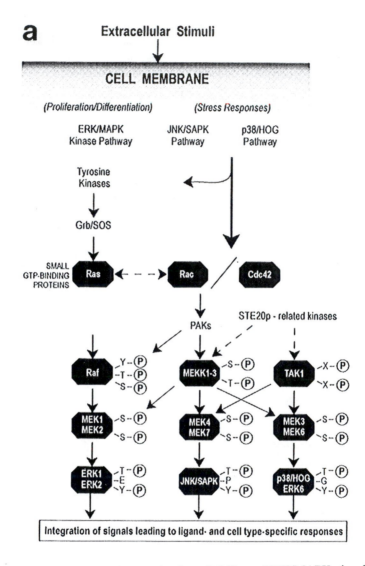

Figure 5a. A scheme of the activation of different ERK/MAPK signaling pathways by different extracellular stimuli.

37

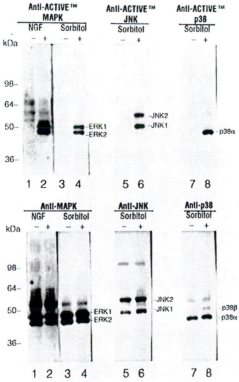

b Western Detection of Coordinated Activation of
MAPK, JNK and p38 Enzymes

Figure 5b. Detection of activated JNK, p38 and MAPK in PC12 cell extracts by Western blotting using Anti-ACTIVE JNK, Anti-ACTIVE p38 and Anti-ACTIVE MAPK polyclonal antibodies. Cells were either untreated, treated with 50 ng/ml of nerve growth factor (NGF) for 5 minutes, or 0.5 M sorbitol for 5 minutes, as indicated. Aliquots of each extract were analyzed by SDS-PAGE (10% gel, under reducing conditions) and transferred to a nitrocellulose membrane. The membranes were probed with either the indicated Anti-ACTIVE antibody (upper part) or with corresponding antibodies that recognize both active and inactive forms of each subfamily of kinases (lower part). Lanes: lanes 1, 3, 5 and 7, 2 μg of unstimulated PC12 cell extract; lanes 2, 4, 6 and 8, 2 μg of stimulated PC12 cell extract.

activation of MAPK, JNK and p38 enzymes by a variety of mitogenic and stress-related stimuli. To begin to characterize the events leading to activation of each ERK/MAPK cascade, we chose stimuli known to potently activate one or more of these signaling cascades (Figure 5b).

Western blotting analysis using the anti-active MAPK antibody clearly illustrated that the MAPK subfamily, ERK1 and ERK2, enzymes were strongly activated by nerve growth factor (NGF) (lane 2) and to a lesser degree (~30%) by osmotic shock using sorbitol (lane 4). In contrast, the JNK1 (46kDa) and JNK2 (54kDa) isoforms were strongly activated by sorbitol (lanes 6) and NGF-withdrawal (data not shown) and, to a lesser degree (~50%), by anisomycin (data not shown). Finally, p38 was potently activated by both sorbitol (lane 8) and anisomycin (data not shown), but not by NGF-withdrawal (data not shown). As an additional control, blots probed with antibodies that recognize total enzyme (active and basal) for each subfamily member (Figure, 5b lower part) resulted in similar signals with both untreated and treated extracts. This illustrates that the differences in signals observed with the corresponding Anti-ACTIVE antibodies were not simply due to differences in the levels of each enzyme. The data demonstrate the utility of these reagents to study the differential activation of endogenous isoforms of the MAPK, JNK and p38 in PC12 cells under different physiological or pathological conditions.

Pardaxin Stimulation of MAPK, JNK and p38 Kinase Pathways in PC12 Cells

To test a possible activation of ERK/MAPK superfamily cascades, PC12 cells were incubated with Nerve Growth Factor (NGF), H_2O_2 and Pardaxin for the given times. The cells were lysed and the resulting lysates were subjected to an immunoblot analysis using anti-phosphorylated (active) MAPK (Figure 6a,b), p38 (Figure 6f) on JNK (Figure 6g) antibodies. Figure 6a shows that the phosphorylated MAPK was detectable within at least 30 min. and disappeared within 60 min., after addition of Pardaxin. The total amounts of protein, detected by using anti- (non phosphorylated) MAPK antibody, did not change with different incubation times (data not shown). Pardaxin stimulation of MAPK phosphorylation activity (both ERK1 and ERK2) is a specific effect since pretreatment of PC12 cells with 2-(2-amino-3-methoxyphenyl-4H-1-benzopyran-4-one (PD 98059, a selective MEK1 inhibitor) completely blocks Pardaxin stimulation of MAPK (Figure 6b). We also observed that the phosphorylated MAPK translocates to the nucleus in PC12 cells treated for 20 min. with Pardaxin (Figure 6c,e). The rapid stimulation of MAPK in PC12 cells by Pardaxin is correlated to the time-course of Pardaxin-induced arachidonic acid and dopamine release. On the other hand, activation of stress-kinases p38 and JNK, by Pardaxin, occurs with a much longer kinetics (Figure 6f,g). This finding and the observation that Pardaxin stimulates p38 and JNK activation to an extent similar to that of anisomycin, H_2O_2, palytoxin, hyperosmotic medium, etc., support a central role of these kinases in pardaxin-induced cytotoxicity. Since it has been demonstrated that certain phospholipases A_2 induce activation of MAPK (*47*), and phospholipase A_2 is activated by MAPK phosphorylation (*48*), it is very tempting to propose the requirement of the MAPK phosphorylation activity in the Pardaxin-induced arachidonic acid and dopamine

Anti - ACTIVE MAPK

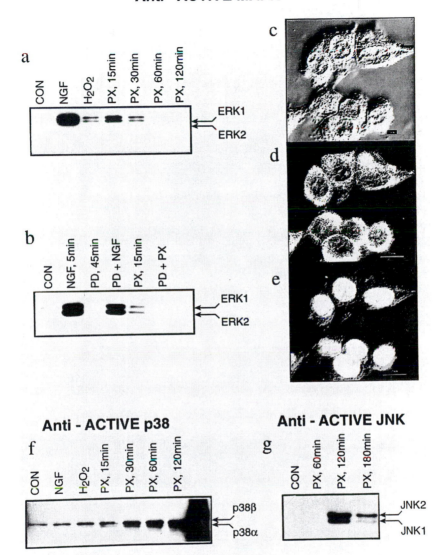

Figure 6. Kinetics of activation of JNK (g), p38 (f) and MAPK (a,b) enzymes and translocation of the phosphorylated MAPK to the nuclei (c,d,e) in PC12 cells after Pardaxin treatment. Effect of 50 µM PD 98059 (a selective MEK inhibitor) on pardaxin-activation of MAPK (b). CON - control, untreated cells; NGF - cultures treated for 5 min with 50 ng/ml nerve growth factor; PX - Pardaxin (5 µM); H_2O_2 - the cultures were treated with 1 µM hydrogen peroxide for 5 min prior lysis. c-e indicates light micrograph of cells untreated (c,d) or treated for 20 min with 5 µM Pardaxin (e), fixed, permeabilized and stained with anti-ACTIVE MAPK antibody.

release from PC12 cells. This aspect is now the major focus of research in our laboratory.

Conclusions

Pardaxins are marine, polypeptide neurotoxins which induce massive release of neurotransmitters from neurons. Recently we reported the isolation of a novel Pardaxin isoform from *Pardachirus marmoratus* which may be useful in studies of structure-function relationships. Using PC12 pheochromocytoma cells as a neuronal model to study secretion, we presented evidence that Pardaxin stimulated dopamine release independent of extracellular calcium. Our results also indicated that Pardaxin stimulated release of arachidonic acid and eicosanoids, independently of calcium. Pardaxin stimulated with different kinetics the extracellular signal-regulated protein kinase (ERK) superfamily of enzymes. We propose that Pardaxin activation of MAPK is involved in Pardaxin stimulation of arachidonic acid and dopamine release while the activation of stress-kinases JNK and p38 is mediating Pardaxin-induced cytotoxicity. Elucidation of the neuronal signal transduction pathways affected by Pardaxin might provide new understandings in synaptic transmission and new targets for therapy of patients affected by neurotoxins.

Acknowledgments

This study was supported in part by David R. Bloom Center for Pharmacy at the Hebrew University, Jerusalem, Israel. Dr. Lazarovici, P. is affiliated with the David R. Bloom Center for Pharmacy at the Hebrew University.

Literature Cited

1. *Toxins and Signal Transduction*. Gutman, Y.; Lazarovici, P. Eds.; Cellular and Molecular Mechanisms of Toxin Action; Harwood Academic Publishers: Amsterdam, The Netherlands, 1998; Vol.1.
2. Jessell, T. M.; Kandel, E.R. *Cell 72 / Neuron*. **1993**, *10*(Suppl.), 1-30.
3. Augustine, G.J.; Charlton, M.P.; Smith S.J. *Annu. Rev. Neurosci.* **1987**, *10*, 633-693.
4. *Neurotoxins in Neurochemistry*. Dolly, J. O., Ed.; Ellis Harwood Ltd.: Chichester, UK, 1998.
5. Lazarovici, P.; Primor, N.; Loew, L. M. *J. Biol. Chem.* **1986**, *261*, 16704-16713.
6. Tachibana, K.; Sakaitanai, M.; Nakanishi, K. *Science.* **1984**, *226*, 703-705.
7. Thompson, S. A.; Tachibana, K.; Nakanishi, K.; Kubota, I. *Science.* **1986**, *233*, 341-343.
8. Bloch-Shilderman, E.; Abu-Raya, S.; Lazarovici, P. In *Toxins and Signal Transduction*; Gutman, Y., Lazarovici P., Eds.; Harwood Academic Publishers: Amsterdam, The Netherlands, 1998; pp. 211-232.
9. Shai, Y.; Fox, J.; Caratsch, C.; Shih, Y. L.; Edwards, C.; Lazarovici, P. *FEBS Lett.* **1988**, *242*, 161-166.

10. Loew, L. M.; Benson, L.; Lazarovici, P.; Rosenberg, I. *Biochemistry.* **1985**, *24*, 2101-2104.

11. Moran, A.; Korchak, Z.; Moran, N.; Primor, N. In *Toxins, Drugs and Pollutants in Marine Animals*; Bolis, L.; Zadunaisky, J.; Gilles, R., Eds.; Springer: Berlin, 1984, pp 13-25.

12. Lazarovici, P.; Edwards, C.; Raghunathan, G. *J. Natural Toxins.* **1992**, *1*, 1-15.

13. Shai, Y. *Toxicology.* **1994**, *87*, 109-129.

14. Shai, Y. *Trends Biochem. Sci.* **1995**, *20*, 460-464.

15. Zagorski, M. G.; Norman, D. G.; Barrow, C. J.; Iwashita, T.; Tachibana, K.; Patel, D.J. *Biochemistry.* **1991**, *30*, 8009-8017.

16. Boman, H. G.; Hultmark, D. *Annu. Rev. Microbiol.* **1987**, *41*, 103-126.

17. Zasloff, M. *Proc. Natl. Acad. Sci. USA.* **1987**, *84*, 5449-5453.

18. Hanke, W.; Methfessel, C.; Wilmsen, H.U.; Katz, E.; Jung, G.; Boheim, G. *Biochem. Biophys. Acta.* **1983**, *727*, 108-114.

19. Shai, Y.; Bach, D.; Yanovsky, A. *J. Biol. Chem.* **1990**, *265*, 20202-20209.

20. Oren, Z.; Shai, Y. *Eur. J. Biochem.* **1996**, *237*, 303-310.

21. Biemann, K. *Methods Enzymol.* **1990,** *193*, 455-479.

22. Fields, G. B.; Noble, R. C. *Int. J. Peptide Protein Res.* **1990**, *35*, 161-214.

23. Schnapp, B.J. *Neuron* **1997**, *18*, 523-526.

24. Bauerfield, R.; Huttner, W.B. *Curr. Opin. Cell Biol.* **1993**, *5*, 628-635.

25. Koenig, J.H.; Ikeda, K. *J. Cell Biol* **1996**, *135*, 797-808.

26. Greene, L. A.; Tischler, A. S. *Proc. Natl. Acad. Sci. USA.* **1976**, *73*, 2424-2428.

27. Burgoyne, R. D. *Biochem. Biophys. Acta.* **1991**, 1071, 174-202.

28. Ahnert-Hilger, G.; Bhakdi, S.; Gratzl, M. *J. Biol. Chem.* **1985**, *260*, 12730-12734.

29. Hochner, B.; Parnas, H.; Parnas, I.; *Nature.* **1989**, *342*; 433-435.

30. Knight, D. E.; Von-Grafenstein, H.; Athayde, C.M. *Trends Neurosci.* **1989**, *12*, 451-458.

31. Lazarovici, P.; Lelkes, P. I. *J. Pharmacol. Exp. Ther.* **1992**, *263*, 1317-1326.

32. Lelkes, P. I.; Pollard, H.B. *J. Biol. Chem.* **1987**, *262*, 15496-15505.

33. Berridge, M.J. *Biochem. J.* **1995**, *312*, 1-11.

34. DiVirgilio, F.; Milani, D.; Leon, A.; Meldolesi, J.; Pozzan, T. *J. Biol. Chem.* **1987**, *262*, 9189-9195.

35. Piomelli, D. *Crit. Rev. Neurobiol.* **1994**, *8*, 65-83.

36. Ray, P.; Berman, J. D.; Middleton, W.; Brendle, J. *J. Biol. Chem.* **1993**, *268*, 11057-11064.

37. Matsuzawa, A.; Murakami, M.; Atsumi, G.; Imai, K.; Prados, P.; Inoue, K.; Kudo, I. *Biochem. J.* **1996**, *318*, 701-709.

38. Barja-Fidalgo, C.; Guimaraes, J. A.; Carlini, C. R. *Toxicon.* **1991**, *29*, 453-459.

39. Harish, O. H.; Poo, M. M. *Neuron.* **1992**, *9*, 1201-1209.

40. Naor, Z.; Kiesel, L.; Vanderhoek, J. Y.; Catt, K. J. *J. Steroid Biochem.* **1985**, *23*, 711-717.

41. Metz, S. A. *Proc. Natl. Acad. Sci.* **1985**, *82*, 198-202.

42.   Robinson, M. J.; Cobb, M.H. *Curr. Opin. Cell Biol.* **1997**, *9*, 180-186.
43.   Cobb, M. H.; Goldsmith, E. J.; *J. Biol. Chem.* **1995**, *270*, 14843-14846.
44.   Gupta, S.; Barrett, T; Whitmarsh, A. J.; Caranagh, J.; Sluss H. K.: Derijard, B.; Davis, R. J. *EMBO J.* **1996**, *15*, 2760-2770.
45.   Molnar, A.; Theodoras, A. M.; Zon, L. I.; Kyriakis, J.M. *J. Biol. Chem.* **1997**, *272*, 13229-13235.
46.   Klokhlatchev, A.; Xu, S.; English, J.; Wu, P.; Schaefer, E.; Cobb, M. H. *J. Biol. Chem.* **1997**, *272*, 11057-11062.
47.   Kinoshita, E.; Handa, N.; Hanada, K.; Kajiyama, G.; Sugiyama, M. *FEBS Lett.* **1997**, *407*, 343-346.
48.   Lin, L.L.; Wartman, M.; Lin, A. Y.; Knopf, J. L.; Seth, A.; Davis, R. J. *Cell.* **1993**, *72*, 269-278.
49.   Adermann, K.; Raida, M.; Paul, Y.; Abu-Raya, S.; Bloch-Shildermann, E.; Lazarovici, P.; Hochman, J.; Wellhoner, H. *FEBS. Lett.* **1998**, *435*, 173-177.

Chapter 4

Anti-Aflatoxigenic Constituents of *Pistacia* and *Juglans* Species

Russell J. Molyneux, Noreen Mahoney, and Bruce C. Campbell

Western Regional Research Center, Agricultural Research Service, U.S. Department of Agriculture, 800 Buchanan Street, Albany, CA 94710

Tree nuts such as pistachios (*Pistacia vera*), almonds (*Prunus dulcis*) and walnuts (*Juglans regia*) are major crops in the state of California, with an aggregate annual value in excess of $1.4 billion. Under certain conditions these nuts may be infected by various strains of *Aspergillus flavus* and *A. parasiticus*, resulting in biosynthesis and accumulation of mycotoxins detrimental to quality and food safety. The carcinogenicity in animals of the primary metabolites, the aflatoxins, has resulted in establishment of tolerance levels. In the U. S. a maximum guidance level limit of 20 ng/g (20 ppb) for nuts intended for human consumption has been set by the Food and Drug Administration, while standards imposed by importing countries are 10 ng/g or less. Phytochemical constituents which are anti-aflatoxigenic have been isolated and identified from pistachios and walnuts, consisting of anacardic acids and naphthoquinones, respectively, and their possible mode of action is described.

Many crops are colonized under certain environmental conditions by the fungi *Aspergillus flavus* and *Aspergillus parasiticus* which are capable of producing the aflatoxins, a group of metabolites that are hepatotoxic and may be hepatocarcinogenic. As a consequence, the levels of aflatoxins in foods and feeds are highly regulated both in the United States and abroad. The occurrence of these mycotoxins in major crops such as corn, peanuts, cottonseed and tree nuts can therefore have a significant economic impact on these commodities for both domestic consumption and export markets. Considerable effort has been devoted to minimizing aflatoxin contamination by agronomic practices and post-harvest sorting but as the analytical techniques for detecting aflatoxins have become exceptionally sensitive the potential for control by such methods has become increasingly difficult to attain and economically impractical. Current research efforts by our group have focused upon aflatoxin elimination through control of vectors of infection, especially insects, investigation of potential biocompetitive organisms such as saprophytic yeasts, and in particular, suppression of *Aspergillus* infection and aflatoxigenesis by phytochemical resistance factors naturally present in tree nuts such as almonds, pistachios and walnuts.

Structure and Toxicity of Aflatoxins

Aflatoxins (1) are a structurally coherent group of acetate-derived fungal metabolites with a core unit consisting of a coumarin unit, fused to a bisfuran moiety, and bearing either a furanone (B series) or pyranolactone (G series) substituent (Figure 1). A double bond is present in the terminal furan ring of the major metabolites, aflatoxins B_1 and G_1; in the minor metabolites, aflatoxins B_2 and G_2, this double bond is saturated.

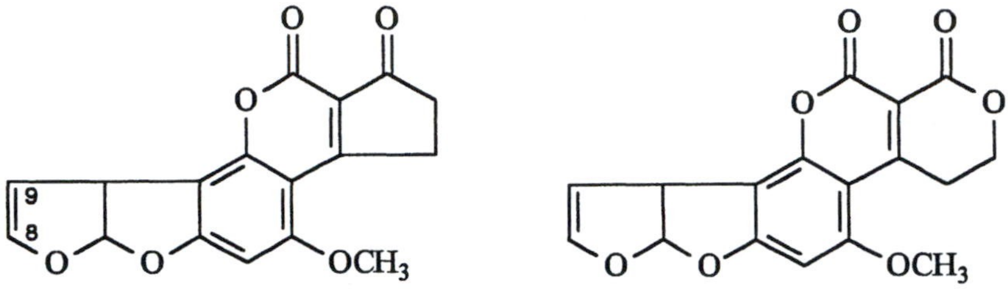

Aflatoxin B$_1$ **Aflatoxin G$_1$**

Aflatoxin B$_2$/G$_2$: 8,9 = dihydro-

Figure 1. Chemical structures of aflatoxin B_1 (AFB$_1$) and aflatoxin G_1 (AFG$_1$). The minor metabolites, aflatoxins B_2 and G_2, are 8,9-dihydro-derivatives.

The aflatoxins are known to be hepatotoxic and may also be hepatocarcinogenic but the toxicity is highly variable between animal species, probably as a consequence of metabolic transformations which lead to either deactivation, conjugation and/or elimination. Nevertheless, the primary mechanism of toxicity of these compounds at the cellular level is well established as cytochrome P450 epoxidation of the double bond at the 8,9-position of the furan ring, giving rise to potently electrophilic species which can alkylate nucleic acids and proteins.

The situation with regard to human toxicity is quite controversial. A recent analysis of epidemiological data (2) has demonstrated that a high risk of liver cancer, as in Africa and Asia, is associated with significant exposure to aflatoxins *via* moldy foodstuffs together with a high endemic rate of hepatitis B infection. In contrast there is a low risk of liver cancer in the U.S. and Europe and the occurrence of this disease correlates primarily with infection by hepatitis C, while exposure to aflatoxins is very limited. A risk assessment for the U.S. indicates that halving the current 20 ppb aflatoxin standard would reduce the cases of liver cancer by 0.001 per 100,000 individuals, *i.e.* a reduction of approximately one case for the entire population. On the basis of this data it appears that the toxicity of aflatoxins is comparable with other natural hepatotoxins and may well be exceeded by the pyrrolizidine alkaloids, which are not so stringently regulated. Nevertheless, the perception of the extreme hazards of aflatoxin has recently received support from news reports that Iraq has admitted filling ballistic warheads and bombs with botulinum, anthrax and aflatoxin (3,4). Even if the human toxicity of aflatoxins is accepted, it is difficult to rationalize use of these compounds as an offensive tactical weapon, because the debilitating effects of human exposure would not be immediate and incapacitating. Possibly these missiles were intended as strategic weapons designed to disperse aflatoxins as a means of economic warfare through contamination of food crops that would then be non-consumable or unexportable.

Economic and Trade Issues of Aflatoxins in Tree Nuts

Economic Value. Almonds, pistachios and walnuts are major crops in California, with an aggregate value which has increased yearly and recently approached $1.5 billion (Table I) (5). As a comparison, this total value is equivalent to or exceeds that of wine grapes, another major agricultural commodity in the State. Essentially 100% of the commercial production of tree nuts is in California and a large proportion is exported annually (6), primarily to Europe and Japan. In 1996 the value of such exports was $1,274 million (7), so that these three nut crops are very important contributors to the agricultural trade balance of the U.S.

Table I. Economic value ($ million) of tree nuts grown in California (1994-1997) and export percentage (1997).

	1994	1995	1996	1997	Exports (1997)
Almonds	965	881	1,009	1,080	55%
Walnuts	239	328	322	142	27%
Pistachios	119	161	122	203	32%
Total	**1,323**	**1,370**	**1,453**	**1,425**	

Aflatoxin Regulatory Levels. Aflatoxins in foods and feeds are regulated at different levels, depending upon the specific item and the country in which it is to be consumed. In the U.S., the Food and Drug Administration has only issued a guidance level for nuts (shells included) of 20 ppb ($20\mu g/kg$) total aflatoxins (8), whereas the Japanese regulations are 10 ppb for aflatoxin B_1. In practice, the level in tree nuts never exceeds 10 ppb, except in cases of serious infection usually caused by improper post-harvest handling.

In contrast, the European Community has proposed a regulatory level of 2 ppb aflatoxin B_1 and 4 ppb total aflatoxins for the edible portion of the nut, based upon incremental sample sizes of 300 grams ranging from 10-100 samples depending upon the weight of the shipment. This regulation will take effect on 1 January 1999 (9). It is likely that a significant proportion of shipments would be unable to conform to a threshold level this low, since a survey of all foodstuffs subject to *Aspergillus* infection produced in the U.S. gave a median level of 4 ppb, with a range of 0-30 ppb (2). It is therefore imperative that practical means be found to control or eliminate aflatoxins from such crops.

Natural Resistance of Tree Nuts to Aflatoxigenesis

One approach to the elimination of aflatoxins in tree nuts is to exploit factors which confer natural resistance to colonization by *Aspergillus* species. The existence of such factors is implicit in the variable susceptibility of different varieties of almonds and walnuts to infection and aflatoxin contamination.

Tree nuts can be considered to have four layers of protection against microbial infection, namely: (1) the hull, consisting of outer and inner layers (epicarp and mesocarp); (2) the shell (endocarp); (3) the seedcoat (pellicle); and, (4) the kernel. The latter two parts comprise the edible portion of the nut, although the seedcoat may be removed by blanching, especially in almonds. The shell is highly lignified and provides a protective barrier which is essentially physical in nature, whereas the hull, seedcoat and kernel are relatively soft and any protection is primarily a consequence of phytochemical constituents. Nevertheless, this combination of physical and chemical barriers does not always confer the protection which might be expected because insects, such as the navel orangeworm and codling moth, can breach them through feeding damage, thus providing avenues for infection.

Physical Barriers. Even when the shell is not penetrated by insect attack or physical damage during harvesting, kernels can become contaminated by aflatoxins. During normal development, the shell of pistachios dehisces from the hull and splits open about a month prior to harvest. However, a significant proportion of the crop does not undergo natural splitting and these closed-shell pistachios must be manually processed and the product marketed as kernels rather than the in-shell product favored by the consumer and the industry. Such processing is generally performed overseas by water-soaking to soften the shells prior to artificial splitting and any nuts sequestering *Aspergillus* spores have the potential to contaminate the batch during this process.

Experiments have shown that fresh, uninoculated closed-shell pistachios do not show any evidence of aflatoxin contamination when incubated at 30°C for up to 6 days. In contrast, rehydrated closed-shell pistachios exhibit levels exceeding the FDA guidance level of 20 ppb after 2-3 days and reach exceptionally high levels after 6 days incubation (10). As shown in Table II, pistachios which were rehydrated by soaking for 3 hours in a bath not inoculated with spores had variable levels of aflatoxin B_1, with one sample also showing the presence of high levels of aflatoxin G_1, indicating that contamination must have arisen from an endemic strain of *Aspergillus*, since this aflatoxin is not biosynthesized by our laboratory strain. As might be expected, pistachios similarly rehydrated in a water-bath inoculated with 100 spores/mL of *A. flavus* NRRL 25347 showed levels of aflatoxin which increased steadily in relation to incubation time, ultimately reaching a level of 87,500 ppb after 6 days.

Table II. Aflatoxin Content of Rehydrated Closed-Shell Pistachios.

Incubation Period (Days)	ppb Aflatoxin B_1		
	Fresh Uninoculated	Rehydrated Uninoculated	Rehydrated Inoculated
0	0	0	0
1	0	0	0
2	0	0	170
3	0	90	15,400
4	0	540 (also 350 ppb AFG_1)	32,700
5	0	160	57,100
6	0	390	87,500

Additional experiments indicated that penetration of the shell by the fungus, and subsequent contamination of the kernel, appears to occur through the attachment point of the stem. This has been shown to be a relatively porous structure and vulnerable to penetration by the stylet-like mouth parts of various heteropteran pests (11), although such damage was not present in the closed-shell pistachios used in this study.

Phytochemical Barriers. The failure of the shell as a physical barrier to prevent contamination of the kernel by aflatoxigenic *Aspergillus* strains renders the discovery of phytochemical resistance factors more significant. Plant constituents which confer resistance to infection can be classified as phytoalexins or phytoanticipins (12). **Phytoalexins** are **inducible** metabolites, formed *de novo* after infection, either by gene derepression or activation of latent enzyme systems. **Phytoanticipins** are **constitutive** metabolites, present *in situ*, either in the active form or easily generated from a precursor. Inducible phytochemicals exhibit a lag-time in their formation, during which period aflatoxin biosynthesis may have already commenced. Also, since their levels may be influenced by the nature of the pathogen and localized to the site of infection, they are inherently less likely to be predictably enhanced by breeding. It therefore appeared more promising to identify, and subsequently manipulate, levels of constitutive factors that are naturally present in various parts of the fruiting bodies of the tree nut species of concern.

Phenolic Lipids of Pistachio Hulls. Members of the plant family Anacardiaceae, to which the pistachio (*Pistacia vera*) belongs, are known to biosynthesize significant levels of phenolic lipids (13). These fall into two groups, cardols (alkyl and alkenyl derivatives of resorcinol) and the anacardic acids (analogous derivatives of salicylic acid) (Figure 2).

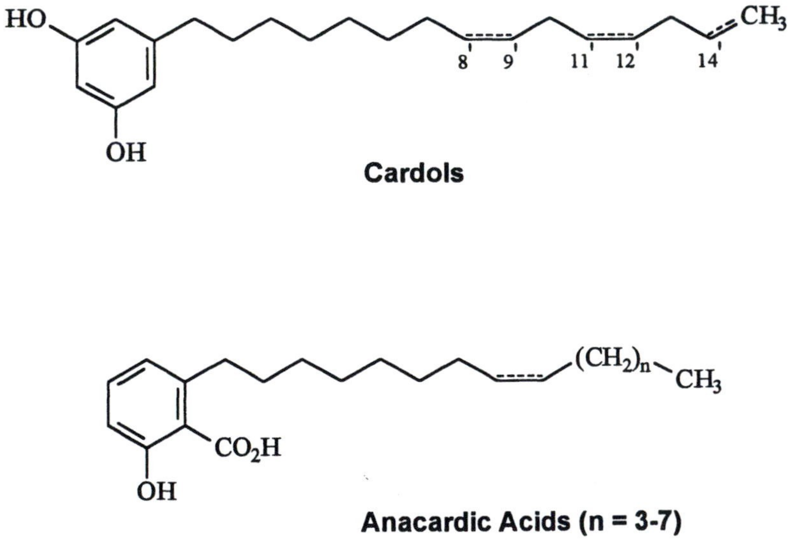

Cardols

Anacardic Acids (n = 3-7)

Figure 2. General Classes of Phenolic Lipid Constituents of the Anacardiaceae. Positions of potential unsaturation are indicated by dotted lines.

Ethyl acetate extraction of pistachio hulls, of the Kerman variety grown in California, gave a mixture of these phenolic derivatives comprising approximately 5% of the dry weight. Analysis of the mixture by gas chromatography - mass spectrometry (GC-MS) of the trimethylsilyl ether derivatives showed the presence of a single cardol, the Δ8,9 alkenyl isomer, together with five anacardic acids. The proportion of each of the individual anacardic acids was established from the GC-MS, with the saturated C_{13} isomer predominating at 51% of the total mixture (Figure 3).

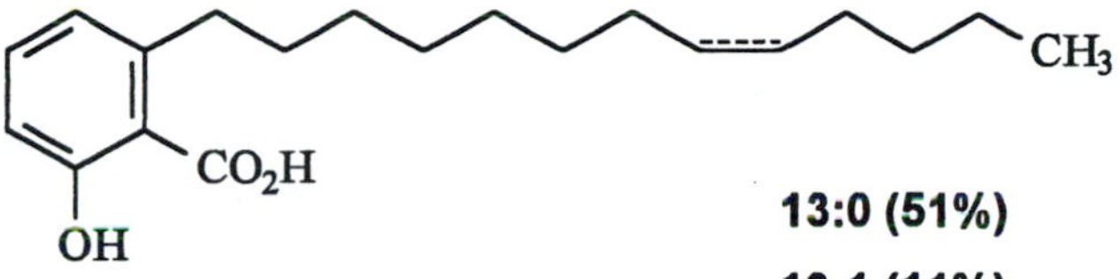

13:0 (51%)

13:1 (11%)

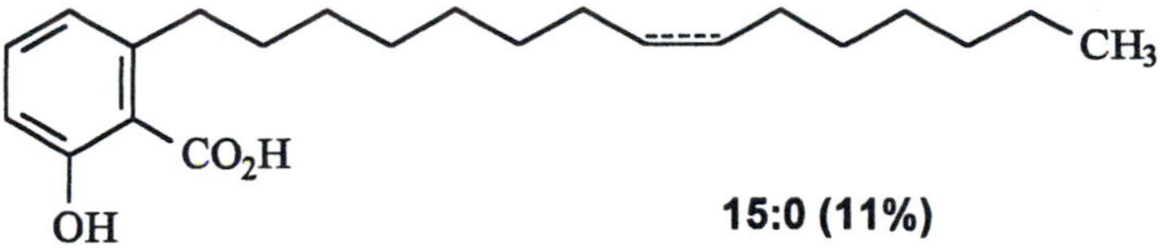

15:0 (11%)

15:1 (4%)

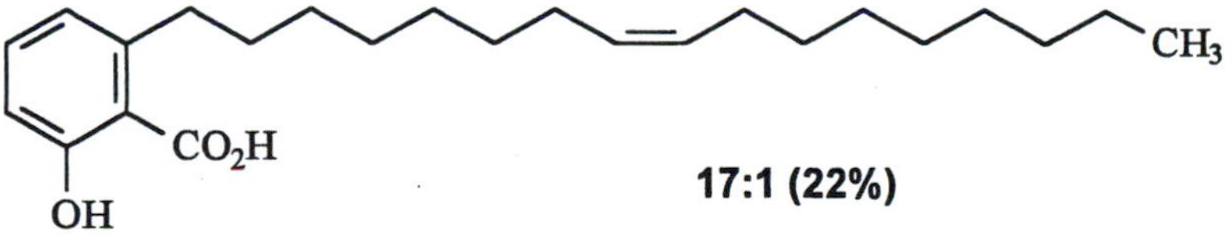

17:1 (22%)

Figure 3. Anacardic Acid Composition of Pistachio Hulls.

The cardol and anacardic acid fractions could be separated by suspension of the total extract in ether, in which the cardol was insoluble. Extraction of the ether-soluble fraction with aqueous sodium hydroxide, followed by acidification and re-extraction with ether, provided the purified anacardic acids. Evaluation of the effect of these two fractions *in vitro* on aflatoxin production by *A. flavus* established that cardol had no significant inhibitory effect. In contrast, the presence of 0.8% anacardic acids in the medium, a concentration approximately equivalent to that in fresh pistachio hulls, reduced aflatoxin production by 67%, relative to controls. Analysis of pistachio varieties or cultivars with higher anacardic acid levels may therefore indicate potential breeding stock with enhanced resistance to aflatoxin biosynthesis.

The mechanism of inhibition of aflatoxin production by these compounds is not established, but it is known that metal ions such as iron, copper and zinc are essential for aflatoxin biosynthesis. Moreover, anacardic acids have been shown to be ionophoric, forming lipophilic complexes with metal ions in the order $Fe^{2+} > Cu^{2+} > Zn^{2+} > Ni^{2+} = Co^{2+} = Mn^{2+}$ (14). These anacardic acid:metal derivatives may be either 1:1 or 2:1 complexes, with binding occurring through the -OH and $-CO_2H$ groups of the salicyclic acid moiety (Figure 4). It is conceivable that this complexation as a lipophilic entity reduces the availability of metal ions in the essentially aqueous environment required for fungal growth and aflatoxin production. Additional experiments with media in which the metal ion content is carefully controlled should establish whether or not this is the mechanism of inhibition.

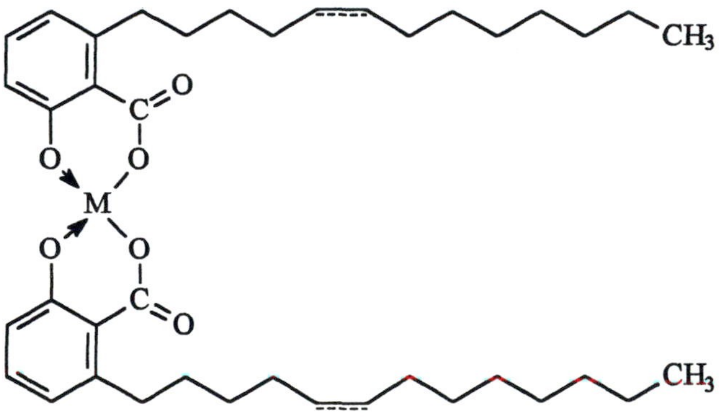

Figure 4. Structure of Lipophilic 2:1 Complexes of Anacardic Acids with Metal Ions.

Naphthoquinones of Walnut Hulls. Walnut hulls are known to contain a series of naphthoquinones (15). The best known of these, juglone (5-hydroxy-1,4-naphthoquinone), has been shown to be responsible for the allelopathic effect of walnut trees towards other plants growing in the vicinity. Juglone has also been isolated from leaves of pecan (*Carya illinoensis*) and found to inhibit mycelial growth of the fungus *Fusicladium effusum*, which is responsible for pecan scab (16).

Evaluation of the antibiotic activity of juglone towards *A. flavus* showed that at 100 ppm in agar, aflatoxin production was completely eliminated, although growth of the fungus was not inhibited. However, preliminary observations indicated that sporulation, which is necessary for initiation of aflatoxin biosynthesis, is significantly delayed. A more detailed evaluation of the inhibition of sporulation by juglone and three structural analogs which occur in walnut hulls showed distinct structure-activity relationships. Thus, 1,4-naphthoquinone required a concentration (170 ppm), approximately twice that of juglone, whereas 2-methyl-naphthoquinone and plumbagin (5-hydroxy-2-methyl-naphthoquinone) were active at 50 ppm or less (Figure 5).

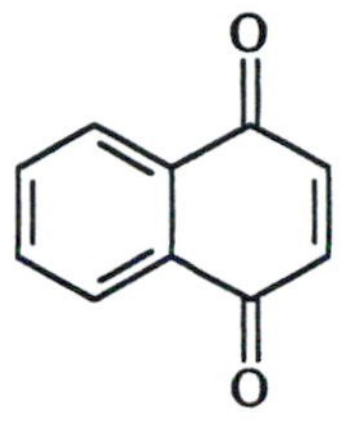

1,4-Naphthoquinone

(170 ppm)

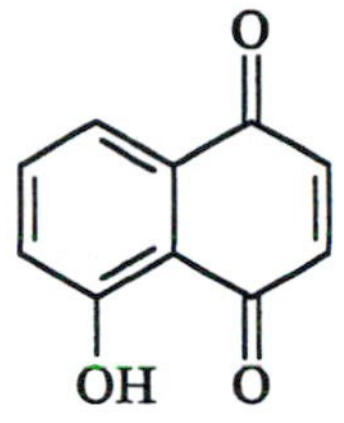

Juglone

(100 ppm)

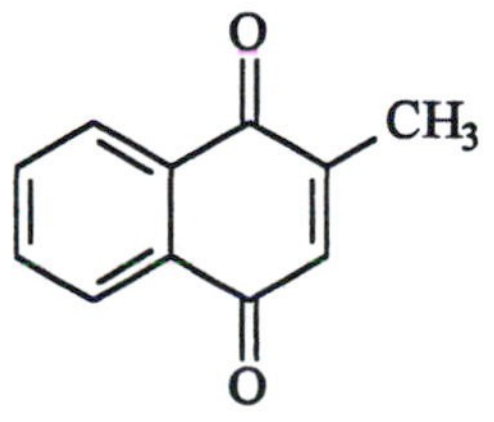

2-Methyl-naphthoquinone

(50 ppm)

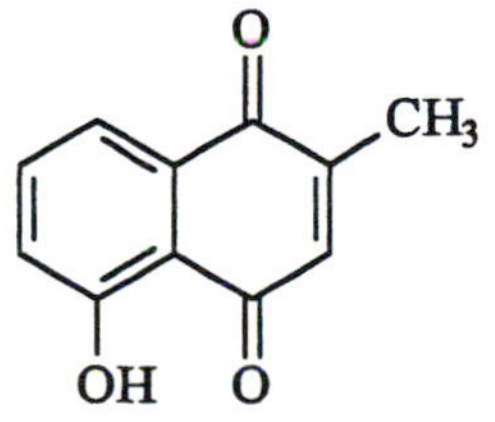

Plumbagin

(40 ppm)

Figure 5.   Naphthoquinones from Walnut Hulls: Concentration Inhibiting Germination of *Aspergillus flavus*.

Recent results have shown that the ED_{50} for germination inhibition by plumbagin and 2-methyl naphthoquinone is about 30 ppm. However, at concentrations between 5-20 ppm there appears to be a stimulatory effect upon aflatoxin production. Experiments are therefore planned to investigate whether or not these compounds directly affect regulatory genes within the aflatoxin biosynthetic pathway gene cluster (17). The levels of individual naphthoquinones will also be correlated with levels in the hulls of various walnut varieties in an attempt to define specific structural features required to prevent aflatoxigenesis.

Constituents of Kernels. Preliminary *in vitro* experiments (18) have shown that the ground kernels of 34 cultivars and breeding lines of almonds (5% in agar), infected with a standardized number of *A. flavus* spores, produce levels of aflatoxin B_1 ranging from 20-192 µg/plate. Under the same conditions, the levels for 20 walnut cultivars of English walnut (*Juglans regia*) were <0.02-28 µg/plate; however, on black walnut (*J. nigra*) 44 µg/plate was produced. It is not possible to obtain similar data for pistachios because California production is essentially a monoculture of the Kerman variety. If such differences are phytochemical in nature then they can be enhanced by conventional breeding or genetic manipulation. In general, the phytochemical constituents of nut kernels have not been thoroughly investigated. Experiments are currently in progress to identify anti-aflatoxigenic compounds through bioassay-directed fractionation. Once this information is available it should be possible to produce new varieties of almonds, pistachios and walnuts which are capable of limiting aflatoxin content to acceptable levels.

Conclusions

The results obtained to date indicate that tree nuts possess phytochemical constituents of considerable structural diversity which are capable of suppressing growth of *A. flavus* and/or inhibiting production of aflatoxins. However, while it is apparent that these constituents can significantly limit aflatoxin contamination, it is unlikely that their enhancement alone will be sufficient to achieve the ultimate goal of entirely eliminating aflatoxins from tree nut crops. This approach is visualized as a major part of a broader strategy of biological control of the problem (19) which will also incorporate semiochemicals such as pheromones and host-plant volatiles to disrupt insect vectors of infection (20), in concert with introduction of microorganisms such as saprophytic yeasts which have been shown to be natural competitors of *A. flavus* in tree nut orchards (21).

Acknowledgements

We thank Drs. Thomas M. Gradziel and Gale McGranahan, Department of Pomology, University of California at Davis, for providing samples of breeding lines and cultivars of almonds and walnuts, respectively.

Literature Cited

1.   Eaton, D. L.; Groopman, J. D., Eds. *The Toxicology of Aflatoxins: Human Health, Veterinary, and Agricultural Significance;* Academic Press: San Diego, CA, 1994.

2.   Henry, S. H. *Proc. Aflatoxin Elimination Workshop*, Memphis, TN. 26-28 Oct. 1997, p. 94.

3.   Wade, N. *New York Times,* 21 Nov. 1997.

4.   Cordesman, A. H. *Iraq's Past and Future Biological Weapons Capabilities;* Center for Strategic and International Studies: Washington, DC, 1997; http://www.csis.org/mideast/reports/iraq_bios.pdf

5.   http://pom44.ucdavis.edu/rank.html

6.   http://www.cdfa.ca.gov/exports/1997_farm.htm

7.   http://www.cdfa.ca.gov/exports/exports_comm.html

8.   Food and Drug Administration. *Compliance Policy Guides Manual;* USFDA: Washington, DC, 1996, Section 555.400, 268; Section 570.500, 299.

9.   Commission of European Communities. *Commission Regulation (EC) No. 1525/98 of 16 July 1998;* Official Journal of the European Communities L201/43 and Annex 1 L201/95, 17 July 1998.

10. Mahoney, N.; Molyneux, R. J. *J. Agric. Food Chem.* **1998,** *46,* 1906-1909.

11. Michailides, T. J. *California Agriculture* **1989,** *43,* 10-11.

12. VanEtten, H. D.; Mansfield, J. W.; Bailey, J. A.; Farmer, E. E. *Plant Cell* **1994,** *9,* 1191-1192.

13. Tyman, J. H. P. *Chem. Soc. Rev.* **1979,** *8,* 499-537.

14. Nagabhushana, K. S.; Shoba, S. V.; Ravindranath, B. *J. Nat. Prod.* **1995,** *58,* 807-810.

15. Binder, R. G.; Benson, M. E.; Flath, R. A. *Phytochemistry* **1989,** *28,* 2799-2801.

16. Hedin, P. A.; Collum, D. H.; Langhans, V. E.; Graves, C. H. *J. Agric. Food Chem.* **1980,** *28,* 340-342.

17. Yu, J.; Chang, P.-K.; Cary, J. W.; Wright, M.; Bhatnagar, D.; Cleveland, T. E; Payne, G. A.; Linz, J. E. *Appl. Environ. Microbiol.* **1995,** *61,* 2365-2371.

18. Mahoney, N.; Molyneux, R. J. unpublished results.

19. Campbell, B. C.; Hua, S.-S. T.; Light, D. M., Molyneux, R. J.; Roitman, J. N.; Merrill, G. B.; Mahoney, N.; Goodman, N.; Baker, J. L.; Mehelis, C. N. *Proc. Aflatoxin Elimination Workshop,* Memphis, TN. 26-28 Oct. 1997, p. 20.

20. Light, D. M., Mehelis, C. N. *Proc. Aflatoxin Elimination Workshop,* Memphis, TN. 26-28 Oct. 1997, p. 32.

21. Hua, S.-S. T.; Baker, J. L.; Grosjean, O.-K.; Flores-Espiritu, M. *Proc. Aflatoxin Elimination Workshop,* Memphis, TN. 26-28 Oct. 1997, p. 82.

Chapter 5

Current Research on Mycotoxins: Fumonisins

W. Thomas Shier[1], Petra A. Tiefel[1], and Hamed K. Abbas[2]

[1]Department of Medicinal Chemistry, University of Minnesota,
Minneapolis, MN 55455
[2]Southern Weed Science Research Unit, Agricultural Research Service,
U.S. Department of Agriculture, Stoneville, MS 38776

Currently, the most active areas of research on established mycotoxin groups (notably aflatoxins, tricothecenes and ochratoxins) involves cloning and characterization of biosynthesis genes, and elucidating the role of mycotoxins in the etiology of important diseases. The most recently discovered group of mycotoxins, first identified in 1988, are the fumonisins, the most abundant of which is FB_1. They are sphingosine analogs which exert all or most of their effects by inhibition of sphingolipid biosynthesis. FB_1 is cytotoxic, phytotoxic and causes equine leucoencephalomalacia and porcine pulmonary edema. However, the major reason for interest in fumonisins is their possible role as environmental tumor promoters in causing human cancer. Because the producing organism, *F. moniliforme*, is a ubiquitous contaminant of corn, and the toxins are stable enough to at least partially survive most food processing techniques, readily detectable amounts contaminate most corn-derived products available for human consumption. Studies on structure-activity relationships among fumonisins indicate that biological activity is retained through most structural changes, including loss of the side chains which constitute up to half the molecular weight. These observations led us to investigate the fate of radiolabeled FB_1 during food processing. Preliminary studies indicate a substantial amount of FB_1 binds covalently to protein and starch in ways that may result in release of an active form upon digestion.

Although the plant, animal and human diseases caused by fungi include some that are of great antiquity, it is only recently that the identity and structures have been determined for the mycotoxins mediating those diseases (1). The study of mycotoxins formally began in the early 1960's with the discovery that aflatoxins produced by *Aspergillus flavus* were the cause of turkey X disease in the U.K. (2). Subsequent research has identified additional aflatoxins and additional structurally distinct classes of mycotoxins, notably the trichothecenes, ochratoxins, zearalenones and most recently the fumonisins. As shown in Table 1, mycotoxins are usually named after the first producing organism identified. Several additional mycotoxins have been identified which do not fit into extended structural groupings, but nevertheless are associated with important agricultural problems, including moniliformin, patulin, citrinin and cyclopiazonic acid (Fig. 1). Novel mycotoxins continue to be discovered. Using any type of toxicity bioassay to examine culture extracts from a wide range of fungi reveals a plethora of unknown toxic substances (3). Clearly, many more mycotoxins remain to be discovered.

Current Research on Mycotoxins

Current research on mycotoxins is not restricted to discovering novel toxins, but also to developing a better understanding of economically important established mycotoxins. Three notably active areas of mycotoxin research can be identified. First, well-studied mycotoxins, such as aflatoxins and trichothecenes, continue to be the focus of active research, particularly through the use of gene cloning and other forms of genetic analysis to understand the regulation of toxin biosynthesis (4,5). This is an area of potentially great practical importance. If stable, non-toxic substances could be found that selectively turn off toxin synthesis genes, they would be very valuable agricultural chemicals. A second active area of mycotoxin research involves ochratoxin, a long known but relatively poorly studied mycotoxin produced by *Aspergillus alutaceus* (formerly known as *Aspergillus ochraceus*) and traditionally associated with Balkan endemic nephropathy (BEN). Ochratoxin A has become the focus of intensive study in Europe and North Africa (6-8). The studies have focused on incidence in processed food products (7), and on the possible role of ochratoxin A or other as yet unidentified mycotoxins in the dramatic but unexplained rise in kidney disease in North Africa (8). A third very active area of mycotoxin research is focused on all aspects of fumonisins (9-12). It will be discussed in more detail below.

Research on the Fumonisins

Fumonisins are the most recently discovered major group of mycotoxins (9,13). The first fumonisin structures were reported in 1988 by Bezuidenhout et al. (14), who used predominantly spectroscopic methods to demonstrate that fumonisins are a series of long chain alkylamines with hydroxylations and methylations on the alkyl chains, including vicinal hydroxyls esterified to two propane-1,2,3-tricarboxylic acid moieties (Fig. 1). The most abundant component is fumonisin B_1 (FB_1), part of a series of fumonisins characterized by a free amino group and varying degrees of hydroxylation. Bezuidenhout et al. (14) also identified a second series, the fumonisin A series with the amino group acetylated. These toxins were immediately recognized to be structural analogs of the AAL-toxins, a series of mycotoxins which had been identified earlier (15,16) and shown to mediate the toxic effects of the host-specific plant pathogen, *Alternaria alternata*. Subsequent research has identified two additional series of fumonisins: the C series, which have no methyl group on the carbon bearing the amino group (17); and the P series, in which the free amino group has been converted to a 3-hydroxypyridinium moiety (18), presumably by reaction with thermal decomposition products of carbohydrates (19). The B and C series exhibit high biological activity, whereas the A series is inactive in mammalian systems and only very weakly active in some plant systems, presumably because plants contain higher levels of amidase activity capable

Table 1. Nomenclature of some major mycotoxin groups

Aflatoxin	*Aspergillus flavus* toxin
Fumonisin	*Fusarium moniliforme* toxin
Ochratoxin	*Aspergillus ochraceus* toxin
Trichothecene	*Trichothecium roseum* toxin in the alkene form
Zearalenone	*Gibberella zeae* resorcylic acid lactone in the alkene and ketone form

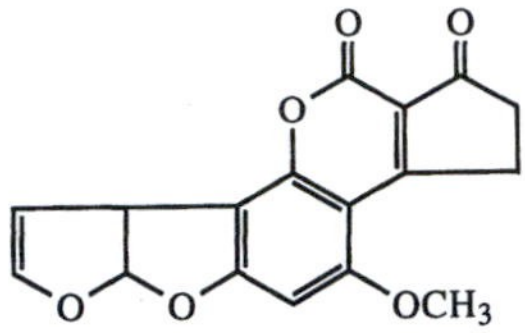

Aflatoxin B_1 (**1**)

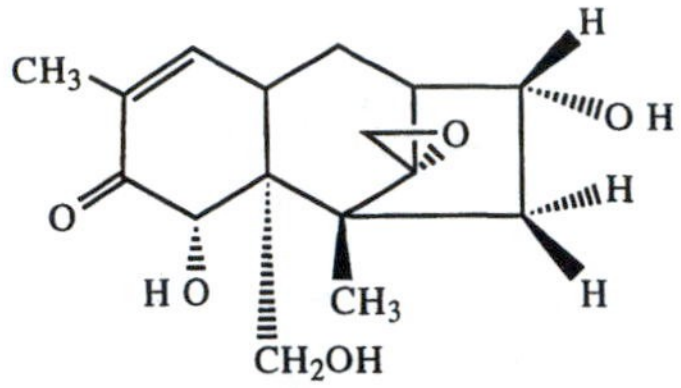

Deoxynivalenol (**2**)

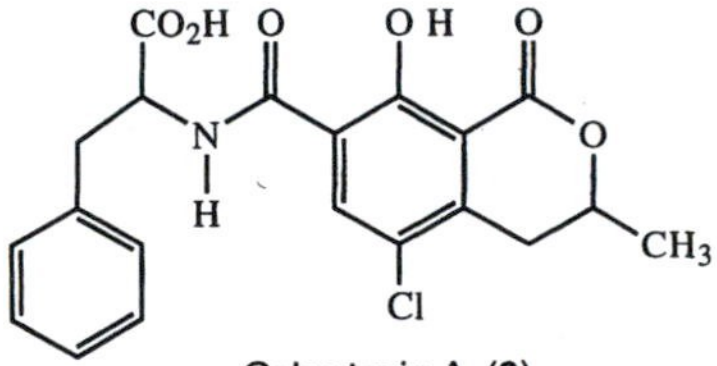

Ochratoxin A (**3**)

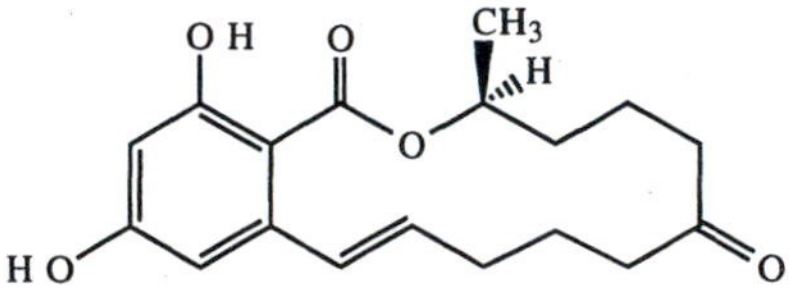

Zearalenone (**4**)

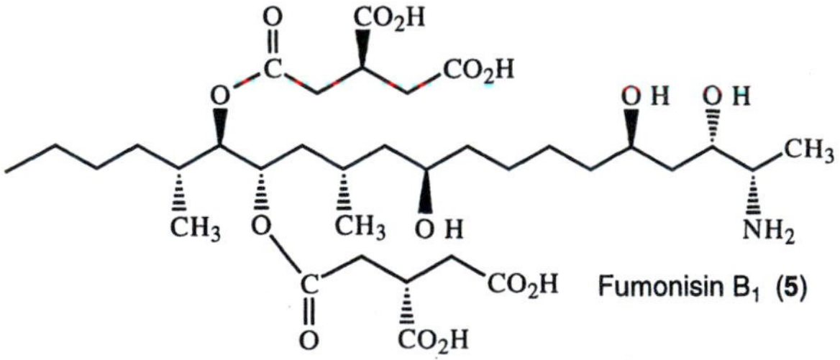

Fumonisin B_1 (**5**)

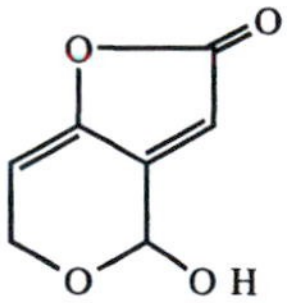

Patulin (**6**)

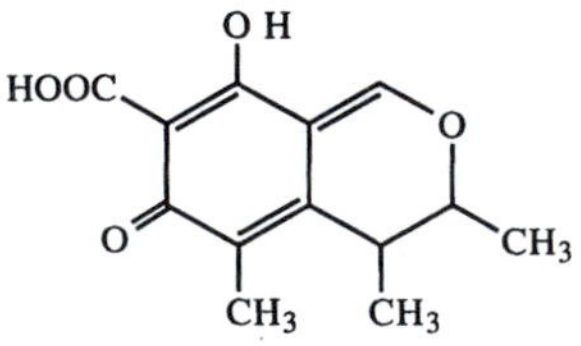

Citrinin (**7**)

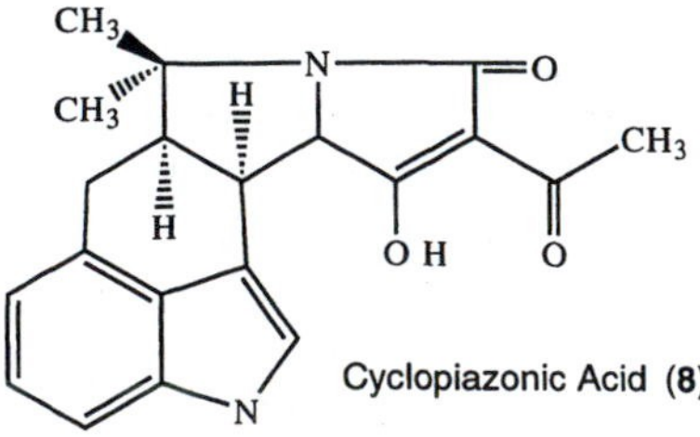

Cyclopiazonic Acid (**8**)

Figure 1. Structures of some mycotoxins (**1** - **8**) of importance to food safety.

of removing the N-acetyl group to generate the corresponding B series toxins. Fumonisin P_1 exhibits only about 1% of the activity of B and C series fumonisins (20).

Fumonisins are produced by *Fusarium moniliforme*, a ubiquitous contaminant of stored corn worldwide and the commonest cause of ear-rot, stalk-rot and root-rot in the United States (21). The fumonisins are believed to be virulence factors secreted by *F. moniliforme* to weaken potential host plants, thereby facilitating invasion of plant tissues by the fungus. Consistent with this suggestion, fumonisins are potent phytotoxins (22). However, they also happen to be toxic in mammalian systems. When purified FB_1 became available, it was demonstrated that two agriculturally-important diseases, equine leucoencephalomalacia and porcine pulmonary edema, known to be caused by consumption of *F. moniliforme*-contaminated feeds, could be reproduced by administration of FB_1 (23). Consumption of foods prepared from *F. moniliforme*-contaminated corn has also been associated with esophageal cancer in native peoples of the Transkei region of South Africa (24) and in China (25). Indeed, the discovery of fumonisins resulted from attempts by Marasas and associates to understand how *F. moniliforme* could cause esophageal cancer. Initial attempts to isolate a carcinogenic mycotoxin using bioassay-guided fractionation with a mutagenesis assay yielded fusarin C (26). However, its toxicity profile differed substantially from that of *F. moniliforme* culture material. Subsequently, use of a "short-term" tumor promotion bioassay yielded the fumonisins (14), which had a toxicity profile with a better match. Extensive studies are underway to evaluate the toxicology of FB_1 in rodents. Preliminary results (27) indicate that the major target organs are liver and kidney. Acute toxicity of FB_1 in the kidney produces dose-dependent non-inflammatory nephrosis with low dose regeneration. The mechanism of cell killing appears to be induction of apoptosis. A major long-term focus of studies on FB_1 toxicity has been the effort to determine if FB_1 is a complete carcinogen, or if it acts only as a tumor promoter (28). This question is not just of academic interest. It is of importance for regulatory purposes, because government regulations on allowable levels of toxins are written for individual toxins with no provisions for additivity, much less synergy, with other toxins (29). Similarly, extensive government regulations have been written for complete carcinogens, but no provisions have been made for environmental tumor promoters. However, from a food safety point of view it is probably not a critical issue whether FB_1 is a complete carcinogen or not, because *F. moniliforme* also produces the potent mutagen fusarin C (26), which would be expected to be an effective cancer initiator.

Studies on the mechanism of action of fumonisins have developed from the recognition by Riley and associates (30) that they are structural analogs of sphingosine [a putative second messenger molecule thought to regulate protein phosphorylation (31)] and sphinganine [an intermediate in the biosynthesis of sphingolipids (32), which are essential components of the lipid phase of biological membranes (33)]. FB_1 was shown (30,34) to inhibit sphingolipid biosynthesis by inhibiting ceramide synthetase, the enzyme which transfers fatty acid residues from an activated form, fatty acyl coenzyme A, to sphinganine to yield ceramide. Treatment with FB_1 results in accumulation of both sphinganine and sphingosine in cultured plant (35) or animal (36) cells and in intact animals (37). Elevated serum sphinganine and sphingosine has been proposed as a useful method for detecting exposure to fumonisins (35). Blockage of sphingolipid synthesis provides a plausible mechanism for most of the acute toxic effects of FB_1, which exhibit a lag of at least 24 hours before the first appearance of toxic symptoms. However, some effects of FB_1 are relatively rapid, such as phosphorylation of mitogen-activated protein kinase (38) and stimulation of phospholipase activity (39). The rapid effects may involve FB_1 acting directly as an analog of sphingosine in its second messenger role (31,40). The mechanism of action of FB_1 as a tumor promoter is not known, in large part because tumor promotion mechanisms are poorly understood in general (41).

Structure-Activity Relationships Among Fumonisins

The "short-term" tumor promotion assay (42) used to guide the original isolation of the fumonisins involved histological examination of liver slices from statistically significant numbers of rats pre-treated with N-ethylnitrosamine, fed culture-extract samples for two three-week periods between which they were given a partial hepatectomy. Clearly, this bioassay is not a practical experimental tool for studies of either mechanism of action or structure-activity relationships (SAR). Bioassays with better time resolution and smaller sample size were needed. For SAR studies of fumonisins we have adapted two assays in culture systems which have the advantages that they use small amounts of test substances, are much more rapid and are semi-quantifiable. The major disadvantage of the culture-system assays is that the relevance of any results obtained to cancer promotion remains to be established.

Cytotoxicity assays in cultured mammalian cell lines were used as indicators of mammalian toxicity (43). Interestingly, it was observed that certain cell lines with differentiated liver and kidney characteristics exhibited unusual sensitivity to B-series fumonisins. This cytotoxicity bioassay was used (44) for the first demonstration that the FB_1 backbone (hydrolyzed FB_1, in which the side chains representing about half the molecular weight of the toxin have been removed) not only retains biological activity but exhibits a broader spectrum of activity, being toxic also to undifferentiated cell lines that are resistant to intact fumonisins. This observation has led to the suggestion (45) that organ-specificity in fumonisin toxic effects may reflect the presence in the most susceptible tissues of esterases able to activate the toxin by removing the side chains. Further, the side chains may have been added to the fumonisins by the producing organism as a detoxification mechanism to protect itself, but not those target cell types expressing esterases able to remove them.

Estimates of phytotoxicity have been made in axenic cultures of duckweed (*Lemna major* L. or *Lemna pausicotata* L.), a small aquatic plant that grows well in culture and exhibits a high degree of sensitivity to fumonisins (46). Fumonisin exposure results in a light-dependent bleaching and death. Toxic effects can be quantitated by measuring leakage of electrolytes or chlorophyll into the culture medium, or by reductions in dry weight gain of the plantlets.

A large family of fumonisin analogs were available for SAR studies from natural fumonisins and AAL-toxins, and from derivatives prepared in the course of determining the absolute configuration of FB_1 (47,48). However, the most useful analogs were a series prepared by Kraus and associates (49) as model compounds to develop methods useful for the total synthesis of fumonisins. These analogs were used to establish that many alterations can be made in the structure of FB_1 without loss of biological activity in either the mammalian or plant toxicity bioassay (44,45,50). The only structural features found to be necessary for biological activity were a free amino group, long alkyl chain and probably the hydroxyl group on the carbon adjacent to the one bearing the amino group. The extensive methylation and hydroxylation of the alkyl chain do not appear to be required for either mammalian or plant toxicity.

Food Safety Considerations with Fumonisins

The recognition that fumonisins have an alkylamine structure made possible the rapid development of sensitive high performance liquid chromatography (HPLC) assays employing a fluorescence detector and pre-column derivatization with fluorescent or fluorogenic reagents (51-53). One of the first uses of these assays was to demonstrate the fumonisins are common contaminants of stored corn worldwide (54-62). In fact, they are so common that it is difficult to obtain a reference sample free of readily detectable amounts of the toxin. The lowest amounts of fumonisins are found in corn grown in cool, dry climates such as northern United States or Canada, with progressively higher amounts in corn grown in warmer, more humid climates (54). Concentrations as high as 130 ppm have been found in corn for animal feed and 330 ppm in corn screenings (55).

The HPLC assays have also been used to examine corn-derived processed food products intended for direct human consumption. Again, fumonisins were widely found (55,56,58-60,63-65). As expected, grinding corn to produce corn meal had little effect on fumonisin levels, but thermal processing resulted in lower levels of extractable, HPLC-assayable fumonisins (66). Thus corn flakes, which are prepared by roasting slices of corn dough with partially gelatinized starch, contain readily detectable amounts of free FB_1, but at levels typically 1/3 to 1/4 that found in the corn meal from which they were made. Stronger heating, as in producing corn chips from similar corn doughs by frying in oil, results in reducing FB_1 levels to near the limit of detection.

Fate of FB_1 During Food Processing

It is not known what happens to FB_1 during thermal food processing. Its disappearance has been generally accepted as a good thing, but a full understanding of the threat to food safety posed by the fumonisins requires that it be established that FB_1 is not merely being reversibly bound to corn component(s), or being converted to another unknown form that retains toxicity. One of the limits of the HPLC assays is that they measure only extractable, unaltered fumonisins (51-53). Either binding in a form that resists extraction or chemical conversion to an altered form that elutes from an HPLC column at a different time will be measured as a reduction of fumonisin levels. Bound fumonisins may be released on digestion of the binding material, and the studies on SAR of fumonisins carried out in these laboratories (44,45,50) suggest that many altered forms of FB_1 may retain biological activity as long as a free amino group has been retained.

A limited amount of research has been carried out to examine the fate of FB_1 during food processing. Scott and Lawrence (67) were the first to consider binding to solid food components as a potential complication in assessing FB_1 content. Hopmans and Murphy (64) have examined corn-derived food products for one expected FB_1 conversion product, hydrolyzed FB_1 (68). They found readily detectable amounts of hydrolyzed FB_1 in nixtmalized (lime-treated) corn products, such as tortillas that have been exposed to alkaline conditions (the lime-treatment step) during production (69). The toxicity of FB_1 is retained or possibly increased by the nixtmalization treatment that produces hydrolyzed FB_1 (70).

A more systematic approach to understanding the fate of FB_1 during thermal food processing has been undertaken in this laboratory. Radiolabeled FB_1 has been prepared by biosynthetic incorporation of [^{3}H-*methyl*]-L-methionine into FB_1 in solid phase cultures of *F. moniliforme* on corn. This procedure labels the methyl side chains on the long alkylamine chain of FB_1 (71-73). Exchange losses of label from this position would not be expected to occur to any significant extent under food processing conditions. Solutions of radiolabeled FB_1 ($[^3H]FB_1$) were absorbed into corn meal or corn kernels, which were then processed in laboratory models of the production of corn flakes (roasting) (74), corn chips (frying) and tortilla chips (nixtmalization) (69). The resulting food products were ground and initially extracted with 50% aqueous acetonitrile under typical HPLC assay extraction conditions (51-66). The insoluble residues were further extracted with water, and then with a 1% (w/v) solution of the detergent sodium dodecylsulfate in water. Analysis of the 50% aqueous acetonitrile extracts by thin layer chromatography with radioactive counting of segments of the chromatograph confirmed observations that roasting and frying conditions eliminate most but not all free FB_1 extractable under standard HPLC assay conditions. Also, it was confirmed that lime treatment (nixtmalization conditions) result in the formation of extractable hydrolyzed FB_1. Two other general observations were made. First, a relatively small percentage of $[^3H]FB_1$ radioactivity was converted to more polar derivatives, which is the type of derivative that would be expected to form by reaction of the free amino group of $[^3H]FB_1$ with Maillard degradation products formed by the decomposition of starch. Second, the fraction from the extract which contained the most radioactivity from $[^3H]FB_1$ after both frying and roasting was the material soluble in 1% aqueous sodium dodecylsulfate, but not soluble in water. This fraction was expected to contain predominantly protein, and it was indeed shown by a series of simple experiments that the radioactivity in the fraction was bound to zein, the major storage protein of corn. That is, the radiolabel was not released on dialysis, but it was released by alkaline treatment in the form of radiolabeled toxin backbone (hydrolyzed $[^3H]FB_1$) identified by thin layer chromatography. Sodium dodecylsulfate polyacrylamide gel electrophoresis with autoradiography indicated the radioactivity co-migrated with a family of proteins having the same molecular weight range as the zein family of proteins (75).

Probable Chemical Reactions of FB_1 During Thermal Processing of Foods

Studies on the reactions undergone by FB_1 during heating in the solid form or dispersed on starch or other matrices have provided some insights into the types of reactions that are likely to occur during thermal processing of FB_1-contaminated corn-derived foods. At the beginning of these studies it was expected that ester pyrolysis would be an important thermal degradation reaction of FB_1 (Fig. 2). Ester pyrolysis was expected to remove one or both side chains generating one or two double bonds in the toxin backbone. The simplest of these degradation products to detect is the diene, **9**. Some other derivatives that were expected to be formed by partial ester pyrolysis of side chains with hydrolysis were of considerable interest for use in SAR studies and as potential biosynthetic intermediates. While it was possible to detect traces of the diene **9** by UV absorption and FAB-MS, only very small amounts were formed under any of a wide range of conditions tested, including varying temperature, atmospheric pressure, exposure time and matrix. It was concluded that if ester pyrolysis occurs at all during thermal food processing, it can be expected to be a minor degradation pathway.

Another type of reaction which was expected to be observed when FB_1 was heated with corn meal or dispersed on a starch matrix was reaction of the free amino group of FB_1 with Maillard products formed by the thermal degradation of the starch (76). This is a common type of reaction in food chemistry. For example, it occurs in non-enzymatic browning, and it is used commercially to produce the coloring agent in cola soft drinks. One example of this reaction was chosen for more detailed

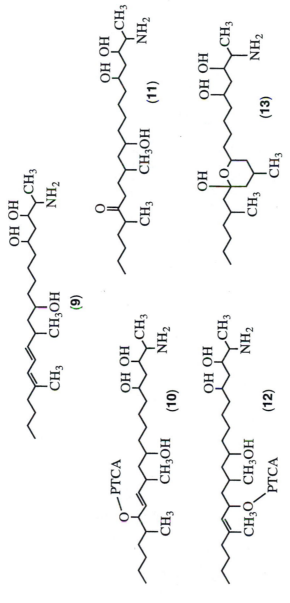

Figure 2. Expected products (**9** - **13**) from ester pyrolysis reactions of fumonisin B_1.

evaluation. It was the reaction of FB_1 with pentose thermal degradation products, which would be expected to yield fumonisin P_1, the 3-hydroxypyridinium derivative, which has been reported to be found at levels as high as 30% of total fumonisins in some studies (18). However, the reaction proceeded very slowly with FB_1 as the amine, producing only traces of FP_1, which was detected chromatographically by a color reaction with ferric chloride and further identified by characteristic pH-dependent shifts in the UV spectrum and downfield signals in the ^{1}H-NMR spectrum. Unfortunately, examination of a wide range of conditions failed to substantially increase yields of FP_1. Thus, it seems unlikely that FP_1 is being formed naturally by vapor phase reaction with thermal degradation products from pentose-containing polysaccharides in corn cobs during drying immediately after harvest. The reactivity of the free amino group of FB_1 may be greater with other Maillard degradation products which have not yet been studied.

The major reaction of FB_1 on heating in air or partial vacuum appears to be loss of water from either side chain to form a cyclic anhydride (Fig. 3). Several observations are consistent with this being the predominant initial reaction that occurs during heating of FB_1 either alone or dispersed on a matrix. First, heating solid [^{3}H]FB_1 under temperature conditions that would be encountered during frying (e.g., 220°C for 20 minutes) converts the bulk of the material to a non-dialyzable polymer. Alkaline hydrolysis of this material releases the toxin backbone (hydrolyzed [^{3}H]FB_1) in good yield. The expected reaction involves initially forming the cyclic anhydride (Fig. 3), then opening the ring by reaction with an amino group on another FB_1 molecule. More direct evidence for formation of FB_1-anhydride intermediate was obtained by heating FB_1 in two forms in which the amino group would not be reactive; i.e., as the perchlorate salt (which maintains the amino group in the unreactive ammonium form) or as the N-acetyl derivative (FA_1). Although the resulting anhydrides were not stable enough to allow rigorous chemical characterization, evidence for formation of the anhydride was obtained by FAB-MS analysis of the residue after heating (spectrum including a species at m/e 704 [M+1]) and by NMR spectroscopy (spectrum including an absorption pattern in the region 2.5 – 3.2 ppm similar to that observed with a model cyclic anhydride prepared from isolated propane-1,2,3-tricarboxylic acid side chains by reaction with acetic anhydride). FB_1 in naturally-contaminated food products would not be expected to react with itself, because the concentration would be much lower than in the solid form or even laboratory models when it is dispersed on matrices at relatively high concentrations. The most abundant nucleophiles in corn meal are expected to be part of the proteins, specifically epsilon amino groups on lysine residues and the amino terminals, or thiol groups, if present. Reactions with amino and possibly thiol groups would result in FB_1 being covalently bound to protein (Fig. 3). Indeed, this reaction is the well-known mixed anhydride reaction commonly used in the development of radioimmunoassays or ELISA assays. The antigens needed to induce the required antibodies to mycotoxins or other relatively small molecules (haptens) can be prepared by covalently coupling amino groups on small molecules to carboxylic acid moieties on proteins using the mixed anhydride reaction (77). This reaction has been modeled by heating a thin film of [^{3}H]FB_1 under frying conditions (220°C for 20 minutes), then cooling and adding a solution of poly(L-lysine) under Schotten-Baumann conditions (pH 10). [^{3}H]FB_1 was covalently bound to poly(L-lysine) in a non-dialyzable form, but could be released as hydrolyzed [^{3}H]FB_1 by alkaline treatment (Fig. 3). Covalent binding of FB_1 to food components in this manner is of great concern for food safety, because it would be expected that the bound FB_1 would be released in the intestinal tract on digestion of the protein either as FB_1 or as hydrolyzed FB_1, which has equal or greater toxicity (70). Another as yet unexplored complication is the possibility that FB_1 covalently

63

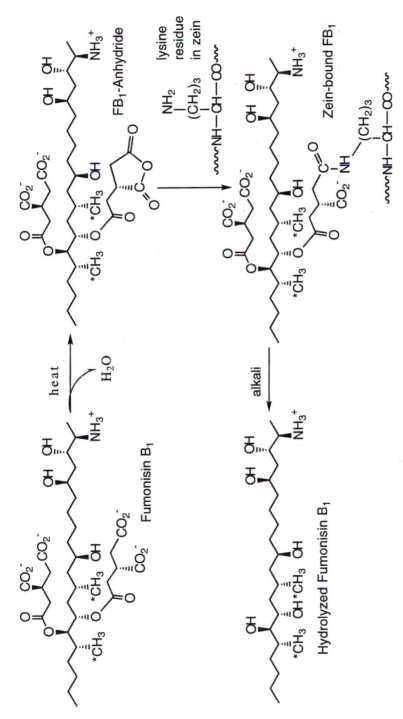

Figure 3. Proposed mechanism for the covalent binding of radiolabeled fumonisin B$_1$ to protein during heating of corn-derived preparations.

bound to a free lysine residue after digestion of the protein part of the complex may be absorbed via amino acid or peptide transporters not available to free FB_1 or hydrolyzed FB_1. If this is true, the bioavailablity of bound FB_1 would be much higher than FB_1 or hydrolyzed FB_1, providing an answer as to how fumonisin toxicity can occur despite its very low observed bioavailability (78-83).

Conclusion

The full extent of the threat to food safety posed by the fumonisins will not be known until there is a full understanding of (i) the chemistry the toxin undergoes during thermal food processing, and (ii) the toxicity and bioavailability of the products formed by those reactions.

Literature Cited

1. Christensen, C. M., *Molds, Mushrooms and Mycotoxins,* University of Minnesota Press: Minneapolis, MN, 1975, pp. 59-85.
2. Palmgren, M. S.; Ciegler, A., *Plant and Fungal Toxins. Handbook of Natural Toxins,* Marcel Dekker, New York, NY, 1983, Vol. 1, pp. 299-323.
3. Abbas, H. K.; Mirocha, C. J.; Shier, W.T. *Appl. Environ. Microbiol.* **1984**, *48*, 654.
4. Bennett, J. W.; Chang, P. K.; Bhatnagar, D. *Adv. Appl. Micro.* **1997**, *45*, 1.
5. Desjardins, A. E.; Hohn, T. M.; McCormick, S. P. *Microbiol. Rev.* **1993**, *57*, 595.
6. Pohland, A. E.; Nesheim, S.; Friedman, L. *Pure Appl. Chem.* **1992**, *64*, 1029.
7. Van Egmond, H. P.; Speijers, G. J. A. *J. Nat. Toxins* **1994**, *3*, 125.
8. Maaroufi, K.; Achour, A.; Zakama, A.; Ellouz, F.; El May, M.; Creppy, E. E.; Bacha, H. *J. Toxicol. Toxin Rev.* **1996**, *15*, 223.
9. Marasas, W. F. O. *Adv. Expt. Med. Biol.* **1996**, *392*, 1.
10. Dutton, M. F. *Pharmacol. Ther.* **1996**, *70*, 137.
11. Riley R. T.; Norred, W. P.; Bacon, C. W. *Annu. Rev. Nutr.* **1993**, *13*, 167.
12. Wood, G. E. *J. Animal Sci.* **1992**, *70*, 3941.
13. Gelderblom W. C. A.; Jaskiewicz, K.; Marasas, W. F. O.; Theil, P. G.; Horak, R. M.; Vleggaar, R.; Kreik, N. P. J. *Appl. Environ. Microbiol.* **1988**, *54*, 1806.
14. Bezuidenhout, S. C.; Gelderblom, W. C. A.; Gorst-Allman, C. P.; Horak, R. M.; Marasas, W. F. O.; Spiteller, G.; Vleggaar, R. *J. Chem. Soc. Chem. Commun.* **1984**, *1984*, 743.
15. Bottini A. T.; Gilchrist D. G. *Tetrahedron Lett.* **1981**, 22, 2719.
16. Bottini, A. T.; Bowen, J. R.; Gilchrist, D. G. *Tetrahedron Lett.* **1981**, *22*, 2723.
17. Plattner, R .D.; Weisleder, D.; Shackelford, D. D.; Peterson, R.; Powell, R. G. *Mycopathologia* **1992**, *117*, 23.
18. Musser, S. M.; Gay, M. L.; Mazzola, E. P.; Plattner, R. D. *J. Nat. Prod.* **1996**, *59*, 970.
19. Pachmayr, O.; Ledl, F.; Severn, T. *Lebensm. Unters. Forsch.* **1986**, *182*, 294.
20. Abbas, H. K.; Shier, W. T.; Seo, J. A.; Lee, W. Y.; Musser, S. M. *Toxicon,* **1998**, in press.
21. Booth, C. *The Genus* Fusarium. Commonwealth Mycological Institute, Kew, Surrey, U.K, 1971.

22. Abbas, H. K.; Gelderblom, W. C. A.; Cawood, M E.; Shier, W. T. *Toxicon* **1993**, *31*, 345.
23. Harrison, L. R.; Colvin, B. M.; Greene, J. T.; Newman, L. E.; Cole, J. R. *J. Vet. Diagn. Invest.* **1990**, *2*, 217.
24. Marasas, W. F. O.; Jaskiewicz, K.; Venter, F. S.; Van Schalkwyk, D. J. *S. Afr. Med. J.* **1988**, *74*, 110.
25. Lin, P.; Tang, W. *J. Cancer Res. Clin. Oncol.* **1980**, *96*, 121.
26. Gelderblom, W. C. A.; Theil, P. G.; van der Merwe, K. J.; Marasas, W. F. O.; Spies, H. S. C. *Toxicon* **1983**, *21*, 467.
27. Bucci, T.J.; Howard, P.C. *J. Toxicol. Toxin Rev.* **1996**, *15*, 293.
28. Gelderblom, W.C.A.; Semple, E.; Marasas, W.F.O.; Farber, E. *Carcinogenesis*, **1992**, *13*, 433.
29. Rosner, H. *Revue Méd. Vét.* **1998**, *149*, 679.
30. Wang, E.; Norred, W. P.; Bacon, C. W.; Riley, R. T.; Merrill, A. H. *J. Biol. Chem.* **1991**, *266*, 14486.
31. Hannun, Y. A.; Linardic, C. M. *Biochim. Biophys. Acta* **1993**, *1154*, 223.
32. Merrill, A. H.; and Jones, D. D. *Biochim. Biophys. Acta* **1990**, *1044*, 1.
33. Rogolsky, M. *Microbiol. Rev.* **1979**, *43*, 320.
34. Yoo, H.-S.; Norred, W. P.; Wang, E.; Merrill, A. H.; Riley, R. T. *Toxicol. Appl. Pharmacol.* **1992**, *114*, 9.
35. Riley, R. T.; Wang, E.; Merrill, A. H. *J. AOAC Int.* **1994**, *77*, 533.
36. Abbas, H. K.; Tanaka, T.; Duke, S. O.; Porter, J. K.; Wray, E. M.; Hodges, L.; Sessions, A. E.; Wang, E.; Merrill, A. H.; Riley, R. T. *Plant Physiol.* **1994**, *106*, 1085.
37. Wang, E.; Ross, P. F.; Wilson, T. M.; Riley, R. T.; Merrill, A. H. *J. Nutr.* **1992**, *122*, 1706.
38. Wattenberg, E. V.; Badria, F. A.; Shier, W. T. *Biochem. Biophys. Res. Commun.* **1966**, *227*, 622.
39. Pinelli, E.; Pipy, B.; Castegnaro, M.; Miller, D. J.; Pfohl-Leszkowizc, A. *Revue Méd. Vét.* **1998**, *149*, 651.
40. Shier W. T. *J. Toxicol.-Toxin Rev.* **1992**, *11*, 241.
41. Vineis, P.; Brandt-Rauf, P. W. *Eur. J. Cancer* **1993**, *29A*, 1344.
42. Tatematsu, M.; Shirai, T.; Tsuda, H.; Miyata, Y.; Shinohara, Y.; Ito, N. *Gann* **1977**, *68*, 499.
43. Shier, W. T.; Abbas, H. K.; Mirocha, C. J. *Mycopathologica* **1991**, *116*, 97.
44. Abbas, H. K.; Gelderblom, W. C. A.; Cawood, M. E.; Shier, W. T. *Toxicon* **1993**, *31*, 345.
45. Shier, W. T.; Abbas, H. K.; Badria, F. A. *J. Nat. Toxins* **1997**, *6*, 225.
46. Tanaka, T.; Abbas, H. K.; Duke, S. O. *Phytochemistry* **1993**, *33*, 779.
47. Hoye, T. R.; Jiménez, J. I.; Shier, W. T. *J. Amer. Chem. Soc.* **1994**, *116*, 9409.
48. Shier, W. T.; Abbas, H. K.; Badria, F. A. *Tetrahedron Lett.* **1995**, *36*, 1571.
49. Kraus, G. A.; Applegate, J. M.; Reynolds, D. *J. Agric. Food Chem.* **1992**, *43*, 2331.
50. Abbas, H. K.; Tanaka, T.; Shier, W. T. *Phytochemistry* **1995**, *40*, 1681.
51. Sydenham, E. W.; Shephard, G. S.; Theil, P. G. *J. Assoc. Off. Anal. Chem.* **1992**, *75*, 313.
52. Bennet, G. A.; Richard, J. L. *J. AOAC Intl.* **1994**, *77*, 501.
53. Rice, L. G.; Ross, P. F.; DeJong, J.; Plattner, R. D.; Coats, J. R. *J. AOAC Intl.* **1995**, *78*, 1002.
54. Thiel, P. G.; Marasas, W. F. O.; Syndenham, E. W.; Shephard, G. S.; Gelderblom, W. C. A.; J. J. Nieuwenhuis, J. J. *Appl. Environ. Microbiol.* **1991**, *57*, 1089.

55. Theil, P. G.; Marasas, W. F. O.; Sydenham, E. W.; Shepard, G. S.; and Gelderblom, W. C. A. *Mycopathologia* **1992**, *117*, 3.
56. Shephard, G. S.; Theil, P. G.; Stockenstrom, S.; Sydenham, E. W. *J. AOAC Int.* **1996**, *79*, 671.
57. Murphy, P. A.; Rice, L. G.; Ross, P. F. *J. Agric. Food Chem.* **1993**, *41*, 263.
58. Ueno, Y.; Aoyama, S.; Sugiura, Y.; Wang, D.-S.; Lee, U.-S.; Hirooka, E. Y.; Hara, S.; Karki, T.; Chen, G.; Yu, S.-Z. *Mycotoxin Res.* **1993**, *9*, 27.
59. Bullerman, L. B.; Tsai, W.- Y. *J. Food Protection*, **1994**, *57*, 541.
60. Doko, M. B.; Visconti, A. *Food Additives and Contaminants* **1994**, *11*, 433.
61. Chamberlain, W. J.; Norred, W. P. *Fd. Chem. Toxicol.* **1993**, *31*, 995.
62. Pittet, A. *Revue Méd. Vét.* **1998**, *149*, 479.
63. Sydenham, E. W.; Shephard, G. S.; Theil, P. G.; Marasas, W. F. O.; Stockenstrom, S. *J. Agric. Food Chem.* **1991**, *39*, 2014.
64. Hopmans, E. C.; Murphy, P. A. *J. Agric. Food Chem.* **1993**, *41*, 1655.
65. Pittet, A.; Parisod, B.; Schellenberg, M. *J. Agric. Food Chem.* **1992**, *40*, 1352.
66. Murphy, P. A.; Hendrich, S.; Hopmans, E. C.; Hauck, C. C.; Lu, Z.; Buseman, G.; Munkrold, G. *Fumonisin in Food*; Plenum Press: New York, NY, 1996.
67. Scott, P. M.; Lawrence, G. A. *J. AOAC Intl.* **1994**, *77*, 541.
68. Scott, P. M.; Lawrence, G. A. *Food Additives and Contaminants* **1996**, *13*, 823.
69. Bressani, R. *Food Rev. Intl.* **1990**, *6*, 225.
70. Hendrich, S.; Miller, K. A.; Wilson, T. M.; Murphy, P. A. *J. Agric. Food Chem.* **1993**, *41*, 1649.
71. Shephard, G. S.; Thiel, P. G.; Sydenham, E. W.; Alberts, J. F.; Gelderblom, W. C. A. *Toxicon* **1992**, *30*, 768.
72. ApSimon, J. W.; Blackwell, B. A.; Edwards, O. E.; Fruchier, A.; Miller, J. D.; Savard, M.; Young, J. C. *Pure Appl. Chem.* **1994**, *66*, 2315.
73. Blackwell, B. A.; Miller, D. J.; Savard, M. E. *J. AOAC Intl.* **1994**, *77*, 506.
74. Rokey, G.J. *Cereal Foods World*, **1995**, *40*, 422.
75. Wilson, C. M. *Cereal Chem.* **1988**, *65*, 72.
76. Huyghues-Despointes, A.; Yaylayan, V. A. *Crit. Rev. Food Sci. Nutr.* **1994**, *34*, 321.
77. Chu, F. S. *Modern Methods in the Analysis and Structural Elucidation of Mycotoxins*; Academic Press: Orlando, Florida, 1986, pp 211-212.
78. Shephard, G. S.; Thiel, P. G.; Sydenham, E. W.; Alberts, J. F.; Gelderblom, W. C. A. *Toxicon* **1992**, *30*, 768.
79. Shephard, G. S.; Thiel, P. G.; Sydenham, E. W.; Savard, M. E. *Natural Toxins* **1995**, *3*, 145.
80. Norred, W. P.; Plattner, R. D.; Chamberlain, W. J. *Natural Toxins* **1993**, *1*, 341.
81. Prelusky, D. B.; Savard, M. E.; Trenholm, H. L. *Natural Toxins* **1994**, *2*, 73.
82. Prelusky, D. B.; Trenholm, H. L.; Savard, M. E. *Natural Toxins* **1995**, *3*, 389.
83. Vudathala, D. K.; Prelusky, D. B.; Ayroud, M.; Trenholm, H. L.; Miller, J. D. *Natural Toxins* **1994**, *2*, 81.

Chapter 6

Bacterial Typing and Identification
by Mass Spectrometry

T. Krishnamurthy[1,5], U. Rajamani[2], P. L. Ross[1], J. Eng[3], M. Davis[4], T. D. Lee[4],
D. S. Stahl[4], and J. Yates[3]

[1]R&T Directorate, U.S. Army Edgewood RDE Center,
Aberdeen Proving Ground, MD 21010
[2]Geo-Centers, Inc., Gunpowder Branch, P.O. Box 68,
Aberdeen Proving Ground, MD 21010
[3]Department of Molecular Biotechnology, University of Washington,
Seattle, WA 98195
[4]Beckman Research Institute, City of Hope, Duarte, CA 91010

Simple, direct, rapid, and accurate mass spectrometric methods have
been developed for identifying hazardous bacteria in unknown samples
and sites. Cellular proteins specific to individual bacterium, were
released by *in-situ* lysis of the intact cells followed by electrospray
ionization (ESI)- or matrix assisted laser desorption ionization
(MALDI)- mass spectrometric analysis of the marker proteins. The
biomarkers specific for individual bacterium were derived from the
acquired mass spectral data. The marker proteins could be used for
very specific bacterial identification, enabling the distinction of
bacterial pathogens from its closely related non-pathogenic
counterparts. The entire analytical process could be completed within
ten minutes and was applicable to both gram-positive and gram-
negative organisms. Algorithms developed for the automated sample
processing enabled the unambiguous identification of bacteria in
unknown samples. This simple but potentially powerful approach has
great application capability in diversified fields such as monitoring and
clean up of hazardous wastes, food processing, and health industrial
sites as well as biological, bioengineering and pharmaceutical research.

[5]Corresponding author. Telephone: (410) 436–5909; Fax: (410) 436–6536;
E-mail: txkrishn@cbdcom-emh1.apgea.army.mil

Routine monitoring of environmental and biological samples for the presence of any kind of toxic substances and microbial contamination is vital for identifying and preventing potential hazards and even human mortality.[1-3]   Recent increase in potential global threat due to biological weapons and terrorist activities makes such technology development even more important.[1-4]   Any successful pursuit for protection against microbial contamination and/or biological agents would initially require reliable identification of the pathogen present in the suspected samples. The methodology should be rapid, specific, sensitive, accurate, reliable and most importantly be applicable for the analysis of any molecule with minor modifications at best. Mass spectrometry (MS), which has been established as a versatile technique, could be used for the analysis of small as well as larger molecules. In addition, the sample preparation prior to MS analysis is simple but effective. Hence, the overall MS approach should be well suited for the investigations and analysis of bacterial pathogens.[1]

Bacterial identification has been accomplished in a limited manner by pyrolyzing the intact bacterial cells and analyzing the basic units of cellular biopolymers.[5,6]   Similarly, mass spectrometric techniques have been applied for distinguishing the bacteria based on the cellular components such as lipids, phospholipids, carbohydrates and DNA.[1, 7-14] None of these methods could be applied for distinguishing the closely related bacteria. The proteins present in greater abundance in any given cell, originating from the DNA of an organism, should provide indirect genetic information about any particular bacterium. In addition, sample preparation for marker proteins, prior to mass spectral analysis, can be accomplished with ease. Hence, during our on-going extensive investigations on bacterial pathogens, we opted to apply cellular protein biomarkers for resolving the problem.

Initially, we isolated and purified the cellular proteins from several strains of γ-radiated *Bacillus anthracis* cells and analyzed by MALDI-TOF-MS methodology.[15] Similarly, MALDI-MS data of protein extracts from *B. thuringiensis, B.cereus, B.subtilis, Brucella melitensis, Francisella tularensis, and Yirsinia pestis* were also obtained.[15] Careful evaluation of the acquired mass spectral data, biomarkers specific for individual genus, species and strains were identified.[15]   In addition, human pathogen *B.anthracis* could easily be distinguished from other non-pathogenic sub-species members, *B.thuringiensis* and *B.cereus,*.[15] Such a distinction between these sub-species organisms was not possible earlier by any other analytical methodology. In addition, identification of other above-mentioned human pathogens was also possible by this procedure.[15]

However, our goal was to identify the pathogenic bacteria present in unknown samples within ten minutes, while the process involving the extraction and purification of cellular proteins took 2-3 hours.   Hence, we induced instantaneous cell lysis by suspending the intact cells in 0.1% aqueous trifluoroacetic acid. The proteins released during the cell lysis were subjected directly to MALDI-MS analyses without further sample processing. Thus, the total sample analysis time was considerably reduced from 2-3 hours to less than ten minutes. This method was also applicable effectively for gram-negative as well as gram–positive bacteria.[16] Numerous common biomarkers were detected during the MS analysis of intact cells and the corresponding protein extracts, even though more proteins were detected during the former process.[16] Protein biomarkers specific for individual genus, species and strains of the investigated bacteria could also be selected easily from the acquired MALDI-MS data of the intact cells.[16] In order to optimize the identification process as well as to reduce the total analysis time, we developed an algorithm for automated spectral library search.[17]

Recently, similar work based on the MALDI-MS investigations of either intact bacterial cells or their corresponding protein extracts have also been reported in the literature.[18-24] Distinction between closely related bacteria and species classification were very clearly demonstrated in most instances.[18-24] Under rigorous cell lysis and sample preparation conditions followed by MALDI-MS analysis, cellular proteins up to 500 kDa have been detected.[23] Recently, protein profiling of mammalian cell preparations, sequencing of neuropeptides from snails, neuropeptide expression patterns, and analyses of lysates from single neuron and mammalian cells have also been accomplished by MALDI-MS analysis.[25-30]

Salts and detergents usually present in excessive quantities in bacterial samples posed a major challenge during some of the MALDI-MS analyses. However, simple and rapid off-line clean-up procedures involving ultrafiltration and electro-microdialysis minimized this limitation.[4,31] However, technology, for direct introduction of liquid samples into a MALDI source and perform tandem mass spectrometric measurements, has not yet been perfected. This could generate a major challenge in accomplishing our ultimate goal, involving automating the bacterial identification in field situations. Hence, we investigated the LC/microspray-MS analysis[32] of the intact cell suspensions in 0.1% aqueous TFA containing 0-20% acetonitrile.[33-35] The observed results are reproducible and comparable with the MALDI-MS data. However, the sensitivity observed during ESI-MS experiments exceeded the detection limits measured during MALDI-MS analysis, by at least two orders of magnitude. In addition, on-line sample preparation, clean up, and concentration during the LC/ESI/MS analysis of samples is currently under investigation.[4] Bacterial pathogens including the individual members of the *Bacillus cereus* sub-species could be easily distinguished from their non-pathogenic counterparts. This would be a potential pathway for field monitoring since the sample preparation and separation units could be linked on-line with the tandem mass spectrometer. Such a system can easily be miniaturized and with adequate software, the entire operation can be performed automatically with minimum human intervention.

MALDI-MS Investigations

All MALDI-MS data were acquired in the linear mode using both bench top and research type time-of-flight mass spectrometers. Some of the earlier data were acquired by the continuous extraction and later by delayed extraction of the ionized molecules.

Protein Biomarkers. Initially, intact bacterial cells were subjected to chemical lysis and the released proteins were extracted and separated from DNA and other biopolymers.[15] The purified proteins were mixed with MALDI matrix and analyzed. On occasion, especially for the thick walled gram-positive bacteria, agitating the cells using micro-tip ultrasonic cell disrupters was required to release the same proteins, especially the low mass markers, in greater abundance (Figure 1).[15] The bacterial cells, regardless of the growth conditions they were generated from, released the same marker proteins.

The *Bacillus cereus* organisms[9] which include *Bacillus anthracis, Bacillus thuringiensis,* and *Bacillus cereus* are genetically very similar[36] and have been difficult to distinguish even by DNA-DNA hybridization.[37] Hence, it has been suggested that the three organisms may actually be members of a sub-species. However, the virulence and pathogenicity of these organisms and hence the exposure risk to humans represented by these species varies drastically. *B.anthracis*, the causative agent of anthrax, is a well-known human and animal pathogen and the infection caused by the organism is generally fatal. *B.cereus* is associated with food poisoning and *B. thuringiensis* used as a biological pesticide is pathogenic only to insects.[38] Hence, it is vital that the presence of *B.anthrax* in unknown samples is unequivocally established in order to ascertain the real danger and minimize the unwarranted panic, caused by incorrect identification. Biomarkers derived from MALDI-MS spectra of the protein extracts from closely related *Bacillus cereus* sub-species group organisms enabled the distinction of the individual bacterium (Figure 2).[15] On careful analysis of the acquired spectra (Figure 2) of different strains of these microorganisms along with harmless *Bacillus subtilis*, protein markers specific for individual genus, species and strains could easily be deduced (Table 1). Additional specific biomarkers were obtained by the proteolytic cleavage of the protein mixtures from individual organism (Table 1). Similar observations were made during the investigations of other pathogens such as *Yersinia pestis, Brucella melitensis* and *Francisella tularensis*.

Direct Intact Cell Analysis. Sample preparation is considerably simple, in comparison with the protein extraction procedure. Bacterial suspension in 0.1% trifluoroacetic acid (TFA) were mixed with the MALDI matrix (sinnapinic acid) and analyzed the mixture directly without further sample treatment. The marker proteins observed must include cell wall components as well as the ones released during cell rupture induced by osmotic pressure generated by buffers and/or by laser shots. The procedure was simple, direct, sensitive, and reproducible and can be applied equally well for gram-positive as well as gram-negative bacterial cells. The entire process can be completed in less than 10 minutes. In addition, more proteins in greater abundance in overall higher molecular weight ranges were observed as a result of direct whole cell experiments. Only some of them were observed during the analysis of extracted proteins and none of them were detected during the analysis of blank experiments. The common ions observed during the MALDI-MS analysis of the whole bacterial cells and their corresponding protein extracts are summarized in Table 2. These similarities indicate that many of the ions generated in the whole cell experiments also arise from proteinaceous components. Optimization of the process was carried

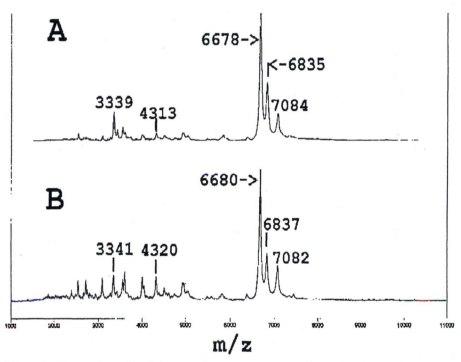

Figure 1. Evaluation of cellular protein extraction procedures. A, chemical lysis; B, ultrasonic disruption prior to chemical treatment. Reproduced with permission from reference 15. Copyright 1996.

72

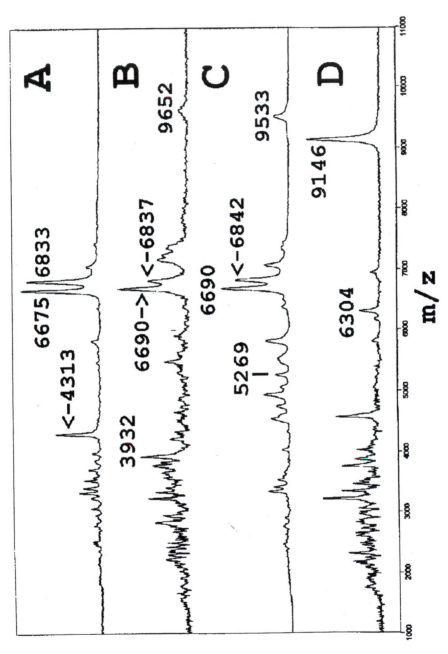

Figure 2. MALDI-MS analysis of protein extracts from *Bacillus* specie. Reproduced from reference 15. Copyright 1996.

Table 1. Extracted Protein Biomarkers for *Bacillus* Species

Organism	Genus Intact	Trypsin[a]	Species Intact	Trypsin[a]	Strain Intact	Trypsin[a]
B. anthracis, vollum	6680, 6837	1192, 1523, 1597, 3359	2385, 3991, 4313	1650	4505	1784, 2784
B. anthracis, sterne	6680, 6837	1192, 1523, 1597, 3359	2385, 3991, 4313	1650	2789	1062, 2278
B. anthracis, zimbabwe	6680, 6837	1192, 1523, 1597, 3359	2385, 3991, 4313	1650	2850	1558, 2697
B. thuringiensis, 4A1	6680, 6837	1192, 1523, 1597, 3359	3932	2424, 2634, 4266	5916	2715, 3116, 3568, 4013
B. thuringiensis, 4A2	6680, 6837	1192, 1523, 1597, 3359	3932	2424, 2634, 4266	4871, 7845	1243, 2310
B. thuringiensis, 4L2	6680, 6837	1192, 1523, 1597, 3359	3932	2424, 2634	2864, 4074, 4548, 5781	2339, 4203
B. cereus, 6E1	6680, 6837	1192, 1523, 1597, 3359	5269, 5537, 7365, 9533	2857, 3049	—	—
B. subtilis, 3A1	6680(?)	1523, 1597	3757, 3871, 6304, 9146	1457, 1880, 2911, 2968	—	—

[a] Tryptic digestion products;
—, Data not available yet.
?, Doubly charged ion, corrected mass/charge ratio

Table 2. Comparison of Ions in MALDI-MS Analysis of Bacterial Protein Extracts and Whole Cells

Organism	Common Markers (m/z)	Unique Whole Cell Markers (m/z)
Bacillus anthracis, Sterne	4340, 6684, 6832	2580, 5171, 6431, 7374, 7772, 9640, 9981
Bacillus anthracis, Zimbabwe	4342, 6686, 6828	2880*, 5178,6436, 7377,7780,9982
Bacillus anthracis, Vollum	4339, 6683, 6837	2580*,5172, 6428, 7379, 7780, 9988
Bacillus cereus, 6E1	2470, 6682, 6827	3015, 4816, 5174, 6437, 9634
Bacillus thuringiensis, 4D5	4335, 6685	5174, 6440, 7378, 9640
Bacillus subtilis	3898, 9130	4340, 6913
Brucella melitensis REV-1	6650, 7044, 7341, 9071	2700, 7403, 9087, 10287, 16100
Yersinia pestis, 195P India		5946, 7288, 9167, 10881
Francisella tularensis,		4725, 6725, 9447, 10280

out, as described below, to ascertain the consistent and maximum release of cellular proteins.

The well-shaken cellular suspensions generated marker ions in greater abundance compared to the centrifuged supernatant (Figure 3). The intact cells, which were not lyzed earlier during the solvent induced lysis, were removed from the supernatant during centrifugation, but were present in the thoroughly mixed cellular suspensions. The laser shot, used for ionizing intact proteins, probably caused additional lysis of the intact cells and subsequent release of marker proteins in greater abundance. Some bacteria such as *Brucella melitensis* and *Bacillus anthracis* lyzed more effectively when suspended in basic buffers such as Tris (pH ~ 8.5). Organisms such as *Yersinia pestis* and *Francisella tularensis* required acidic buffers such as 0.1% TFA. However, in most instances, the experiments were conducted using the TFA buffer since both gram-positive and gram-negative bacteria generated reasonable signal and in addition, it provides means to analyze all types of microorganisms by a single experiment. Ultrasonic disruption of liquid suspensions containing the intact cells improved the sensitivity considerably. However, when bacterial samples were disrupted, the pH of the buffer used for suspending the cells, had no significant effect on the release of the biomarkers. Any variations introduced in the sample preparations had no noticeable effect in the acquired spectrum. In addition, the lyophilized disrupted cells can be stored in the freezer for prolonged period of time. Reconstitution of the lyophilized material can be carried out effectively prior to analysis. These processes can be repeated few times without the disintegration of the marker proteins.

During the delayed extraction MALDI-MS investigations, all of the mass measurements were carried out using external standards. Generation and mass measurements of biomarkers for most of these bacteria were carried out consistently at four different sites and five analysts using six different instruments. *Francisella tularensis* grown in four different media was analyzed and the results indicated that regardless of the growth conditions, most of the observed biomarkers were identical. The slight differences detected between the measured molecular masses of the same marker proteins in different samples are due to the anticipated errors, while assigning masses with external mass calibrants (Figure 4).[39] Most of the measured marker proteins for an individual bacterium, both in its virulent and non-virulent forms, are also identical and hence the bacterium could be identified at least up to the species level. However, the biomarkers released from the vegetative and sporulated cells of a single bacterium were quite different (Figure 5). The results observed during the analysis of over 15 different organisms in their corresponding vegetative and sporulated states, emphasized that the identification of biomarkers released from different growth states of any single bacterium is essential for the accurate analysis of unknown samples.

Ionic profiles obtained during whole-cell MALDI-MS analysis of four well-known human and animal pathogens are easily distinguishable even by visual inspection (Figure 6). Profiles obtained from several major classes of human pathogens are distinct, as are comparative spectra from closely related *Bacillus* and

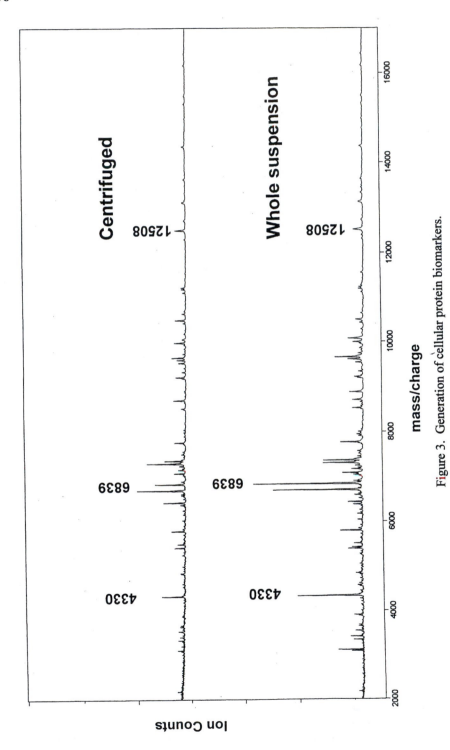

Figure 3. Generation of cellular protein biomarkers.

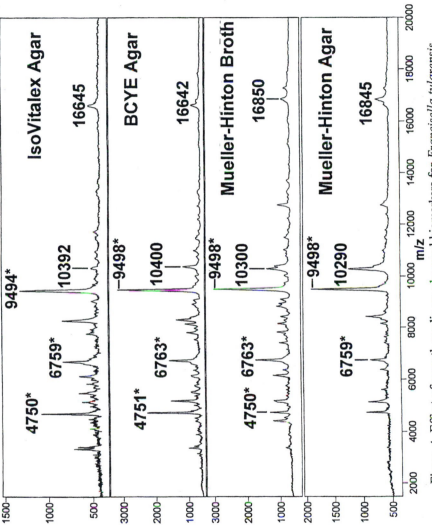

Figure 4. Effect of growth media on observed biomarkers for *Francisella tularensis*. Reproduced with Permission from Reference 39. Copyright 1996.

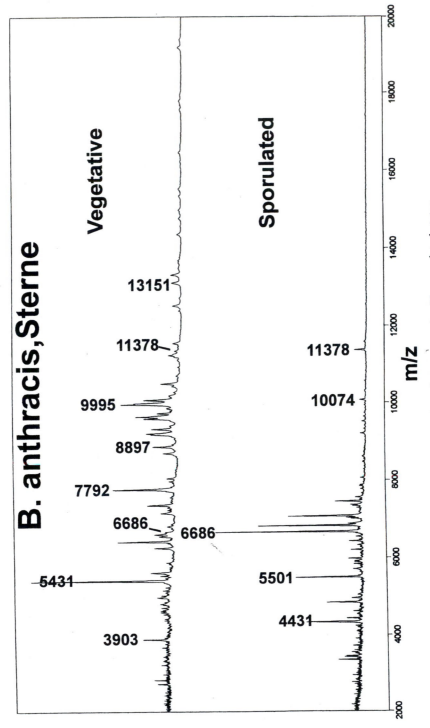

Figure 5. Biomarkers from vegetative Vs sporulated states.

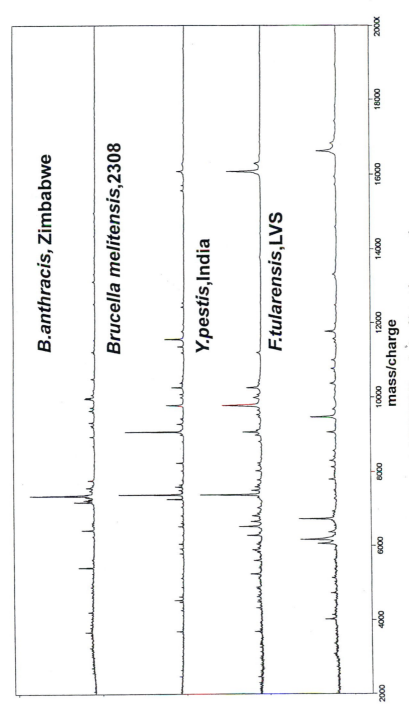

Figure 6. MALDI-MS analysis of intact human pathogens

Brucella species. More importantly, a number of intense and consistently occurring ions can be identified which are characteristic of the genus and the species.

Biomarkers. Results obtained from the whole cell MALDI-MS analysis of four *Bacilli* species are shown in Figure 7. As expected, the most notable ions observed for *B. cereus* group organisms were totally absent for *B. subtilis* (Table 3). Common ion detected in all four of the *Bacillus* species (Table 3) has been designated the specific marker for genus, *Bacillus.* Similar observations were made in the case of other pathogens as well (Table 3). A series of different strains of the four pathogens thus far investigated and other related species of the same genus were analyzed. The species specific biomarkers derived from the mass spectra are also listed in Table 3. As mentioned earlier, the genetically similar *Bacillus cereus* organisms,[7] which include *Bacillus anthracis, Bacillus thuringiensis,* and *Bacillus cereus,* differ in their virulence and pathogenicity. The presence of human pathogen, *B.anthracis,* must be rapidly and distinctly established in samples in order to detect the hazardous exposure to humans and avoid mortality.[2, 3, 36-38] MALDI-MS spectra of the intact cells of *Bacillus cereus* sub-species group organisms enabled the distinction of the individual organisms much more rapidly than the earlier methods (Figure 7; Table 3).

Distinction of a bacterium at the most basic level involves the identification of the individual strains. This could be very challenging since there could be numerous wild-type strains for each microorganism. The numbers of the pathogenic strains may increase due to natural variations in environment as well as induced changes in the genome. In addition, many known wild-type pathogenic strains, grown in the laboratory after removal of portions of DNA and adjustment of growth conditions, could be converted into their corresponding non-virulent and non-pathogenic strains. In fact, such closely related, non-pathogenic strains are commonly used in vaccines for livestock and humans. The MALDI-MS analysis of intact cells of representative virulent (Sterne) and non-virulent (d-Sterne) strains of *B.anthracis* is illustrated in Figure 8. Even though, most of the observed biomarkers in both instances are the same, some biomarkers present in the virulent strain are either considerably reduced or totally absent in its corresponding non-virulent form (Figure 8). There is significant information in the recorded MALDI-MS spectra of any microorganism, which could be used to identify the bacterium up to its strain level as well as distinguish the virulent form of a specific strain from the corresponding non-virulent form.

Analysis of Bacterial Mixtures. Bacteria commonly occurring in the background as well as multi-component bacterial mixture will pose a challenge in identifying the microorganism(s) present in the sample. The developed methodology should be applicable to identify various bacteria present in the sample as well as background material(s). Numerous artificial mixtures containing 2-4 bacterial pathogens along with *E.coli* were prepared and analyzed directly, without any sample purification. Known biomarkers, present in abundance, assigned for individual bacterium could be easily detected in the MALDI-mass spectra of the mixtures. The identification of the individual components in a bacterial mixture is illustrated in Figure 9. This demonstrates a great potential for identifying the individual components in a bacterial mixture.

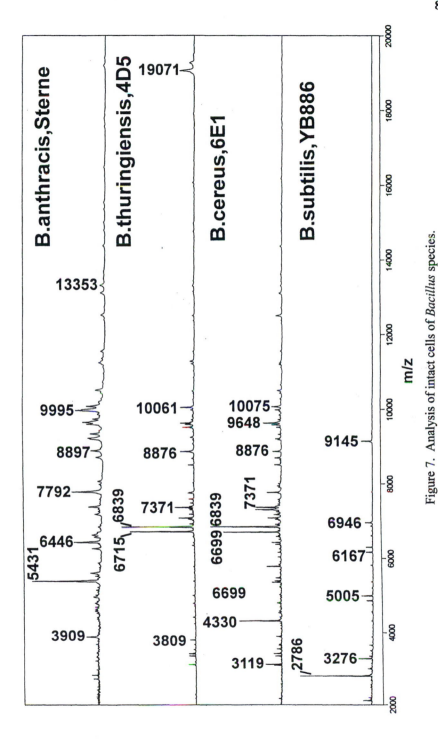

Figure 7. Analysis of intact cells of *Bacillus* species.

Table 3. Biomarkers from MALDI-MS Investigations

Organism	Genus	Species
B. anthracis	6845,7189,7389,7793,8877 9674,11573,12551	2715,2821,5560,5638,6632, 8880,10100
B.thuringiensis	6840,7187,7371,7755,8876 9647,11559,12525	3809,4438,8876,19070
B.cereus	6840,7168,7371,7773,7773 9648,11559,12514	4331,5798,6380
B.subtilis	7170,7780,11570	2786,2993,6167,7170,7780, 9146

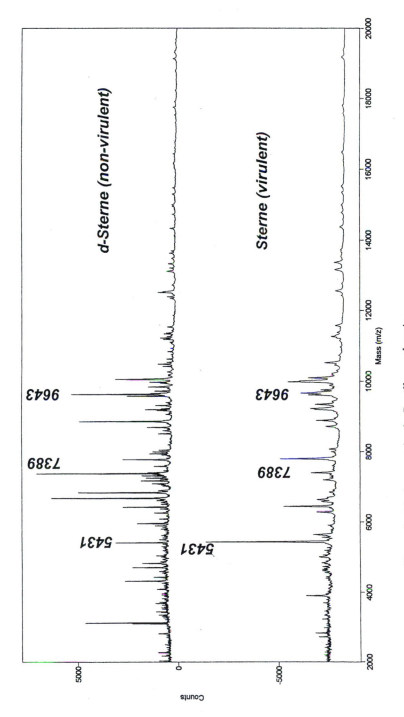

Figure 8. Marker proteins in *Bacillus anthracis*.

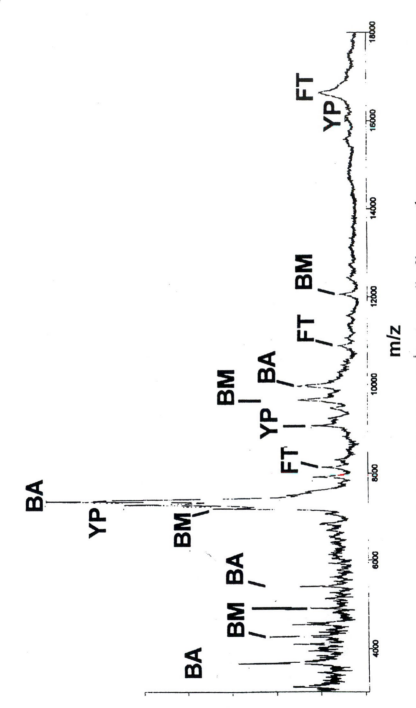

Figure 9. Analysis of mixture containing intact cells of human pathogens.

Automated MALDI-MS Data Processing.[17] In order to automate the process, an algorithm was developed both for the storage of acquired standard spectra in a specified MS library and comparison of the sample spectrum with the library spectra. The individual acquired mass spectral file was converted into the corresponding ASCII file. Relevant information such as the genus, species, strain, growth conditions and the cellular state of the investigated bacteria, mass range of the generated spectra and instrumental (MS) parameters are included. The full spectrum is read, reduced by selecting the maximum intensity peaks at every 10 amu range, and the reduced spectrum along with the annotations was stored. Sample input file, generated by a similar process is subjected to comparison with the library spectra. During the process, the sample spectrum loops through each library file and cross correlation against each library file is performed and the individual scores are stored and sorted. Ten library spectra with highest scores assigned during matching are displayed with the corresponding scores. Comparative display of the sample spectrum and the library spectrum with the highest score is also accomplished. In most instances, the genus and the species of the bacteria present in the samples were correctly identified. Comparative displays of the sample and library spectra origination during the searches are illustrated in Figure 10. All of the library spectra were generated in the linear delayed extraction mode using a research type mass spectrometer. In order to test the process more rigorously, crude sample spectra were using a bench-top Voyager-DE mass spectrometer. Despite this major variation, identifications were accurate in most instances (~90%) at least up to the species level. Optimization of the search routines is presently underway to enable the identification of the bacteria up to its strain level.

LC/Electrospray-MS Method[34]

Despite the success of the MALDI-MS methodology, we investigated an alternate method for the identification of bacterial pathogens. within the targeted time frame of 10-15 minutes. This approach should not only lead to the future developments involving the on-line sample preparation and separation capabilities, but also to more specific tandem mass spectrometric methods. During this investigation, we injected (1 µl) of the bacterial suspension into a fused silica capillary reverse phase (C_8) micro-HPLC column. The separated components were introduced on-line into the ESI source and the resulting multiply charged ions were focused and detected in an Ion-trap mass spectrometer. A mixture of specific peptides and proteins were detected when individual bacteria were analyzed. It was deduced that the TFA present in the solvent was sufficient to cause the cell lysis and the release of the cellular proteins along with the peptides. When the sample was lysed in 20% acetonitrile in 0.1% TFA, only proteins were detected (Figure 11). Thus, we could selectively utilize either protein and/or peptide biomarkers for our investigations. A series of specific biomarkers for some pathogenic as well as non-pathogenic gram positive and negative bacteria were generated from the acquired LC/ESI-MS data (Table 4).

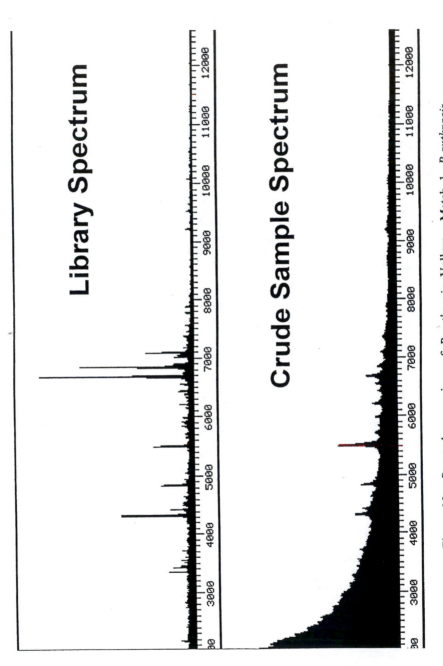

Figure 10. Spectral comparison of *B.anthracis*, Vollum. Match 1, *B.anthracis*, Vollum; score 10,746. Match 2, *B.anthracis*, ANR2; score 6,462.

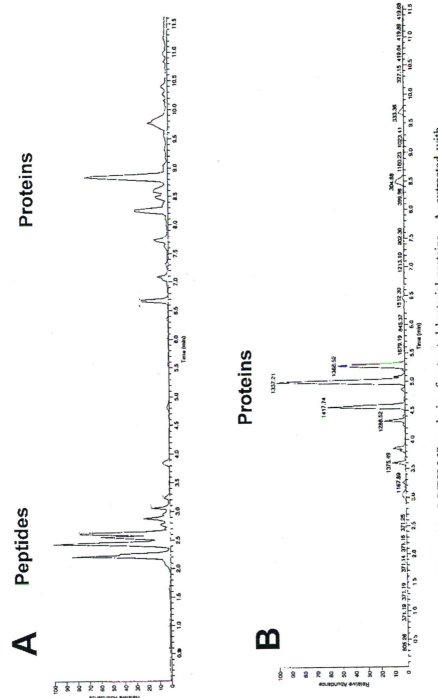

Figure 11. LC/ESI-MS analysis of extracted bacterial proteins. A, extracted with 0.1% aqueous TFA. B, extracted with 20% acetonitrile in 0.1% aqueous TFA.

Table 4. Biomarkers from ESI Investigations

Organism	Genus	Species
B.anthracis	6429,7775,9188,9336, 9996	4231,4942,5835,5913, 6201,6683,7017,7558, 8465,8846,9524,10071, 11376
B.thuringiensis	7774,9160,9335,9994	2790,2805,3441,3798, 3850,5578,5616,6334, 6611,9670,9734
B.cereus	6429,7771,9187,9332, 9992	3937,5458,5549,6253, 6905,10507,11234
B.subtilis	7771,9187	6511,7301,7470,9062, 9891,11140,15201

This methodology clearly depends on observing a set of few particularly abundant proteins, with better ionization properties under the ESI conditions, which are specific to a given organism. The biomarker set used to determine an organism's identity could be created without the knowledge of the genome or the structure of the individual proteins. Even though the observed biomarkers were dependent upon extraction conditions, cell disruption played a significant role in releasing the protein biomarkers from microorganisms reproducibly as well as in greater abundance. In addition, cell disruption greatly improved the stability of the bacterial suspensions.

All of these biomarkers could be generated reproducibly on repeated analyses. However, these biomarkers differed from the ones observed during MALDI-MS investigations (Table 5). Some proteins have been known to ionize under either ESI or MALDI conditions. Even though thousands of proteins are expressed in any given cell, only a fraction of the proteins released from the cells are identified during any of these mass spectrometric methods. Although the sets of biomarkers observed in the ESI-MS analyses are different from those obtained by MALDI-MS methods, biomarkers representing the genus, species and strain of individual organism can be generated reproducibly based on the adopted method. Hence, differences between measured biomarkers under ESI and MALDI conditions should not have any undesirable effect on the bacterial identification process. An artificial mixture containing equal amounts of *B.anthracis*, *B.thuringiensis*, *B.cereus*, and *B.subtilis* was made and analyzed by the same procedure (Figure 12). The identification of the organisms was accomplished by comparing the acquired sample spectrum with the corresponding standard spectrum (Figure 13).

Real-time Identification of Bacteria.[4,33] An automated real-time detection of bacteria present in unknown samples was accomplished as follows. LC/MS data of disrupted and filtered bacterial suspensions were acquired. As mentioned earlier, numerous peptides and proteins with molecular masses ranging from 3 to 30 kDa were detected for each of these microorganisms. The ESI- mass spectrum of individual protein has specific pattern of charge states which could be converted into a one dimensional Boolean array (Figure 14) . A set of biomarkers for one or more organisms can be stored in two-dimensional Boolean array to generate a biomarker library. During the analysis, "logical AND" operation is performed to match the spectral patterns with the biomarker library entries (Figure 15). A match is identified and presented in real-time when sufficient ions corresponding to the charge states of one or more biomarkers are present. At the end of the analysis, results for all of the collected spectra are displayed and the length of the bar in the final display is a measure of the area under the peak for a given biomarker. This approach is very robust and can be used to analyze very crude samples. Co-eluting components can be correctly identified despite noisy spectra. When a mixture of *B.anthracis* and *Brucella melitensis* was analyzed, both components were correctly identified (Figure 15).

Conclusions.

The protein biomarkers measured by either MALDI-MS or ESI-MS methods could be

Table 5. Comparison of Biomarkers from MALDI- & ESI- MS Data

Organism	Genus		Species	
	ESI	MALDI	ESI	MALDI
B.anthracis	6429,7775, 9188,9336, 9996	6845,7189,7389, 7793,8877,9674, 11573,12551	4231,4942,5835, 5913,6201,6683, 7017,7558,8465, 8846,9524, 10071,11376	2715,2821, 5560,5638, 6632,8880, 10100
B.thuringiensis	7774,9160, 9335,9994	6840,7187,7371, 7755,8876,9647, 11559,12525	2790,2805,3441, 3798,3850,5578, 5616,6334,6611, 9670,9734	3809,4438, 8876,19070
B.cereus	6429,7771, 9187,9332, 9992	6840,7168,7371, 7773,7773,9648, 11559,12514	3937,5458,5549, 6253,6905, 10507,11234	4331,5798, 6380
B.subtilis	7771,9187	7170,7780, 11570	6511,7301,7470, 9062,9891, 11140,15201	2786,2993, 6167,7170, 7780,9146

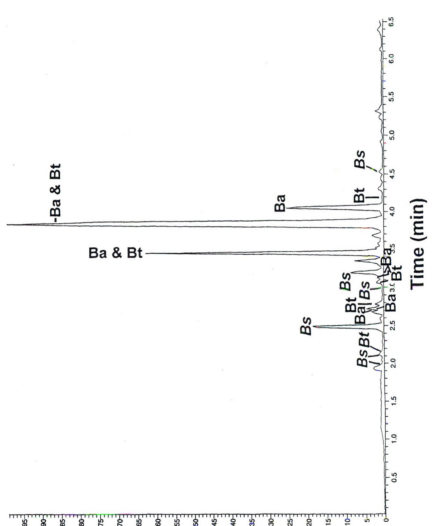

Figure 12. LC/MS analysis of a bacterial mixture. Ba, *B.anthracis*; Bs, *B.subtilis*; Bt, *B.thuringiensis*.

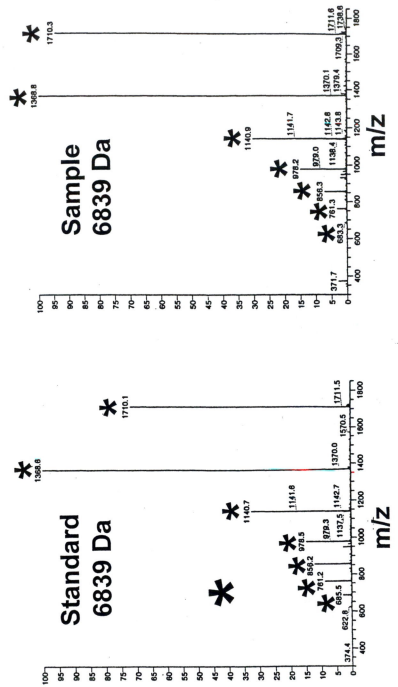

Figure 13. ESI-mass spectra of a biomarker in standard and sample spectra of *B.anthracis*.

Identification of
Bacterial (Protein) Biomarkers

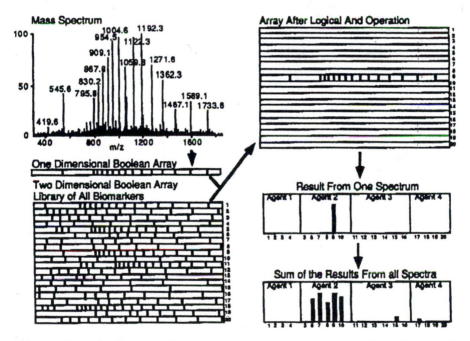

Figure 14. On-line identification of bacterial pathogens. Reproduced with permission from reference 40. Copyright 1997.

94

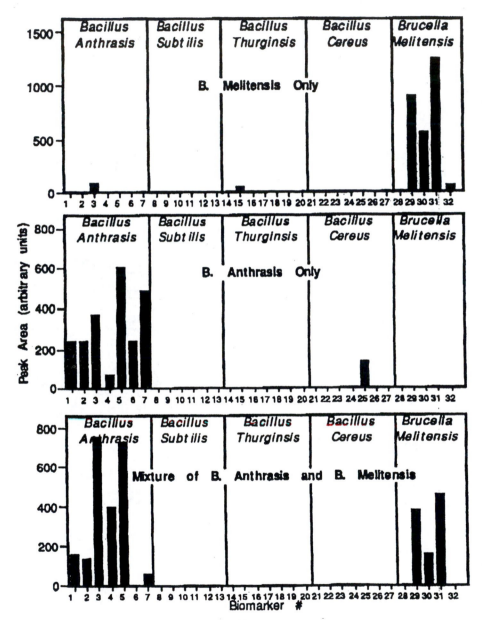

Figure 15. Automated library search results from ESI-MS spectral data. Reproduced with permission from reference 40. Copyright 1997.

used to identify a single or multiple organisms present in a sample. In both cases, the total analysis time was less than 10 minutes and the results were accurate and reproducible. Only off-line clean up of the sample could be performed prior to MALDI-MS analysis. Where as, this process could be performed on-line during the ESI-MS operations. In addition, tandem mass spectrometry could be applied to introduce additional stages of MS analysis for increasing sensitivity as well as specificity. Automated data reduction is possible in both instances. There is a great potential in the near future for development of fully automated MS/MS method(s) for rapid, reliable, and highly sensitive identification of one or more microorganisms . The application potential of such simple yet powerful methodologies in monitoring, industrial, and research fields is immeasurable.

Acknowledgments

The authors gratefully acknowledge Dr. Randolph Long (US Army Edgewood RDE Center) and Dr. Hari Nair (Geo-centers, Inc., APG, MD) for their valuable suggestions during the preparation of this manuscript. In addition, they also acknowledge the US Army laboratories at Dugway Proving Ground, Utah, USA for their generous supply of gamma radiated bacterial pathogen samples.

References

1. Mass Spectrometry for the Characterization of Microorganisms, Fenselau, C., (Ed.), ACS Symposium Series, Washington, DC, **1994**, Volume 541.
2. Tournabene, T. G., CBIAC Internal Report on the Development of Biochemical Data Base for the Bacterial Pathogens, prepared for U.S. Army Edgewood RD&E Center, Aberdeen Proving Ground, MD **1995**.
3. Handbook of New Bacterial Systematics, Goodfellow, M and O'Donnell, A.G., (Ed.), Academic Press, San Diego, CA **1993**.
4. Proceedings of First Joint Services Workshop on Biological Mass Spectrometry, **1998**, Aberdeen Proving Ground, MD, USA.
5. Meuzelaar, H. L. C., Kistemaker, P. G., Eshius, W., and Engel, H. W. B., In: Rapid Methods and Automation in Microbiology, Newsom, S.W.B., Johnston, and H. H., Eds., **1976**, 225-230.
6. Snyder, A. P., McClennen, W. H., Dworzanski, J. P., and Meuzelar, H. L. C., Anal. Chem., **1990**, 62, 2565.
7. Black, G. E., Fox, A., Fox, K., Snyder, A. P., and Smith, P. B. W., Anal. Chem. **1994**, 66, 4171-4176.
8. Cole, M. J., and Enke, C.G., Anal. Chem., **1991**, 63, 1032-1038.
9. Wunschel, D. S., Fox, K. F., Fox, A., Bruce, J. E., Muddiman, D. C., and R. D. Smith, Rapid Commun. Mass Spectrom., **1996**, 10, 29-35.
10. Hurst, G. B., Doktycz, M. J., Vass, A. A., and Buchanan, M. V., Rapid Commun, Mass Spectrom., **1996**, 10, 377-382.

96

11. Schleiffer, K. H., Ludwig, W., and Amann, R., Handbook of New Bacterial Systematics, Goodfellow, M., and O'Donnell, A.G., (Ed.), Academic Press, San Diego, CA, **1993**, pp 463-510.
12. Saiki, R. K., Gelfund, P. H., Stoffel, S., Scharf, S. J., Higuchi, R., Horn, G. T., Mullis, K. B., and Erlich, H. A., Science, **1988**, 239, 487-491.
13. Starnach, M. N., Falkow, S. and Tompkins, L. S., J. Clin. Micro., **1989**, 27, 1257-1261.
14. Hance, A. J., Grandchamp, B., Levy-Frebault, V., Lecosier, D., Rauzier, J. , Bocart, D. and Gicquel, B., Molec. Microbiol., **1989**, 3, 843-849.
15. Krishnamurthy, T., Ross, P. L., and Rajamani, U., Rapid Commun. Mass Spectrom., **1996**, 10, 883-888.
16. Krishnamurthy, T., and Ross, P. L., Rapid Commun. Mass Spectrom., **1996**, 10, 1992-1996.
17. Krishnamurthy, T., Eng, J., Yates, J., and Rajamani, U., Proceedings of the 45th ASMS Conference on Mass Spectrometry and Allied Fields, Palm Springs, **1997**, 60.
18. Liang, X., Zheng, K., Qian, M.G., and Lubman, D. M., Rapid Commun. Mass Spectrom., **1996**, 10, 1219-1226.
19. Holland, R. D., Wilkes, J. G., Sutherland, J. B., Persons, C.C., Voorhees, K. J., and Lay, J. O. Jr., Rapid Commun. Mass Spectrom., **1996**, 10, 1227-1232.
20. Arnold, R. J., and Reilly, J. P., Rapid Commun. Mass Spectrom., **1998**, 12, 630-636.
21. Claydon, M. A., Davey, S. N., Edwards-Jones, V., and Gordon, D. B., Nature Biotechnol., **1996**, 14, 1584-1586.
22. Welham, K. J., Domin, M. A., Scannell, D. E., Cohen, E., and Ashton, D. A., Rapid Commun. Mass Spectrom., **1998**, 12, 176-180.
23. Chong, B. E., Wall, D. B., Lubman, D. M., and Flynn, S. J., Rapid Commun. Mass Spectrom., **1997**, 11, 1900-1908.
24. Wang, Z., Russon, L., Li, L., Roser, D. C., and Long, S. R., Rapid Commun, Mass Spectrom, **1998**, 12, 456.
25. Li, K. W., Geraerts, W. P. M., in Methods in Molecular Biology, Vol. 72: Neurotransmitter Methods, Rayne, R. C., (Ed), Humana Press, Inc., Totowa, NJ, **1996**, pp, 219-223.
26. Jimenez, C. R., Li, K. W., Dreisewerd, K., Mansvelder, H. D., Brussaard, A. B., Reinhold, B. H., Van der Schors, R. C., Karas, M., Hillenkamp, F., Burbach, J. P. H., Costello, C. E., and Geraerts, W. P. M., Proc. Natl. Acad. Sci. USA, **1996**, 94, 9481-9486.
27. van Adrichem, J. H. M., Bornsen, K. O., Conzelmann, H., Gass, M. A. S., Eppenberger, H., Kresback, G. M., Ehrat, M., and Leist, C. H., Anal. Chem., **1998**, 70, 923-930.
28. Dreisewerd, K., Kingston, R., Geraerts, W. P. M., Li, K. W., Int. J. Mass Spectrom and Ion Processes, **1997**, 169/170, 291-299.
29. van Golen, F. A., Li, K. W., de Lange, R. P. J., Jespersen, S., and Geraerts, W. P. M., J. Biol. Chem., **1995**, 270, 28487-28493.
30. Li, K. W., Kingston, R., Dreisewerd, K., Jimenez, C. R., van der Schors, R. C., Bateman, R. H., and Geraerts, W. P. M., Anal. Chem., **1997**, 69, 563-565.
31. Krishnamurthy, T., and Nair, H., Unpublished results.

32. Davis, M. T., and Lee, T. D., J. Amer. Soc. Mass Spectrom., **1998**, 9, 194-201.
33. Davis, M. T., Stahl, D. C., Lee, T. D., and Krishnamurthy, T., 46th ASMS Conference on Mass Spectrometry and Allied Fields, **1998**, Orlando, Fl, USA
34. Davis, M. T., Stahl, D. C., Lee, T. D., and Krishnamurthy, T., Rapid Communications in Mass Spectrom., **1999**, 13. (In print)
35. Krishnamurthy, T., Eng, J., Yates, J., Davis, M., Lee, T. D., Rajamani, U., Wang, S., Heroux, K., Aabshire, T., Ezzell, J., Henchal, E., Proceedings of the First Joint Services Workshop on Biological Mass Spectrometry, Baltimore, MD, **1997**, 31.
36. Kaneko, T., Nozati, R., Aizawa, K., Microbiol. Immunol., **1978**, 22, 639-641.
37. Ash, C., Farrow, J. A. E., Dorsch, M., Stackebrandt, E., Collins, M. D., Int. J. Syst. Bacteriol., **1991**, 41, 343-346.
38. Hofte, H., and Whitely, H. P., Micro. Rev., **1989**, 53, 242-255.
39. Krishnamurthy, T., and Ross, P. L., In Proceedings of the Seventh National Symposium on Mass Spectrometry, Aggrawal, S. K., and Jain, H. C., (Eds.), Indian Society for Mass Spectrometry, Mumbai, India, **1996**, pp 105-122
40. Davis, M. T., Stahl, D. C., Lee, T. D., and Krishnamurthy, Proceedings of the First Joint Services Workshop on Biological Mass Spectrometry, Baltimore, MD, **1997,** pp 133-137

PLANT TOXINS

Chapter 7

Pyrrolizidine Alkaloids: Physicochemical Correlates of Metabolism and Toxicity

Ryan J. Huxtable and Roland A. Cooper

Department of Pharmacology, College of Medicine,
University of Arizona, Tucson, AZ 85724–5050

Pyrrolizidine alkaloids occur in the Leguminosae, Asteraceae and Boraginaceae families. Toxicity is dependent on hepatic oxidation by cytochrome P450 to reactive pyrrolic dehydroalkaloids (DHAs). These alkylate nucleophilic sites on cell metabolites and macromolecules. DHAs undergo hydrolysis at rates ranging from 2.2 sec^{-1} to 0.1 sec^{-1} (i.e. with half-lives ranging from 0.3 sec to 5.3 sec). Both the site of toxicity (liver, lung, central nervous system, etc.) and degree of toxicity (lethality or LD_{50}) are related to the pattern of metabolism. This pattern, in turn, is related to the rate of hydrolysis. This is greater for DHAs having acyclic esters than for cyclic diester DHAs. The rate also appears to be greater for cyclic diester DHAs with smaller substituents at the C14 position. We postulate that the rate of hydrolysis of a given DHA and hence the degree and kind of toxicity expressed by the parent PA is strongly dependent on steric hindrance at the C7 ester linkage.

Pyrrolizidine alkaloids are widespread in plants, being found in more than 400 species (*1-4*). Those having significant toxicological interest are found in *Crotalaria* (Leguminosae), *Senecio* (Asteraceae), and *Symphytum*, *Heliotropium* and *Trichodesma* (Boraginaceae). Pyrrolizidine-containing plants are found throughout the inhabited world and are public health and economic problems in many regions (*5; 6*).

Pyrrolizidine alkaloids are quintessential hepatotoxins, producing veno-occlusive disease leading to ascites, hepatomegaly and cirrhosis (*5*). Some alkaloids are also hepatic carcinogens (*7*). A select number of alkaloids additionally produce extrahepatic toxicity, including pulmonary arterial hypertension and right ventricular hypertrophy (*8; 9*), neurotoxicity (*10*), and damage to other organ systems (*11*).

Veno-occlusive Disease in Humans

The most widespread human exposure to pyrrolizidine alkaloids comes from contamination of food grain with the seeds of pyrrolizidine-containing plants. The grain is processed into food (bread, tortillas, chappaties, etc.) resulting in chronic low-level exposure of those consuming the food (e.g. South Africa (12; 13)). Toxicity results from the cumulative exposure over a period of time (Table 2 in (6)). Pyrrolizidine-containing plants are intentionally consumed as vegetables (14; 15), herbs and bush teas (16; 17). Such exposure can lead to intoxication. In industrialized countries, there are increasing reports of pyrrolizidine poisoning due to the use of herbal supplements prepared from alkaloid-containing plants (18-24). Table I contains a partial list of reported pyrrolizidine poisonings.

An outbreak in Afghanistan in the mid-1970s is an example of chronic low-level contamination of food supplies. Approximately 7800 people suffered pyrrolizidine poisoning after contamination of wheat with *Heliotropium* seeds (25). This outbreak occurred in a region that was suffering economic and social disruption because of the Soviet invasion. This is not an uncommon occurrence in underdeveloped regions of the world, a similar phenomenon occurring in Tadjikistan following the break-up of the Soviet Union (26).

Direct consumption of pyrrolizidine-containing plants in the form of "bush teas" was the cause of a long-lasting epidemic in the West Indies (17). Investigation of this epidemic led to the recognition of veno-occlusive disease as a clinical entity (16). Although often described as being caused by the use of leaves of *Crotalaria*, the original literature supports both *Senecio* and *Crotalaria* as causative agents (5).

Cases of pyrrolizidine poisoning resulting from the use of herbs or herbal supplements have been reported from the US and other industrialized nations, primarily from *Senecio* (23; 27; 28) (or plants such as *Adenostyles* that contain also contain *Senecio* alkaloids (29)) and *Symphytum*-containing products (18; 19; 22; 30; 31).

Structure and Toxicity of Pyrrolizidine Alkaloids

Typical structures of toxic pyrrolizidine alkaloids are shown on Figure 1. A common structural feature is the presence of two fused five-membered rings with a bridgehead nitrogen and a double bond in the 1,2 position. The alkaloids differ in the size of the macrocyclic diester ring, the *Crotalaria* type alkaloids having an 11-membered ring and the *Senecio* type a 12-membered ring. The alkaloids also differ in the pattern of substitution, degree of unsaturation and level of hydroxylation in the diester ring. Acyclic diester alkaloids also occur, such alkaloid being responsible for the toxicity of *Symphytum* (comfrey) and *Heliotropium* species.

Of the alkaloids in Figure 1, monocrotaline was one of the causative agents in the epidemic of veno-occlusive disease of the liver that plagued the Caribbean for so long (17). The structurally similar trichodesmine was largely responsible for an epidemic of poisoning in Uzbekistan in the 1950s. This outbreak was associated with vertigo, vomiting, and death from respiratory failure (15). Retrorsine was a major alkaloid involved in the gordolobo poisonings detected in Arizona in the 1970s (32). Seneciphylline is another constituent of gordolobo (*Senecio longilobus*) (32; 33). It produces lung disease and cor pulmonale in young rats under appropriate conditions (34) and it is also mutagenic (35; 36).

Structural differences in the diester ring are associated with marked differences in toxicity. Pyrrolizidine alkaloids differ not only in the degree of toxicity (e.g. the LD_{50}; Table II) but in the type of toxicity (e.g. the organ system affected). It is the

Table I. Some Cases of Pyrrolizidine-Induced Veno-Occlusive Disease and Other Conditions in Humans

Location	Plant Involved	Year Reported	Number of Cases	References
Tadjikistan	*Heliotropium*	1993	3906	*(26; 56)*
New Zealand	*Symphytum*	1990	1	*(18)*
New Zealand	*Symphytum?*	1989	1	*(31)*
New York	*Symphytum*	1989	1	*(19)*
Switzerland	*Tussilago farfara; Petasites officinalis*	1988	1	*(57; 58)*
England	*Symphytum*	1987	1	*(59)*
Switzerland	*Senecio*	1985	2	*(27)*
Massachussetts	*Symphytum*	1985	1	*(22)*
Hong Kong	*Heliotropium lasciocarpum*	1983	4	*(60-62)*
India (Bihar and Madhya Pradesh)	*Crotalaria nana*	1981	67	*(63)*
Arizona	*Senecio*	1980	4	*(64)*
Northwest India	*Heliotropium*	1978	7	*(65)*
India	*Heliotropium*	1978	6	*(66)*
Arizona	*Senecio longilobus*	1977	2	*(23; 24)*
Afghanistan	*Heliotropium popovii*	1976	7200	*(25)*. 14
Central India	*Crotalaria* species	1976	67	*(25)*
India	Unknown	1976	25	*(67; 68)*
Britain	Unknown	1976	1	*(69)*
Ecuador	*Crotalaria juncea (?)*	1976	1	*(70)*
USSR (Asia)	*Heliotropium*	1965	61	*(71)*
India	Unknown herbs	1963	2	*(72)*
Barbados	*Crotalaria*	1956	Several	*(73)*
West Indies	*Crotalaria, Senecio*	1954	Several hundred	*(16; 17)*
USSR (Asia)	*Heliotropium*	1952	28	*(74)*
Jamaica	*Crotalaria, Senecio*	1951	137	*(75)*
Uzbekistan	*Trichodesma incanum*	1950	>200	*(15)*
South Africa	*Senecio*	1950	12	*(13)*
Barbados	*Heliotropium angiospermum; H. indicum*	1949	?	*(76)*
South Africa	*Senecio ilifolius; S. burchelli*	1920	Many	*(12)*

Adapted from (6). This list is not intended to be exhaustive. For more complete literature coverage, see e.g. (5; 77; 78).

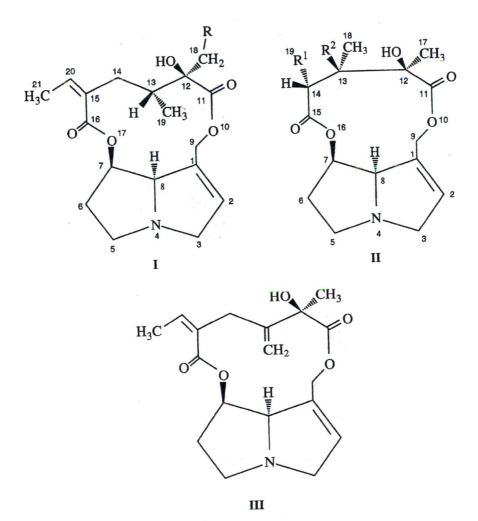

Figure 1. Typical macrocyclic diester pyrrolizidine alkaloids with numbering system: **I:** R=H Senecionine; R=OH Retrorsine; **II:** R^1=CH$_3$, R^2=OH Monocrotaline; R^1=CH$_3$, R^2=OH Trichodesmine; **III:** Seneciphylline

purpose of this chapter to review our work on how minor structural differences affect the metabolism of pyrrolizidine alkaloids, and how metabolism affects toxicity, both qualitatively and quantitatively.

Metabolism of Pyrrolizidine Alkaloids

It has been well established that pyrrolizidine alkaloids must undergo metabolism before toxicity is expressed. To date, metabolism has only been demonstrated in the liver, efforts to show metabolism in other tissues being unsuccessful. However, it is possible that low levels of metabolism may be found at extrahepatic sites, and that such metabolism may be of toxicological consequence.

In 1968, Mattocks (37) proposed that pyrrolizidine alkaloids were metabolized to pyrrolic intermediates, or dehydroalkaloids (Figure 2). Although chemically this is a simple dehydrogenation, it has the consequence of changing stable diester alkaloids into rapidly hydrolyzed products incapable of existence in aqueous solution for more than a few seconds. Pyrroles exhibit such instability because generation of a cation at C7 by an S_N1 process is favored by resonance stabilization by the adjacent pseudoaromatic ring (and, of course, resonance stabilization of the departing carboxylate anion) (Figure 3). The cation is discharged by a nucleophile. If this is water, the resulting product is 6,7-dihydro-7-hydroxy-1-hydroxymethyl-5H-pyrrolizine (DHP), a relatively stable pyrrole (formation of a cation at C7 from this compound would release a nonresonance stabilized hydroxide anion). If the nucleophile is an amine or sulfhydryl on a cell macromolecule, alkylation of the macromolecule results. This presumably provides the molecular basis of the toxicity of pyrrolizidine alkaloids.

Mattocks' hypothesis was so attractive that it was quickly adopted, although proof of the actual nature of the initial metabolite was lacking. The production of DHP following *in vitro* incubation of pyrrolizidine alkaloids with rat liver microsomes (38) is presumptive evidence of the intermediacy of a dehydroalkaloid. The involvement of cytochrome P450 in this conversion is suggested by observations that pretreatment of rats with phenobarbital leads to increased DHP production by liver microsomes while pretreatment with SKF-525A decreases DHP production (39; 40). In the intact rat, metabolism of monocrotaline to pyrrolic metabolites (41) was increased by pretreatment with phenobarbital, but decreased by treatment with SKF-525A, again suggesting involvement of cytochrome P450. At least in the case of senecionine, the P450 species has been established as IIIA4 (42).

Perfusion of monocrotaline through the isolated rat liver with the bile duct cannulated resulted in the appearance of a pyrrolic metabolite in the perfusate and two pyrroles in the bile (41; 43; 44). Tandem perfusion of liver perfusate into the isolated lung led to inhibition of serotonin transport by the lung (41). The same phenomenon had earlier been observed in monocrotaline-exposed rats *in vivo* (45). In the tandem organ perfusion, the cytochrome P450 inhibitor, SKF525A, blocked inhibition of serotonin transport. Subsequently, the metabolite released into perfusate was confirmed to be dehydromonocrotaline (46), in keeping with the hypothesis of Mattocks (37). Dehydromonocrotaline was shown to be generated by incubation of monocrotaline with rat liver microsomes, perfusion of isolated liver with monocrotaline, or superfusion on rat liver slices with monocrotaline. Dehydromonocrotaline generated by all these systems was identified by trapping it in a

Table II. Pyrrolizidine Alkaloid Toxicity in Rats

Alkaloid	Structural type of Ester	LD_{50} (mg/kg)
Senecionine	12-member macrocycle	50
Seneciphylline	12-member macrocycle	77
Trichodesmine[a]	11-member macrocycle	25
Retrorsine	12-member macrocycle	34
Monocrotaline	11-member macrocycle	109
Heliotrine	9-monoester	296
Echimidine	Acyclic 7,9-diester	ca. 200
Lycopsamine[b]	9-monoester	>1000

[a]Data from (79). [b]Data from (80). Other data from (1)

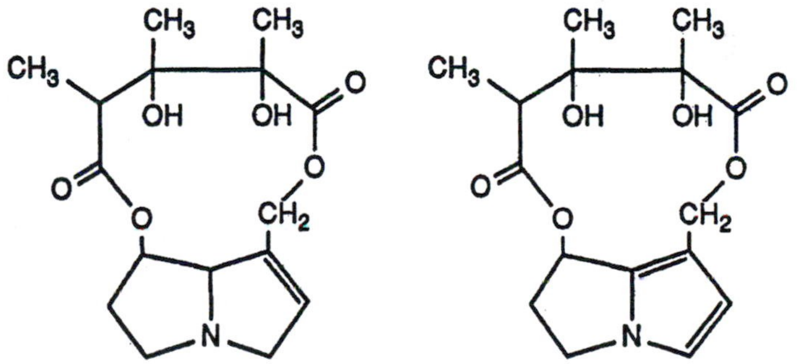

Figure 2. Relationship between pyrrolizidine alkaloids, exemplified by monocrotaline (Left) and their pyrrolic dehydroalkaloid metabolites, exemplified by dehydromonocrotaline (Right).

stable form on thiopropylsepharose resin. The resin acts as a nucleophile, covalently binding diester pyrroles, but not DHP, via a thioether linkage. The thioester can be solvolyzed with acidified silver nitrate. When solvolysis is performed in ethanol, a stable ethyl ether is formed, the ether position indicating the site at which the dehydromonocrotaline had reacted with the thiopropylsepharose resin (46). The principle underlying this procedure is shown on Figure 4. The ethyl ether was identified by mass spectral comparison with authentic material.

This procedure unequivocally confirms the hepatic metabolism of pyrrolizidine alkaloids to dehydroalkaloids. The trapped intermediate is demonstrated to be a pyrrole. In addition, it has to be a pyrrole capable of alkylating thiopropylsepharose at C7. Direct comparison with authentic dehydroalkaloids, generated in nonaqueous solvents by chemical dehydrogenation of the parent alkaloid (47), show that these are capable of doing this. Other putative intermediates, such as DHP or N-oxides, are not.

In collaboration with Mattocks, we identified the major biliary metabolite as 7-glutathionyl-6,7-dihydro-1-hydroxymethyl-5H-pyrrolizine (GSDHP) (43; 47). The other trace metabolite was DHP, the hydrolytic product of dehydroalkaloids. GSDHP is formed, in varying amounts, by every pyrrolizidine alkaloid we have examined (48). *In vivo* experiments show that GSDHP is less toxic than monocrotaline, and thus cannot contribute to monocrotaline-induced veno-occlusive disease or pulmonary arterial hypertension. Regardless of the parent pyrrolizidine alkaloid, GSDHP formation is a function of GSH content of the liver. Agents that decrease GSH concentration also decrease GSDHP formation (48).

The pattern of metabolism of pyrrolizidine alkaloids can now be summarized as shown on Figure 5.

Comparative Metabolism of Pyrrolizidine Alkaloids. In the isolated, perfused liver, we now have the tools for measuring dehydroalkaloid release into perfusate, GSDHP release into bile, and covalent binding of pyrroles to liver tissue (presumably the biochemical basis for toxicity). This allows a "check book" comparison of metabolic patterns in a variety of pyrrolizidine alkaloids of varying toxicity (Table III) (49).

Several points are apparent: (i) the lower the LD_{50} (i.e. the higher the lethality), the greater the amount of dehydroalkaloid released into the perfusate; (ii) the primarily hepatotoxic *Senecio* alkaloids, retrorsine and seneciphylline, show higher levels of pyrroles covalently bound to liver compared to the alkaloids having marked extrahepatic effects (however, both the *Senecio* alkaloids can produce pneumotoxicity under certain conditions); (iii) The *Senecio* alkaloids show higher biliary release of GSDHP than the *Crotalaria*-type alkaloids; and (iv) *Crotalaria*-type alkaloids showing marked extrahepatic toxicity have a higher ratio of perfusate release of dehydroalkaloid to biliary release of GSDHP. These findings suggest that both the quantitative and qualitative aspects of toxicity relate to varying patterns of metabolism.

Half-Lives of Hydrolysis of Dehydropyrrolizidines. The instability of dehydroalkaloids has long been recognized. The first successful approach to estimating rates of hydrolysis involved a delay line between the introduction of dehydromonocrotaline into an aqueous stream and the "trapping" of residual dehydroalkaloid by thiopropylsepharose resin downstream (50). However, the method is slow and inconvenient and it is difficult to assess accuracy and reproducibility.

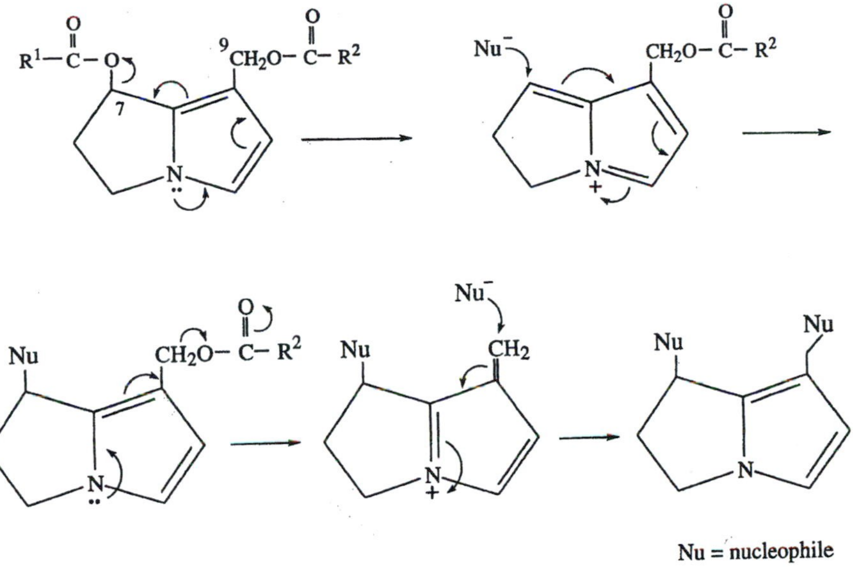

Figure 3. Stabilization of anion at C7 of a dehydroalkaloid during nucleophilic substitution by an S_N1 mechanism. Substitution at C7 is favored over C9 due to greater stabilization of the secondary carbonium ion at C7.

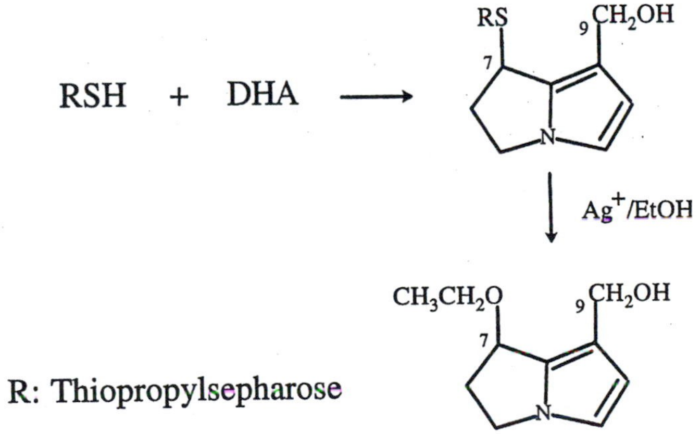

Figure 4. "Trapping" of dehydroalkaloid generated from incubation of a pyrrolizidine alkaloid with rat liver microsomes or perfusion of isolated liver with a pyrrolizidine alkaloid (46; 49). The dehydroalkaloid (DHA) reacts covalently with thiopropylsepharose® resin to form a stable thioether adduct. The adduct is solvolyzed by silver nitrate in ethanol yielding a stable ethylether identified by TLC and mass spectrometry. The ether linkage indicates the site of thioether formation on the resin. DHP or the parent alkaloids do not react with the resin.

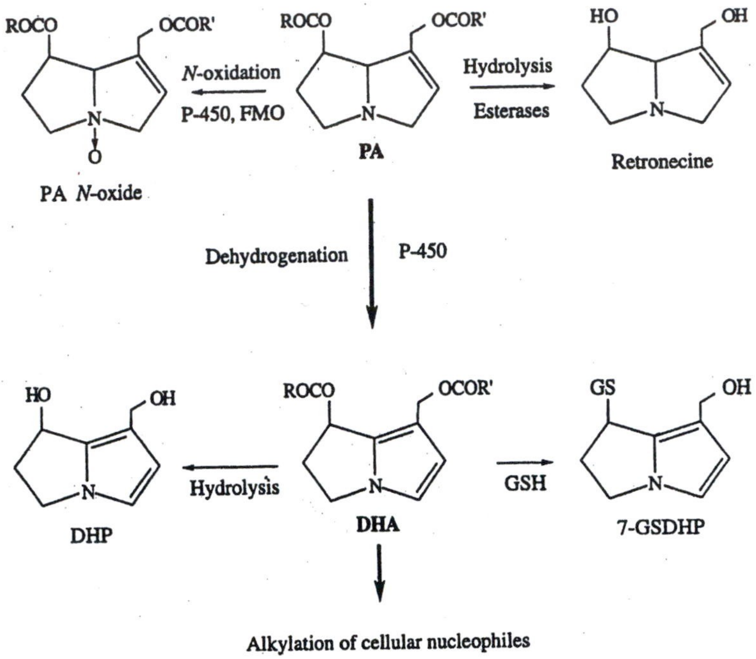

Figure 5. Hepatic metabolism of pyrrolizidine alkaloids. The initially formed dehydroalkaloid (DHA) is an alkylating agent, this presumably being the basis of its toxicity. It can alkylate protein and nucleic acids in the liver, be hydrolyzed to the relatively nontoxic 6,7-dihydro-7-hydroxy-1-hydroxymethyl-5H-pyrrolizine (DHP), or be conjugated with glutathione to the polar, detoxified metabolite, 7-glutathionyl-6,7-dihydro-1-hydroxymethyl-5H-pyrrolizine (GSDHP). The latter is released into bile. Depending on the parent alkaloid, DHA may also survive long enough to be released into the circulation to damage organs downstream from the liver.

Table III. Comparative Metabolism of Pyrrolizidine Alkaloids in the Isolated Perfused Rat Liver

Alkaloid	LD_{50} ($\mu mol/kg$)	Pyrrole in Perfusate	Binding in Liver	Bile GSDHP	DHA/GSDHP
			(nmol/g liver)[a]		
Trichodesmine	57	688	65 ± 7	80 ± 7	5.85
Retrorsine	89	539	197 ± 23	881 ± 45	0.19
Seneciphylline	231	397	91 ± 5	404 ± 33	0.41
Monocrotaline	335	257	57 ± 8	180 ± 16	0.64

DHA: Dehydroalkaloid. Data from (49)

Hydrolysis of the ester linkages in dehydroalkaloids is accompanied by the release of protons from the carboxylate functions generated. We took advantage of this to develop a rapid and accurate potentiometric method for the determination of hydrolysis rates (Table IV) (*51*). Among the dehydroalkaloids examined, rates varied by about 1.5 orders of magnitude.

Hydrolysis is be initiated by S_N1 cleavage of the carbon-oxygen bond at C7. The driving force for this step is the generation of a resonance-stabilized carboxylate anion in the one moiety, and a secondary carbonium ion in the other moiety that is stabilized by delocalization onto the pseudoaromatic ring. Reaction is completed by the discharge of the carbonium ion by a water molecule, with release of a proton. A similar process occurs at C9, with the intermediacy of a less stable primary carbonium ion. Alkylation by cell nucleophiles occurs by a similar process, the nucleophile (GSH or macromolecule) taking the place of the water.

Given this process, what factors modify hydrolysis rates? Initial S_N1 cleavage occurs at C7 rather than C9 due to the greater stability of a secondary compared to a primary carbonium ion (Figure 6). The carbonium ion can be discharged by one of two routes: an incoming nucleophile can complete the lysis of the ester bond, or the carboxylate group – still part of the molecule due to the patency of the ester linkage at C9 – can reverse the reaction and reform the initial dehydroalkaloid (Figure 6). We reasoned on this basis that (i) nonmacrocylic esters should hydrolyze faster than macrocylic diesters (on carbonium ion formation, the departing carboxylate moiety diffuses away); and (ii) increased steric hindrance in the vicinity of C7 would decrease access for an incoming nucleophile and increase the chance of ester reformation, leading to an overall decrease in rate of hydrolysis.

The data in Table IV support this reasoning. The acyclic diester, 7-acetyldehydrolycopsamine, has a hydrolysis rate three times faster than that of the C9-monoester, dehydrolycopsamine (Figure 7). Its hydrolysis rate is similarly three times faster than that of the most rapidly hydrolyzing macrocyclic diester, dehydroretrorsine (Figure 1). Thus, the range in rates of hydrolysis between dehydroretrorsine and dehydrotrichodesmine must reflect steric hindrance around the C7 position.

Rates of Hydrolysis of Pyrroles and Comparative Metabolism of Pyrrolizidine Alkaloids. When metabolic patterns in the isolated perfused liver are compared with rates of hydrolysis of dehydroalkaloid metabolites, several associations appear (Table V): The faster the rate of hydrolysis, (i) the lower the proportion of pyrroles released as dehydroalkaloid into the perfusate; (ii) the greater the proportion of pyrroles released as GSDHP into the bile; (iii) the greater the covalent binding of pyrroles to liver macromolecules; (iv) the lower the likelihood of extrahepatic toxicity; (v) the greater the ratio of covalent binding of pyrrole to liver macromolecules to perfusate release of dehydroalkaloid; and (vi) the greater the ratio of GSH conjugation to perfusate release of dehydroalkaloid. In other words, the more reactive the dehydroalkaloid, the more "reaction" (binding, hydrolysis or conjugation) is favored over release.

Rates of Alkylation. Dehydroalkaloids react with 4-(*p*-nitrobenzyl)pyridine under pseudo-first order conditions to yield the corresponding dihydropyridine adduct at C7 of the pyrroline nucleus (*52*; *53*). The incoming nucleophile in this case is the nitrogen of the pyridine ring. Rate constants measured in our laboratory are shown on Table IV. The

110

Table IV. Reaction Rates of Dehydroalkaloids

Dehydroalkaloid	Rate Constant for hydrolysis[a] (sec⁻¹)	Half-life of hydrolysis[a] $t_{1/2}$ (sec⁻¹)	Rate Constant for alkylation[b] (min⁻¹)
Dehydro-7-acetyllycopsamine	1.77 ± 0.09	0.39 ± 0.02	0.152 ± 0.011
Dehydrolycopsamine	0.68 ± 0.12	1.02 ± 0.22	0.076 ± 0.003
Dehydroretrorsine	0.65 ± 0.10	1.06 ± 0.20	0.226 ± 0.063
Dehydroseneciphylline	0.43 ± 0.06	1.60 ± 0.28	0.041 ± 0.002
Dehydromonocrotaline	0.20 ± 0.06	3.39 ± 0.99	0.026 ± 0.002
Dehydrotrichodesmine	0.13 ± 0.02	5.36 ± 0.92	0.013 ± 0.001

[a]Data are means $\pm$ SD (from (51)); [b]Data are means $\pm$ SE (Cooper and Huxtable, unpublished observations). All rates are statistically different at $p < 0.01$ except for hydrolysis rates of dehydrolycopsamine and dehydroretrorsine, which are not statistically

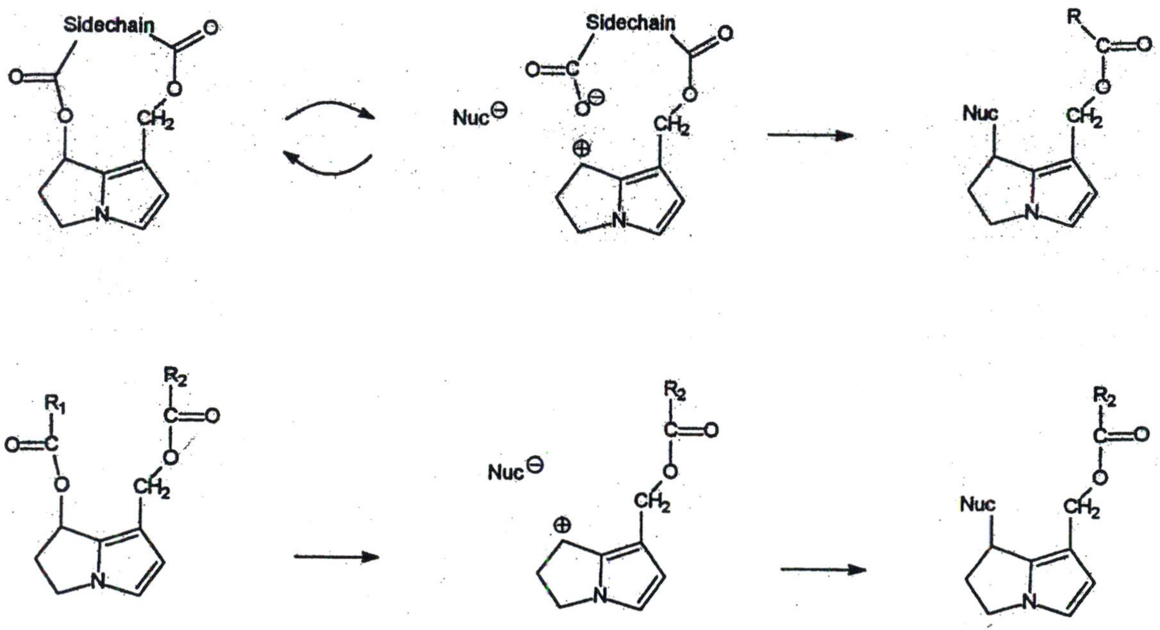

111

Figure 6. Solvolysis of dehydroalkaloids by S_N1 reaction. Top: A carbonium ion formed at C7 of a macrocyclic diester pyrrole remains a single molecular entity. The ion may be discharged either by reaction with an incoming nucleophile (Nuc⁻) or by reformation of the ester linkage at C7. The more hindrance faced by an incoming nucleophile, the more favored is reversal compared to nucleophile substitution. Bottom: An acyclic ester has a much lower rate of reformation of the ester linkage as carbonium ion formation cleaves the molecule into separate moieties.

Figure 7. R^1=H, R^2=H, R^3=OH Lycopsamine; R^1=COCH$_3$, R^2=H, R^3=OH 7-Acetyllycopsamine

Table V. Relationship Between Rates of Hydrolysis of Dehydroalkaloids and Pyrrolizidine Metabolism in the Isolated, Perfused Rat Liver

Alkaloid	Half-Life of Hydrolysis (sec)	GSDHP (% Total Pyrroles)	DHA (% Total Pyrroles)	Liver Binding (nmol/g liver)
Retrorsine	1.06	54.4	10.2	197
Seneciphylline	1.60	45.3	18.5	91
Monocrotaline	3.39	36.4	23.4	57
Trichodesmine	5.36	9.6	56.2	65

Data from (*49*; *51*)

rank order of reaction rates for the macrocyclic diester dehydroalkaloids examined is the same as for hydrolysis. Furthermore, the acyclic diester, 7-acetyldehydrolycopsamine, has a faster rate of reaction for both alkylation and hydrolysis than the 9-monoester, dehydrolycopsamine. These findings substantiates the conclusions above reached from the rates of hydrolysis. However, the reason for the high reactivity of dehydroretrorsine compared to dehydroseneciphylline is not obvious to us.

Trichodesmine and Monocrotaline. The *Crotalaria*-type alkaloids, monocrotaline and trichodesmine, differ only in the substitution of an isopropyl group at C14 in the latter compound for a methyl group in monocrotaline (Figure 1). However, this seemingly minor difference is associated with a marked difference in the pattern of metabolism (Table VI). In the isolated rat liver, perfusion with trichodesmine releases four times more dehydroalkaloid into perfusate compared with perfusion with monocrotaline. There is correspondingly lower biliary release of GSDHP with trichodesmine.

There are differences in both the quantitative and qualitative toxicity of these structurally similar alkaloids. In rats, the LD_{50} of trichodesmine is around 1/6th that of monocrotaline. The major extrahepatic effects of monocrotaline are pulmonary arterial hypertension and right ventricular hypertrophy. In the *Trichodesma* intoxications in Uzbekistan, however, trichodesmine was neurotoxic, the clinical description being encephalitis with associated vertigo and nausea proceeding to delirium and death from respiratory depression (*15*).

Our metabolic studies illuminate the basis for the toxicological differences between these alkaloids. First, with trichodesmine much more dehydroalkaloid is released from the alkaloid-perfused liver. This is the toxin responsible for downstream organ damage. Secondly, dehydrotrichodesmine hydrolyses more slowly than dehydromonocrotaline (Table IV). Although the difference may not seem profound, the two factors together result in 54 times more dehydrotrichodesmine surviving in the circulation after 5 sec. Thus, greater amounts of dehydrotrichodesmine than of dehydromonocrotaline are delivered to the brain. We have shown that trichodesmine is more liposoluble than monocrotaline (*54*). Hence, penetration of the brain is presumably greater for dehydrotrichodesmine. This is in agreement with our demonstration that greater covalent binding of pyrrole residues is found in brains of trichodesmine-exposed compared to monocrotaline-exposed rats (*55*). Thus, the increased size of the isopropyl group at C14 in trichodesmine compared to the methyl group in monocrotaline (Figure 1) presumably stabilizes the corresponding dehydroalkaloid by steric hindrance, resulting in slower rates of hydrolysis, alkylation (Table IV) and conjugation with GSH, and greater release of toxic metabolite into the circulation (Table V). The increased lipophilicity of the isopropyl group presumably allows greater delivery of toxin into the brain, a largely lipid compartment. The increased lipophilicity may also relate to the greater rate of P450 dehydrogenation of trichodesmine.

Acknowledgements

At various times, Mark Lafranconi, Sue Maxwell Long, Pam Shubat, Stacie Wild and others have also contributed from my laboratory to the studies on which the

Table VI. Comparison of Metabolism of Trichodesmine and Monocrotaline in the Isolated Perfused Liver

	Monocrotaline	Trichodesmine
	nmole/g liver	
Disposition of Parent Alkaloid		
Delivered in perfusate	5090 ± 250	5260 ± 180
Remaining in perfusate	2150 ± 110	2340 ± 230
Extracted into liver	2940 ± 220	2920 ± 280
Released in bile	14.4 ± 1.9	11.9 ± 2.5
Disposition of Pyrrolic Metabolites		
Dehydroalkaloid in perfusate	116 ± 20	468 ± 64*
GSDHP (bile)	180 ± 16	80 ± 7*
DHP (perfusate)	141 ± 11	220 ± 19*
Tissue-bound in liver	57 ± 8	64 ± 7
Total pyrroles detected	495 ± 40	833 ± 66*

Isolated livers were perfused for 1 hr in a recirculating system with an initial concentration of 0.5 mM alkaloid. *$p < 0.05$. Data are means ± SD from (*49*)

above discussion is based. We also acknowledge the contributions of A. Robin Mattocks, who has provided so much methodology and so many of the tools used in this field.

References

1. Mattocks, A.R. (1986) *Chemistry and Toxicology of Pyrrolizidine Alkaloids*, 1 Ed., 1-393, London, Academic Press.
2. Robins, D.J. *Fortschr.Chem.Org.Naturst* **1982,** *41*, 115-203.
3. Smith, L.W.; Culvenor, C.C.J. *J.Nat.Prod.* **1981,** *44*, 129-152.
4. Rizk, A.-F.M. In *Naturally Occurring Pyrrolizidine Alkaloids*; Rizk, A.-F.M., Ed., C.R.C. Press:.Boca Raton, Florida, 1991, pp 6-23.
5. Huxtable, R.J. In *Toxicants of Plant Origin, Vol I: Alkaloids*; Cheeke, P.R., Ed., CRC Press:.Boca Raton, Florida, 1989, pp 41-86.
6. Huxtable, R.J. In *Comprehensive Toxicology, Vol. 9. Hepatic and Gastrointestinal Toxicology*; McCuskey, R.S.; Earnest, D.L., Eds.; Pergamon:.1997, pp 423-431.
7. Hirono, I. In *Genetic Toxicology of the Diet*; Knudsen, I., Ed., Alan R. Liss, Inc., New York:.1986, pp 45-54.
8. Huxtable, R.J.; Paplanus, S.; Laugharn, J. *Chest* **1977,** *71S*, 308-310.
9. Lafranconi, W.M.; Duhamel, R.C.; Brendel, K.; Huxtable, R.J. *Biochem.Pharmacol.* **1984,** *33*, 191-197.
10. Ismailov, N.I.; Madzhidov, N.M.; Magrupov, A.L.; Makhkamov, G.M.; Mukminova, S. *Meditsina (Tashkent,Uzbek SSR)* **1970,** 85.
11. Hooper, P.T. *J.Comp.Pathol.* **1975,** *85*, 341-349.
12. Willmot, F.C.; Robertson, G.W. *Lancet* **1920,** 848-849.
13. Selzer, G.; Parker, R.G.F. *Am.J.Pathol.* **1950,** *27*, 885-907.
14. Hirono, I.; Mori, H.; Culvenor, C.C.J. *Gann Monograph.Cancer Res.* **1976,** *67*, 125-129.
15. United Nations Environment Programme, International Labour Organization, and World Health Organization (1988) *Environmental Health Criteria 80: Pyrrolizidine Alkaloids*, 1 Ed., 1-345, Geneva, World Health Organization.
16. Bras, G.; Jelliffe, D.B.; Stuart, K.L. *Arch.Pathol.* **1954,** *57*, 285-300.
17. Stuart, K.L.; Bras, G. *Quart.J.Med.* **1957,** *26*, 291-315.
18. Yeong, M.L.; Swinburn, B.; Kennedy, M.; Nicholson, G. *J.Gatroenterol.Hepatol.* **1990,** *5*, 211-214.
19. Bach, N.; Thung, S.N.; Schaffner, F. *Amer.J.Med.* **1989,** *87*, 97-99.
20. Ridker, P.M.; McDermott, W.V. *Lancet* **1989,** *1*, 657-658.
21. Huxtable, R.J.; Luthy, J.; Zweifel, V. *N.Engl.J.Med.* **1986,** *315*, 1095-1095.
22. Ridker, P.M.; Ohkuma, S.; McDermott, W.V.; Trey, C.; Huxtable, R.J. *Gastroenterology* **1985,** *88*, 1050-1054.
23. Fox, D.W.; Hart, M.C.; Bergeson, P.S.; Jarrett, P.B.; Stillman, A.E.; Huxtable, R.J. *J.Pediatr.* **1978,** *93*, 980-982.
24. Stillman, A.E.; Huxtable, R.; Consroe, P.; Kohnen, P.; Smith, S. *Gastroenterology* **1977,** *73*, 349-352.

116

25. Tandon, H.D.; Tandon, B.N.; Mattocks, A.R. *Am.J.Gastroenterology* **1978,** *70,* 607-613.
26. Mayer, F.; Lüthy, J. *Lancet* **1993,** *342,* 246-247.
27. Margalith, D.; Heraief, E.; Schindler, A.M.; Birchler, R.; Mosimann, F.; Aladjem, D.; Gonvers, J.J. *J.Hepatol.* **1985,** *1: Supp.* 2, S280.
28. Stillman, A.E.; Huxtable, R.J.; Fox, D.W.; Hart, M.C.; Bergeson, P.S. *Arizona Med.* **1977,** *34,* 545-546.
29. Sperl, W.; Stuppner, H.; Gassner, I.; Judmaier, W.; Dietze, O.; Vogel, W. *Eur.J.Pediatr.* **1995,** *154,* 112-116.
30. Abbott, P.J. *Med.J.Australia* **1988,** *149,* 678-682.
31. Jones, J.G.; Taylor, D.E. *Ann.Rheum.Dis.* **1989,** *48,* 791-791.
32. Huxtable, R.; Stillman, A.; Ciaramitaro, D. *Proc.Western Pharmacol.Soc.* **1977,** *20,* 455-459.
33. Adams, R.; Looker, J.H. *J.Amer.Chem.Soc.* **1951,** *73,* 134-136.
34. Ohtsubo, K.; Ito, Y.; Saito, M.; Furuya, T.; Hikichi, M. *Experientia* **1977,** *33,* 498-499.
35. Candrian, U., Luthy, J., Graf, U., and Schlatter, Ch. (1984) Mutagenic activity of the pyrrolizidine alkaloids seneciphylline and senkirkine in *Drosophila* and their transfer into rat milk. *Chem.Toxic.* **22,** 223-225.
36. Rubiolo, P.; Pieters, L.; Calomme, M.; Bicchi, C.; Vlietinck, A.; Vanden Berghe, D. *Mutat.Res.Lett.* **1992,** *281,* 143-147.
37. Mattocks, A.R. *Nature (London)* **1968,** *217,* 723-728.
38. Jago, M.V.; Edgar, J.A.; Smith, L.W.; Culvenor, C.C.J. *Mol.Pharmacol.* **1970,** *6,* 402-406.
39. Mattocks, A.R.; White, I.N.H. *Chem.-Biol.Interact.* **1971,** *3* , 393-396.
40. Tuchweber, B.; Kovacs, K.; Jago, M.V.; Beaulieu, T. *Res.Comm.Chem.Pathol.Pharmacol.* **1974,** *7,* 459-480.
41. Lafranconi, W.M.; Huxtable, R.J. *Biochem.Pharmacol.* **1984,** *33,* 2479-2484.
42. Miranda, C.L.; Reed, R.L.; Guengerich, F.P.; Buhler, D.R. *Carcinogenesis* **1991,** *12,* 515-519.
43. Mattocks, A.R.; Croswell, S.; Jukes, R.; Huxtable, R.J. *Toxicon* **1991,** *29,* 409-415.
44. Lafranconi, W.M.; Ohkuma, S.; Huxtable, R.J. *Toxicon* **1985,** *23,* 983-992.
45. Huxtable, R.J.; Ciaramitaro, D.; Eisenstein, D. *Mol.Pharmacol.* **1978,** *14,* 1189-1203.
46. Glowaz, S.L.; Michnika, M.; Huxtable, R.J. *Toxicol.Appl.Pharmacol.* **1992,** *115,* 168-173.
47. Mattocks, A.R.; Jukes, R.; Brown, J. *Toxicon* **1989,** *27,* 561-567.
48. Yan, C.C.; Huxtable, R.J. *Toxicol.Appl.Pharmacol.* **1995,** *130,* 132-139.
49. Yan, C.C.; Cooper, R.A.; Huxtable, R.J. *Toxicol.Appl.Pharmacol.* **1995,** *133,* 277-284.
50. Mattocks, A.R.; Jukes, R. *Chem.-Biol.Interact.* **1990,** *76,* 19-30.
51. Cooper, R.A.; Huxtable, R.J. *Toxicon* **1996,** *34,* 604-607.
52. Karchesy, J.J.; Deinzer, M.L. *Heterocycles* **1981,** *16,* 631-635.
53. Karchesy, J.J.; Arbogast, B.; Deinzer, M.L. *J.Org.Chem.* **1987,** *52,* 3872.

54. Huxtable, R.J.; Yan, C.C.; Wild, S.; Maxwell, S.; Cooper, R. *Neurochem.Res.* **1996,** *21,* 141-146.
55. Yan, C.C.; Huxtable, R.J. *Toxicon* **1995,** *33,* 627-634.
56. Chauvin, P.; Dillon, J.-C.; Moren, A.; Talbak, S.; Barakaev, S. *Lancet* **1993,** *341,* 1663-1663.
57. Roulet, M.; Laurini, R.; Rivier, L.; Calame, A. *J.Pediatr.* **1988,** *112,* 433-436.
58. Spang, R. *Journal of Pediatrics* **1989,** *115,* 1025-1025.
59. Weston, C.F.M.; Cooper, B.T.; Davies, J.D.; Levine, D.F. *British Medical Journal* **1987,** *295,* 183-183.
60. Culvenor, C.C.J.; Edgar, J.A.; Smith, L.W.; Kumana, C.R.; Lin, H.J. *Lancet* **1986,** *i,* 978-978.
61. Kumana, C.R.; Ng, M.; Lin, H.J.; Ko, W.; Wu, P.-C.; Todd, D. *Lancet* **1983,** *ii,* 1360-1361.
62. Kumana, C.R.; Ng, M.; Lin, H.J.; Ko, W.; Wu, P.-C.; Todd, D. *Gut* **1985,** *26,* 101-104.
63. Arora, R.R.; Pyarelal Ghost, T.K.; Mathur, K.K.; Tandon, B.N. *J.Commun.Dis.(India)* **1981,** *13,* 147-151.
64. Huxtable, R.J. *Perspect.Biol.Med.* **1980,** *24,* 1-14.
65. Aikat, B.K.; Bhusnurmath, S.R.; Datta, D.V.; Chhuttani, P.N. *Indian J.Pathol.Microbiol.* **1978,** *21,* 203-211.
66. Datta, D.V.; Khuroo, M.S.; Mattocks, A.R.; Aikat, B.K.; Chhuttani, P.N. *J.Assoc.Physicians India* **1978,** *26,* 383-*.
67. Tandon, H.D.; Tandon, B.N.; Tandon, R.; Nayak, N.C. *Ind.J.Med.Res.* **1977,** *65(5),* 679-684.
68. Tandon, R.K.; Tandon, B.N.; Tandon, H.D.; Bhatia, M.L.; Bhargava, S.; Lal, P.; Arora, R.R. *Gut* **1976,** *17,* 849-855.
69. McGee, J.O.; Patrick, R.S.; Wood, C.B.; Blumgart, L.H. *J.Clin.Path.* **1976,** *29,* 788-794.
70. Lyford, C.L.; Vergara, G.G.; Moeller, D.D. *Gastroenterol.* **1976,** *70,* 105-108.
71. Braginskii, B.M.; Bobokhodzaev, I. *Soviet Medicine* **1965,** *28,* 57.
72. Gupta, P.S.; Gupta, G.D.; Sharma, M.L. *British Medical Journal* **1963,** *1,* 1184-1186.
73. Stuart, K.L.; Bras, G. *West Indian Med.J.* **1956,** *5,* 33-36.
74. Savvina, K.L. *Arkh.Patol.* **1952,** *14,* 65.
75. Hill, K.R.; Rhodes, K.; Stafford, J.L.; Aub, R. *West Indian Med.J.* **1951,** *1,* 49-63.
76. Baylen, I. *J.Barbados Mus.Hist.Soc.* **1949,** *16,* 103-112.
77. Huxtable, R.J. In *The Vulnerable Brain and Environmental Risks, Vol. I: Malnutrition and Hazard Assessment*; Isaacson, R.L.; Jensen, K.F., Eds.; Plenum Publishing:.New York, 1992, pp 77-108.
78. Huxtable, R.J. *Drug Safety* **1990,** *5 (Suppl. 1),* 126-136.
79. Huxtable, R.J.; Wild, S.L. *Proc.Western Pharmacol.Soc.* **1994,** *37,* 109-111.
80. Fowler, M.E.; Schoental, R. *J.Am.Vet.Med.Assoc.* **1967,** *150,* 1305.

Chapter 8

Transfer of Pyrrolizidine Alkaloids into Eggs: Food Safety Implications

John A. Edgar and Leslie W. Smith

CSIRO Animal Health, Private Bag 24, Geelong, Victoria 3213, Australia

The maximum permitted concentration of pyrrolizidine alkaloids (PAs) in herbal medicines on sale in Germany is 0.1 micrograms per daily dose. These regulations provide a basis for assessing the safety of foods contaminated by PAs. PAs occur widely in agricultural production systems throughout the world and can enter the human food chain as contaminants. Grain, milk and honey are among the products that can sometimes be contaminated by PAs. A natural episode of PA poisoning in chickens caused by a lapse in the quality control of the grain component of their feed, has provided evidence that PAs can also be transferred into eggs. The concentrations of PAs detected in the eggs exceeded the levels deemed tolerable for herbal medicines in Germany. Even higher concentrations of PAs have previously been recorded in milk and honey. While there is no evidence of chronic health problems caused by PAs in these products, they and other products, including grain and meat, warrant monitoring for PAs to ensure safe levels of dietary intake are not being exceeded.

In 1992, following a comprehensive risk assessment study, the German Federal Health Bureau established regulations specifying that 0.1 micrograms was the maximum amount of hepatotoxic pyrrolizidine alkaloids (PAs) and their N-oxides allowed in a daily dose of herbal medicines incorporating PA-containing plants or plant extracts (1). One microgram per day is allowed if intake is limited to 6 weeks per year. Prescription of herbal medicines containing PAs is prohibited in the case of pregnant and lactating women.

While the harmful effects of PAs have been recognized for many years (2) the establishing of these regulations appears to be the first time PAs have been subjected to a formal risk assessment process leading to the determination of a maximum tolerable daily intake for PAs. As well as providing a measure of the safety of herbal preparations, the German drug regulations provide a basis on which to judge the risk to public health and safety from the presence of PA contaminants in food.

The term risk as it applies to the determination of safety of chemicals in food reflects not only the toxicity or hazard of the chemicals but also the likely intake of the chemical in a normal diet (3). Thus while the hazardous nature of a chemical may be well recognized the actual risk it poses to public health and safety may be negligible if the quantity in a normal diet is below the threshold of toxicity for that substance. Exceptions include genotoxic carcinogens such as aflatoxins and the pyrrolic metabolites of PAs (1, 3). Genotoxic carcinogens theoretically have no threshold of toxicity (1, 3). A single molecule of a genotoxic carcinogen has the potential to induce a genetic change that may ultimately be fatal. It has been suggested therefore that genotoxic substances in food should be reduced to as low as reasonably achievable (3).

While PAs cause acute poisoning when consumed in relatively large amounts, it is their potential to cause chronic, long-term health effects such as cirrhosis and possibly cancer that the German regulators were most concerned with when they established a maximum permitted intake of PAs in herbal preparations (1).

The literature contains many examples of poultry being poisoned by PAs in their feed (4-9) but eggs have not previously been investigated for PA contamination. A natural outbreak of PA poisoning in layer birds in a small-scale egg production system provided an opportunity to investigate the transfer of PAs into eggs. It also enabled consideration of the food safety implications of any PA contamination that was found in light of the German maximum permitted concentration for PAs in herbal products.

Transfer of PAs into eggs

The poisoning episode. The likely cause of PA poisoning among three flocks of layer chickens was considered to be *Heliotropium europaeum* seeds present in wheat at an estimated concentration of 0.6% by weight. Also present as contaminants in the grain, but in much smaller quantities, were seeds of yellow Burr weed (*Amsinckia* spp.) and Sheepweed (*Buglossoides arvensis*) which are also possible sources of PAs.

The farmer involved was a small-scale egg producer who had purchased grain from a neighbour. As a result the grain was not subject to normal quality assurance testing or cleaning before being incorporated into the feed.

Rather surprisingly, the oldest flock (750 birds) was the first affected after the contaminated feed was introduced. They were taken out of production within 12 days of being given the contaminated feed when their egg production was badly affected. Egg production in the second flock (620, 12 month-old birds) dropped significantly over 4 -5 weeks after the contaminated grain was incorporated into the feed. The third flock of 1179, 22 week-old birds was also badly affected over the same period of exposure to the contaminated feed. After withdrawal of the contaminated wheat, egg production in both remaining flocks continued to fluctuate for several months.

Samples examined and methods of analysis. The contaminated wheat and two lots of eggs, designated A and B, were analyzed for PAs.

The A eggs are believed to have come from the oldest flock, and were collected soon after the first signs of poisoning had been seen. The B eggs were collected from the remaining flocks after the contaminated wheat had been withdrawn but when the birds were still badly affected.

Standard procedures for alkaloid isolation were followed, including reduction of PA N-oxides to free bases before analysis (*10*). Alkaloids were characterised and identified by fast atom bombardment-mass spectrometry (FAB-MS) and gas chromatography-electron impact mass spectrometry (GC-EIMS) using standard procedures (*10-12*). Samples for GC-EIMS were run both as acetyl and combined acetyl-methylboronate derivatives (*12*).

The content of individual alkaloids in the total alkaloid fraction from each sample was estimated from GC-EIMS peak areas using a standard curve generated with authentic lasiocarpine. The total PA content was obtained by summation of the concentrations of the individual alkaloids.

Analysis of the grain. The FAB spectrum of the total alkaloid extract from the grain (Figure 1) showed [M+H]$^+$ ions corresponding to *H. europaeum* alkaloids: supinine, heleurine, heliotrine, europine and lasiocarpine. Exact mass measurements of the [M+H]$^+$ ions were in good agreement with the expected elemental compositions for these alkaloids.

The three main alkaloids (heliotrine, europine, and lasiocarpine) were also detected by GC-EIMS. Comparison of EI spectra and GC retention times with those of authentic samples confirmed the identity of the alkaloids. The concentration of these alkaloids in the wheat was estimated to be 26 parts per million (ppm, milligrams per kilogram) made up of 6.7 ppm heliotrine, 9.5 ppm europine and 9.8 ppm lasiocarpine.

Analysis of the eggs. The A and B eggs were each divided into two samples and both sub-samples were separately extracted to give extracts A1 and A2 and B1 and B2.

The FAB spectrum of a total alkaloid extract of the A1 eggs is shown in Figure 2. A very similar result was seen with the A2 egg extract. Alkaloids indicated by [M+H]$^+$ ions were: 7-angelyl retronecine, lycopsamine/intermedine, heliotrine, europine, uplandicine, echiumine, echimidine, lasiocarpine and acetyl-echimidine, or isomers of these. Only heliotrine, europine and lasiocarpine are typical of *H. europaeum*. FAB-MS detected only *H. europaeum* alkaloids in the B eggs.

GC-MS retention times and EI mass spectra confirmed the identifications of the principal alkaloids indicated in the FAB spectra, i.e. heliotrine, europine, uplandicine, angelylretronecine, lasiocarpine and acetylechimidine. The total ion chromatogram obtained for the A1 egg total base fraction is shown in Figure 3. Location of the GC peaks of alkaloids indicated by FAB-MS was greatly facilitated against considerable background of co-extractives by selecting fragment ions of m/z 180 and 220 (Figure 3). An intense fragment ion at m/z 180 is characteristic, after acetylation, of lycopsamine, intermedine, heliotrine, europine and uplandicine and an intense ion at m/z 220 is found in the mass spectra of 7-angelylretronecine, echiumine, echimidine, lasiocarpine and acetyl echimidine.

Quantitation of the total alkaloids present in the egg extracts is shown in Table I. The highest levels of PAs were found in the A eggs collected while birds were still being exposed to the contaminated feed. Consumption of a single A or B egg would have resulted in a PA exposure hundreds of time the tolerable daily intake specified by the German regulations.

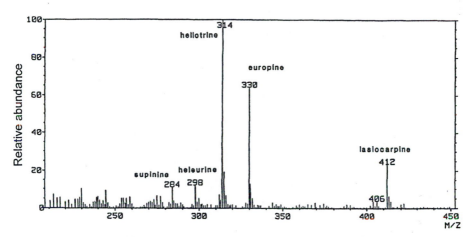

Figure 1. Fast atom bombardment mass spectrum of an alkaloid extract of contaminated wheat considered to be the cause of PA poisoning in chickens, showing the presence of PAs typical of *Heliotropium europaeum viz.* supinine, [M+H]$^+$ m/z 284; heleurine, [M+H]$^+$ m/z 298; heliotrine, [M+H]$^+$ m/z 314; europine, [M+H]$^+$ m/z 330; lasiocarpine, [M+H]$^+$ m/z 412.

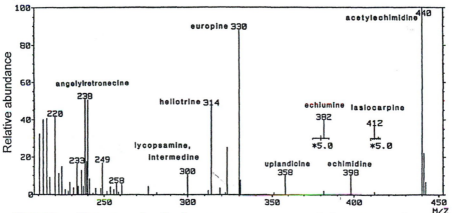

Figure 2. Fast atom bombardment mass spectrum of the extract of eggs collected at an early stage of the poisoning episode. The eggs, designated sub-sample A1, were collected from the oldest birds while they were receiving the suspect feed. PAs indicated by [M+H]$^+$ ions (and confirmed in most cases by GC-EIMS comparison with authentic standards) are: angelylretronecine, [M+H]$^+$ m/z 238; lycopsamine/intermedine, [M+H]$^+$ m/z 300; heliotrine, [M+H]$^+$ m/z 314; europine, [M+H]$^+$ m/z 330; uplandicine, [M+H]$^+$ m/z 358; echiumine, [M+H]$^+$ m/z 382; echimidine, [M+H]$^+$ m/z 398; lasiocarpine, [M+H]$^+$ m/z 412; acetylechimidine, [M+H]$^+$ m/z 440.

122

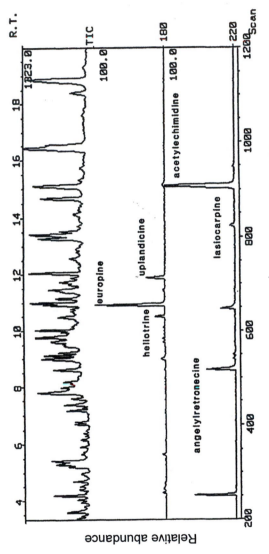

Figure 3. Total (top), m/z 180 (middle) and m/z 220 (bottom) ion chromatograms obtained on GC-EIMS of the extract from sub-sample A1 eggs after acetylation and methylboranate derivatization. The selected ion chromatograms indicate the locations of the principal PAs present in the total ion chromatogram. Comparison of retention times and EI mass spectra with those of authentic standards confirmed the identity of the PAs indicated.

Table I. Quantitation of pyrrolizidine alkaloids (PAs) present in the egg extracts.

Egg sample	No. of eggs Extracted	Wt. of eggs (g)	Total PAs (µg)	PAs per egg (µg)[a]	ppb (µg/kg)
A1	4	230	38.75	9.7	168
A2	4	227	33.0	8.2	145
B1	4	235	4.6	1.15	19
B2	4	267	13.2	3.3	49

German regulations for herbal preparations permit a maximum daily intake of PAs of 0.1 microgram (0.1µg).

While the expected *H. europaeum* alkaloids are present in both A and B eggs the A eggs also contained PAs normally found, albeit in different proportions, in *Echium plantagineum* (Paterson's curse, Salvation Jane) (*12, 13*) and similar to those in related Boraginaceae such as *Amsinckia* spp. (*14*) and *Symphytum* spp. (Comfrey) (*10*).

The FAB-MS spectrum shown in Figure 4 is typical of that produced by a total alkaloid extract of *E. plantagineum*, the seeds of which are known to occasionally contaminate grain, and this is considered the most likely source of the additional alkaloids seen in the A eggs. Similar alkaloids may occur in *Buglossoides arvensis*, seeds of which were found as contaminants in the wheat. *Amsinckia* spp., the seeds of which were also present, have been shown to contain intermedine, lycopsamine, sincamidine and echiumine but no echimidine (*14*). These two PA-containing plants cannot therefore be ruled out as possible sources of some of the unexpected PAs detected in the A eggs.

The high proportion of acetylechimidine, the very low level of lycopsamine and intermedine and the absence of their acetyl derivatives, distinguish A egg alkaloids from the typical mixture of alkaloids found in *E. plantagineum*. Despite these differences, *E. plantagineum* seems to be the most likely source of the alkaloids found in the A eggs. The atypical mixture of *Echium* alkaloids found in the eggs may reflect a selective transfer of alkaloids from feed into the eggs. This is to be expected on the basis of significant variation in lipid solubility of the alkaloids concerned and the ease with which 7-acetyllycopsamine and 7-acetylintermedine would be expected to be hydrolysed *in vivo*.

Whatever the source of the unexpected alkaloids, it seems that they were present in the feed in substantial amounts at an early stage of the poisoning episode, probably in feed pre-dating introduction of the *H. europaeum*-contaminated wheat. This would help to explain why the older birds were the first to be affected even though the younger birds could be expected to be more susceptible to poisoning. *Echium* seeds were not found in the wheat sample provided for analysis so that some other component of the diet or an earlier grain source is implicated.

124

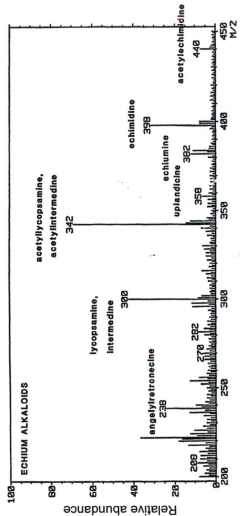

Figure 4. A typical FAB-MS spectrum obtained from a total alkaloid extract of *Echium plantaginaeum*. [M+H]⁺ ions corresponding to the PAs present in *E. plantaginaeum* are: angelylretronecine, [M+H]⁺ m/z 238; lycopsamine, [M+H]⁺ m/z 300; intermedine, [M+H]⁺ m/z 300; acetyllycopsamine, [M+H]⁺ m/z 342; acetylintermedine, [M+H]⁺ m/z 342; uplandicine, [M+H]⁺ m/z 358; echiumine, [M+H]⁺ m/z 382; echimidine, [M+H]⁺ m/z 398; acetylechimidine, [M+H]⁺ m/z 440.

Food safety and pyrrolizidine alkaloids

It is unusual for PA-containing plants to be intentionally used as foods. Examples of such limited use include Comfrey (*Symphytum* spp.) that is sometimes eaten in salads (*10*) and Borage (*Borago officinalis*) which is occasionally used as a cucumber-flavored garnish (*15*).

The principal route of human exposure to PAs in food is via contaminated staple foods (*2*). There have, for example, been a number of cases in which very large numbers of people have been acutely poisoned by PA contamination of grain (*16-26*).

It should be noted however that most of the recorded cases of acute PA poisoning from contaminated grain have involved people in economically disadvantaged countries and under socially stressful circumstances or where a local source of a contaminated grain, not subject to quality assurance, has been consumed.

Most economically advanced countries in which PA-containing plants are known to be present in agricultural systems have regulations or marketing standards for grain contamination that are likely to prevent acute PA poisoning (*27*). In addition grain for human consumption is normally freed of visible contamination by PA-containing seeds before milling. However the adequacy of the existing quality assurance measures and grain cleaning to prevent chronic effects of PAs is open to debate in light of the German risk assessment outcome and the genotoxic nature of PA metabolites.

In the case of grain, quality assurance usually takes the form of a tolerance for a certain number of PA-containing seeds per volume or weight of grain (*27, 4*). While counting seeds as a measure of contamination may prevent acute PA poisoning it is inadequate for preventing levels of contamination that could, according to the German risk assessment, lead to chronic effects such as cirrhosis and possibly cancer. As an example it has been calculated that the 0.1% tolerance for *H. lasiocarpum* seeds in stored grain allowed in the former USSR could result in 1820 micrograms of PAs per kg of grain (*27*). The regulations, guidelines and marketing standards set by other countries, if based on tolerance of a certain number of PA-containing seeds rather than an actual concentration of PAs, are also likely to result in levels of PAs much higher than would be deemed tolerable according to the German standards for herbal preparations. Indeed the presence of the seeds of PA-containing plants in grain is only a visible sign that such plants were growing in the crop at the time of harvest. The dust from these plants also adheres to the grain and can be another source of PAs (data not shown here). Grain cleaning that normally precedes milling is likely to remove foreign seeds but not PA-containing dust.

PA contamination of other foods. Chronic and acute PA poisoning of food-producing livestock occurs in many countries (*28, 29*), indicating considerable exposure to PAs. There is therefore potential for even low level retention to result in animal products contaminated by PAs in excess of the German standard. PAs have for example been shown to pass into the milk of cows, goats and rats (*30-37*). In the case of cows, experimental exposure to a PA-containing plant, *Senecio jacobaea*, has been reported to lead to levels of PAs between 470 and 835 micrograms per liter of milk (*32*). Honey has also been found to contain PAs (*38-41*). Honey from *Senecio jacobaea* has been shown to contain between 300 and 3200 micrograms of PAs per kilogram (*38, 39*). *Senecio vernalis* honey is reported to contain between 500 and 1000 microgram of PAs per kilogram (*40*) while *E. plantagineum* honey has been shown to contain between 270 and 950 micrograms per kilogram (*41*).

As expected from a consideration of the levels of PAs present in these products relative to the amount required to cause acute toxicity, there have been no confirmed cases of acute human poisoning from these sources. There is however an unpublished report of infants in Egypt showing typical PA liver damage attributed to the consumption of contaminated goats milk (cited in *42*). Consumption of milk and honey, eggs and grain will sometimes result in the ingestion of PAs in excess of the level of 0.1 micrograms PAs per day deemed the maximum permitted by the German regulations and thus could expose consumers to the risk of long-term chronic health effects.

Conclusions

The investigation described here shows that PAs can enter the human food chain in eggs. It adds to previous knowledge that grain, milk and honey are items of food that can be contaminated by PAs at levels sometimes considerably in excess of that specified as tolerable in German regulations for herbal medicines.

In countries where effective quality assurance standards are in place, the level of PAs in food is likely to be very low and may not pose an unacceptable risk to health. However to fully assess the risk that PA-contaminated food poses for public health requires, among other things, determination of the level of these natural toxicants in a normal diet and consensus on what constitutes an acceptable level of exposure.

Given the unpredictable occurrence of low-level dietary exposure to PAs and the likelihood that exposure will differ from country to country and individual to individual, there is a need to develop biochemical methods for quantifying individual human exposure to PAs. Methods for quantifying PA pyrrolic metabolite-DNA or protein adducts in tissues could provide such a measure. This approach has been used to measure exposure to other genotoxic carcinogens in the environment such as polycyclic aromatic hydrocarbons and aflatoxins, the metabolites of which also produce DNA adducts *in vivo* (*43, 44*).

Long-term monitoring of populations that have been acutely poisoned by PAs in food, such as the victims of the recent Tadjikistan poisoning (*25, 26*) and their contemporaries, could also contribute to defining the risk to human health from dietary exposure to PAs.

It should be possible, using these approaches, to establish normal levels of PA exposure in various populations and to confirm or cast doubt on the appropriateness of the maximum tolerable daily intake for PAs specified in the German regulations.

References

1. Bundesgesundheitsamt (German Federal Health Bureau). *Bundesanzeiger* **1992**, *111*, 4805v, cit. *Pharm. Ztg* **1992**, *137*, 2088; *Dtsch. Apoth. Ztg* **1992**, *132*, 1406.
2. Anon. *Pyrrolizidine Alkaloids*; Environmental Health Criteria 80, WHO: Geneva, Switzerland, 1988.

3. Anon. *Application of Risk Analysis to Food Standards Issues*. Report of the Joint FAO/WHO Expert Consultation (WHO/FNU/FOS/95.3) WHO: Geneva, Switzerland, 1995.

4. Sippel,W. L. *Ann. NY Acad. Sci.* **1964**, *111*, 562.

5. Bierer, B. W.; Vickers, C. L.; Rhodes, W. H.; Thomas, J. B. *J. Am. Vet. Med. Assoc.* **1960**, *136*, 318.

6. Golpinath, G.; Ford, E. J. H. *Br. Poult. Sci.* **1977**, *18*, 137.

7. Hooper, P. T.; Scanlan, W. A. *Aust. Vet. J.* **1977**, *53*, 109.

8. Pass, D. A.; Hogg, G. G.; Russell, R. G.; Edgar, J. A.; Tence, I.M.; Rikard-Bell, L. *Aust. Vet. J.* **1979**, *55*, 284.

9. Gaul, K.L.; Gallagher, P.F.; Reyes, D.; Stasi, S.; Edgar, J.A. (1994). *Plant-associated Toxins: Agricultural, Phytochemical and Ecological Aspects.*, CAB International: Oxon.,UK, 1994; pp 137.

10. Culvenor, C. C. J.; Edgar, J. A., Frahn, J. L.; Smith, L. W. *Aust. J. Chem.* **1980**, *33*, 1105.

11. Edgar, J.A.; Lin, H.J.; Kumana, C.R.; Ng, M.M.T. *Amer. J. Chinese Med.* **1992**, *20*, 281.

12. Edgar, J.A. *Plant Toxicology*; Proceedings of the Australia – USA Poisonous Plants Symposium, Brisbane, Australia, May 14-18, 1984; Queensland Poisonous Plants Committee: Brisbane, Queensland, Australia, 1984 pp. 227.

13. Culvenor, C. C. J.; Jago, M. V.; Peterson, J. E.; Smith, L. W.; Payne, A. L.; Campbell, D.G.; Edgar, J. A.; Frahn, J. L. *Aust. J. Agric. Res.* **1984**, *35*, 293.

14. Culvenor, C. C. J.; Smith, L. W. *Aust. J. Chem.* **1966**, *19*, 1955.

15. Mabberley, D.J. *The Plant-book*; Cambridge University Press: Cambridge, UK, 1987, pp. 75.

16. Steyn, D.G. *Onderstepoort J. Vet. Sci. and Animal Industry* **1933**, *1*, 219.

17. Dubrovinskii, S. B. *J. Sov. Prot. Health* **1946**, *6*, 17.

18. Khanin, M. N. *Arch. Pathol. USSR* **1948**, *1*, 42.

19. Mohabbat, O.; Srivastava, R. N.; Younos, M. S.; Sediqu, G. G.; Merzad, A. A.; Aram, G. N.; *Lancet* **1976**, *2*, 269.

20. Tandon, B. N.; Tandon, H. D.; Mattocks, A.R. *Ind. J. Med. Res.* **1978**, *68*, 84.

21. Tandon, B. N.; Tandon, H. D.; Mattocks, A.R. *Am. J. Gastroenterol.* **1978**, *70*, 607.

22. Tandon, R. K.; Tandon, B. N.; Tandon, H. D.; Bhatia, M. L.; Bhargava, S.; Lal, P.; Arora, R.R. *Gut* **1976**, *17*, 849.

23. Tandon, B. N.; Tandon, H. D.; Tandon, R. K.; Nerandranathan, M; Joshi, Y. K. *Lancet* **1976**, *2*, 271.

24. Krishnamachiari, K. A. V. R.; Bhat, R. V.; Krishnamurthy, D.; Krishnaswamy, K.; Nagarajan, V. *Ind. J. Med. Res.* **1977**, *65*, 672.

25. Chauvin, P.; Dillon, J-C.; Moren, A.; Tablak, S.; Barakaev, S. *Lancet* **1993**, *341*, 1663.

26. Mayer, F.; Lüthy, J. *Lancet* **1993**, *342*, 246.

27. Anon. *Pyrrolizidine Alkaloids Health and Safety Guide*; IPCS Health and Safety Guide No. 26; WHO: Geneva, Switzerland, pp17.

28. Peterson, J. E.; Culvenor, C. C. J. *Plant and Fungal Toxins*; Handbook of Natural Toxins; Marcel Dekker, Inc.: New York, NY, 1983, Vol.1; Chapter 19, pp 637.

29. Cheeke, P. R. *Toxicants of Plant Origin*; CRC Press Inc.: Boca Raton, Florida, 1989, Vol. 1; Chapter 1, pp1.

30. Schoental, R. *J. Pathol. Bacteriol.* **1959**, *77*, 485.

31. Johnson, A.E. *Am. J. Vet. Res.* **1976**, *37*, 107.

32. Dickinson, J. O.; Cooke, M. P.; King, R. R.; Mohamed, P. A. *J.Am. Vet. Med. Assoc.* **1976**, *169*, 1192.

33. Dickinson, J. O.; King, R.R. *Effects of Poisonous Plants on Livestock* Academic Press: New York, NY, 1978; pp. 201.
34. Groeger, D. E.; Cheeke, P. R.; Schmitz, J. A.; Buhler, D. R. *Am. J. Vet. Res.* **1982**, *43*, 1631.
35. Deinzer, M. L.; Arbogast, D. R.; Buhler, D. R.; Cheeke, P. R. *Anal. Chem.* **1982**, *54*, 1811.
36. Lüthy, J; Heim, T.; Schlatter, C. *Toxicol. Lett.* **1983**, *17*, 283.
37. Candrian, U.; Lüthy, J.; Graf, U.; Schlatter, Ch. *Fd. Chem. Toxic.* **1984**, *22*, 223.
38. Deinzer, M. L.; Thompson, P. A.; Burgett, D. M.; Isaacson, D. L. *Science* **1977**, *195*, 497.
39. Crews, C.; Startin, J.R.; Clarke, P.A. *Food Additives and Contaminants* **1997**, *14*, 419.
40. Roeder, E. *Pharmazie* **1995**, *50*, 83.
41. Culvenor, C. C. J.; Edgar, J. A.; Smith, L. W. *J. Agric. Food Chem.* **1981**, *29*, 958.
42. Hippchen, Von C.; Entzeroth, R.; Roeder, E; Greuel, E. *Der Praktische Tierarzt* **1986**, *67*, 322.
43. Randerath, K.; Sriran, P.; Moorthy, B.; Aston, J. P.; Baan, R. A.; van den Berg, P. T. M.; Booth, E. D.; Watson, W. P. *Chem-Biol. Interact.* **1998**, *110*, 82.
44. Harrison, J.C.; Carvajal, M; Garner, R.C. *Environ. Health Perspec.* **1993**, *99*, 99

Chapter 9

Calystegine Alkaloids in the Potato and Other Food Plants

A. A. Watson[1], D. R. Davies[1], N. Asano[2], B. Winchester[3], A. Kato[4], R. J. Molyneux[5], B. L. Stegelmeier[6], and R. J. Nash[1]

[1]Institute of Grassland and Environmental Research, Aberystwyth, United Kingdom
[2]Hokuriku University, Kanazawa 920–11, Japan
[3]Institute of Child Health, University College London, London, United Kingdom
[4]Toyama Medical and Pharmaceutical University, Toyama 930–01, Japan
[5]Western Regional Research Center, Agricultural Research Service, U.S. Department of Agriculture, Albany, CA 94710
[6]Poisonous Plant Research Laboratory, Agricultural Research Service, U.S. Department of Agriculture, Logan, UT 84341

The calystegines are polyhydroxylated *nor*tropane alkaloids first reported from the Convolvulaceae in 1988 but they have also now been found in most genera of the Solanaceae and in one *Morus* species (Moraceae). We have found them in over 70 varieties of potatoes and also other edible species of the Solanaceae and Convolvulaceae such as *Solanum melongena* and *Ipomoea batatas*. They survive cooking and are stable enough to be found in processed potato products. This group of alkaloids is of interest in that most of them are potent inhibitors of glycosidase enzymes. Storage of potatoes at 5°C increases the proportions of the 4-O-α-D-galactoside of calystegine B_2 and the trihydroxylated calystegine A_3. Mice treated with calystegine A_3 show vacuolation of Kupffer cells with minimal vacuolation in other histocytic cells. The microflora in rumen fluid removed from sheep previously fed hay reduced calystegines B_1 and B_2 to undetectable levels but the concentrations of calystegine A_3 and the control compound swainsonine were not affected. There was no effect on the overall respiratory rate of the microbial population by any of these alkaloids.

Glycoalkaloids such as solanine are well established as toxic components of the potato. Their presence in potatoes is monitored routinely world-wide to ensure that concentrations dangerous to humans are not reached in commercial varieties (*1*). Nutritional components in potatoes such as starch and vitamins are also studied by many laboratories. It was, therefore, a great surprise to discover in 1993 two new alkaloids as major components in potato tubers which were not related to glycoalkaloids and which had not been discovered in edible tubers by other laboratories (*2*). The new alkaloids had actually been reported for the first time in 1988 by Tepfer *et al.* (*3*) who were studying microbial associations with the bindweed *Calystegia sepium* and named the new class of *nor*tropane alkaloids

129

130

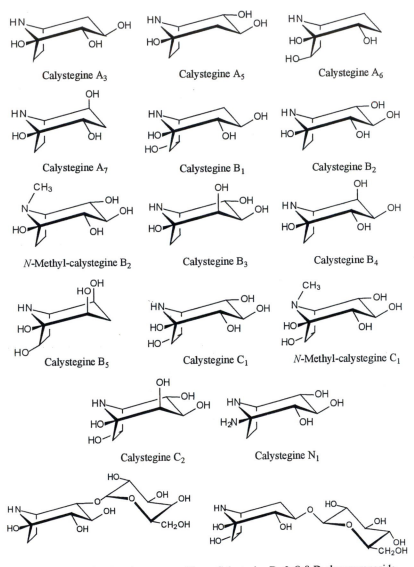

Calystegine A₃

Calystegine A₅

Calystegine A₆

Calystegine A₇

Calystegine B₁

Calystegine B₂

N-Methyl-calystegine B₂

Calystegine B₃

Calystegine B₄

Calystegine B₅

Calystegine C₁

N-Methyl-calystegine C₁

Calystegine C₂

Calystegine N₁

Calystegine B₂-4-O-α-D-galactopyranoside

Calystegine B₁-3-O-β-D-glucopyranoside

"calystegins". They showed that only one rhizosphere bacterium (*Rhizobium meliloti* 41) of 42 tested could catabolise and utilise them as their sole sources of carbon and nitrogen. Since 1993 eleven calystegines have been purified that can be grouped into classes A, B, or C depending on whether they have three, four or five hydroxyl groups, respectively, a amino derivative (calystegine N₁) has also been reported. The chemistry of the calystegines was reviewed extensively by Molyneux *et al.* (*4*) but subsequently a number of new structures have been isolated, most notably the *N*-methylcalystegines from *Lycium chinense* (Solanaceae) (*5*). The calystegines possess an unusual aminoketal functionality which generates a tertiary hydroxyl group at the bicyclic ring bridgehead. Their distribution seems to be largely restricted to species of the Solanaceae (but not *Nicotiana* or *Petunia*, Nash *et al.*, unpublished) and the Convolvulaceae, although they have also been reported in *Morus alba* and *M. bombycis* (Moraceae) (*6*).

The calystegines form one class of a group of glycosidase-inhibiting alkaloids known to have a wide range of potential pharmaceutical uses, including as therapeutic agents in the treatment of cancer and viral diseases. As well as the *nor*tropane alkaloids, this group contains polyhydroxylated pyrrolidine (e.g. DMDP), piperidine (e.g. deoxynojirimycin), indolizidine (e.g. swainsonine) and pyrrolizidine (e.g. alexine) alkaloids. Glycosidases catalyse the hydrolysis of the glycosidic bonds in oligosaccharides and glycoconjugates and as such are involved in a wide range of metabolic functions such as digestion, intracellular catabolism and the biosynthesis of the oligosaccharide portions of glycoproteins. Most of the polyhydroxylated alkaloids are potent and specific competitive inhibitors of these processes and as such have aroused great interest in a wide range of potential therapeutic areas (*7*).

DMDP   Deoxynojirimycin

Swainsonine   Alexine

However, ingestion by livestock of plants producing glycosidase inhibitors can lead to severe metabolic disruption. For example, in the USA and Australia the α-mannosidase inhibitor swainsonine has long been known to induce a lysosomal storage disorder in cattle grazing certain legumes (*Astragalus* and *Oxytropis* species in the USA and *Swainsona* species in Australia) that gives rise to neurological dysfunction in severe cases (*8,9*). In general, the calystegines are inhibitors of glucosidases and galactosidases (*10,11*) but there is evidence to suggest that these alkaloids could also induce lysosomal storage disorders in cattle. For instance, *Solanum dimidiatum* in west Texas causes a disease known to ranchers as "Crazy Cow Syndrome" (*12*) and *S. kwebense* causes "Maldronksiekte" in South Africa (*13*). Both disorders are characterised by neurological signs of cerebellar dysfunction, including staggering and incoordination and there is severe cellular vacuolation and degeneration of Purkinje cells in the brain. Both plants contain high concentrations of a range of calystegines (2). Therefore, it seems likely that these syndromes are also lysosomal storage disorders caused by glycosidase inhibition produced by the calystegines. Symptomatically, these disorders could be considered to be similar to Gaucher's and Fabry's diseases in humans which are caused by the genetic absence or diminution of activity of β-glucosidase and α-galactosidase, respectively (9). Certain *Ipomoea* species in the Convolvulaceae have been associated with instances of animal poisonings in Australia and Africa and these have been found to contain both calystegines and swainsonine (8,14). The Solanaceae contains a large number of important crop species (e.g. tomatoes, eggplants, and peppers in addition to potatoes) and sweet potatoes belong to the closely related family Convolvulaceae. Therefore, the levels of the calystegines found in the edible parts of these plants, plus any other polyhydroxylated alkaloids, needed to be assessed in order to evaluate their potential influence on human as well as animal health.

We studied over 70 varieties of potato available in the UK and every variety contained the calystegines in the tubers (for GC-MS detection method see ref. 4). The highest levels were found just beneath the skin (periderm) of the tubers (up to 1mg/g fresh weight) and the lowest were detected in the centres of the tubers. The varieties all had a distinctive ratio of the two major calystegines (B_2:A_3 = 2:1). Low levels of the other "B" calystegines were also detected in potatoes, but the structures and levels of these varied considerably between the varieties analysed. One cultivar (Désirée) also produced the polyhydroxylated pyrrolidine alkaloid DMDP. Unlike the glycoalkaloids, the concentrations of the calystegines are not altered by greening of the potato tubers (green sprouts do, however, contain a high concentration of calystegines), but the common practice of long-term cold storage of tubers at approximately 5°C alters the ratios by decreasing the content of calystegine B_2 and raising the levels of calystegine A_3. The 4-O-α-D-galactoside of calystegine B_2 is also synthesised during cold storage and other glycosides of the calystegines have also been reported (15,16). We detected the calystegines in other edible species of the Solanaceae, Convolvulaceae and Moraceae (Table I). Variability was observed in both the type and concentration of the calystegines found in potatoes, eggplants and sweet potatoes originating from Japan and the UK but this could have been due to varietal or climatic differences. However, calystegine B_2 was present in all of the species analysed and its concentration was generally higher than those of the other calystegines (when present). Calystegines A_3 and B_1 were detected in approximately half of the species analysed but the presence of calystegine C_1 was restricted to sweet potatoes, sweet peppers and cape gooseberries (11). Swainsonine was not detected in any of the sweet potatoes analysed.

In order to assess the likely level of exposure to calystegines in the human diet a series of cooking trials was undertaken using potatoes. After boiling or oven-roasting the potatoes the total calystegine content had fallen to approximately 15% of the starting value whereas 20% remained after microwave preparation and deep fat frying. The calystegines are also stable enough to be found in processed potato products and interestingly, certain processing techniques actually appear to enhance the calystegine content (Table II).

Table I. Calystegine Content of Edible Fruits and Vegetables.

Plant	Means of Acquisition[a]	Calystegines Detected (μg/g fresh weight)			
		A_3	B_1	B_2	C_1
Solanaceae					
Capsicum annum var. angulosum (sweet peppers)	A	None	12.0	37.0	4.0
Capsicum frutescens (chilli peppers)	B	0.235	Trace	0.269	None
Cyphomandra betaceae (tomatillo)	C	None	None	0.002	None
Lycopersicum esculentum (tomatoes)	B	1.1	None	4.5	None
Physalis peruviana (cape gooseberries)	C	0.003	0.038	0.048	0.005
Physalis exocarpa (tamarillo)	C	None	None	0.002	None
Solanum melongena (eggplants)	A	None	7.0	73.0	None
	C	0.312	None	0.473	None
Solanum scabrum (huckleberries)	B	None	None	0.007	None
Solanum tuberosum (potatoes)	A	None	None	7.0	None
	B	1.17	None	2.22	None
Convolvulaceae					
Ipomoea batatas (sweet potatoes)	A	None	16.0	19.0	9.0
	C	0.11	2.37	1.12	0.61
Moraceae					
Morus alba (mulberries)	A	None	None	7.0^b	None

[a] A, Shop-bought in Japan. B, Grown at Inst. of Grassland & Env. Research, UK.
C, Shop-bought in UK,
[b] μg/g dry weight.
SOURCE: Adapted with permission from reference 11. Copyright 1997 Oxford University Press.

Table II. Calystegine Content of Commercially Available Potato Products.

Product and Manufacturer	Concentration of Calystegines (μg/g)	
	A_3	B_2
Potato Waffles (Bird's Eye) [a]	3.06	4.04
Oven ready chips (Somerfield, UK) [a]	1.32	1.13
Microwave fries (Lamb Weston) [a]	15.16	19.53
Instant mashed potato granules (Somerfield, UK) [a]	8.96	22.66
Crisps made from whole potatoes (Walker)	4.89	20.97
Crisps made from freeze dried potato granules (Pringles)	0.21	0.30
Hula Hoops (KP)	1.10	5.09

[a] Product cooked according to manufacturer's instructions prior to analysis.

Screens were conducted to assess the range of mammalian digestive and lysosomal glycosidases inhibited by the major calystegines detected in food plants (Tables III and IV) (*11*). The IC_{50} values of calystegines toward rat digestive glycosidases are shown in Table III (*17*). Calystegines B_1, B_2 and C_1 potently inhibited lactase and calystegines A_3, B_2 and B_4 were good competitive inhibitors of the intestinal trehalase activity. The introduction of a hydroxyl group at C6*exo* in calystegines A_3 and B_2 to give B_1 and C_1, respectively, markedly enhanced the inhibition of β-glucosidases in the gut (lactase and cellobiase) and also the mammalian liver lysosomal β-glucosidase and β-xylosidase activities (Table IV). This suggests that the hydroxylated pyrrolidine ring is the key structural determinant for inhibition of these enzymes. In contrast, the addition of this group lowered the affinity for rat intestinal trehalase (Table III) and abolished the inhibition of bovine, human, and rat liver α-galactosidase by calystegine B_2 and of human liver α-galactosidase by calystegine A_3 (*11*). It is difficult to understand the structural basis for the selective inhibition of α-galactosidase by calystegine B_2 but the configuration at carbons 1, 4, and 5 is the same as in carbons 1, 4, and 5 of α-galactose although the substituents at C2 and C3 are in the wrong configuration.

Potent glycosidase inhibitors do not need to be present at high concentrations for toxic effects. In the case of the α-mannosidase inhibitor swainsonine, the low concentration ingested is made more effective by its ability to permeate the plasma and lysosomal membranes freely, but once inside the lysosomes it gradually accumulates because it is protonated due to the low pH and this reduces its membrane permeability (*18*). A threshold of toxicity is, therefore, difficult to establish but Molyneux *et al.* (*8*) suggested that levels of swainsonine in the diet in excess of 0.001% of the dry weight would be sufficient to produce intoxication in livestock. It is clear that the presence of calystegines in foods such as tomatoes, potatoes, peppers, eggplants, and sweet potatoes poses questions about the possible effects these compounds might have on humans. It has been established that there are differences between mammalian groups in the susceptibility of their liver glycosidases to inhibition by calystegines (*11*) (Table IV). Although calystegines B_1 and C_1 potently inhibited human liver lysosomal β-glucosidase, preliminary experiments have indicated that they do not cause additional lysosomal storage in human fibroblasts in culture for one week in the presence of 1mM of the calystegine (B. Winchester, unpublished observations). Feeding experiments with mice have, however, shown that calystegine A_3 (but not calystegine B_2) causes vacuolation in Kupffer cells in the liver (Figure 1) (Stegelmeier *et al.*, manuscript in preparation). Interestingly, macrophages in other tissues were not affected. This may have been due to the concentrations used in the study as it has been shown with swainsonine that concentration can occur in the liver to 10 times the serum levels in sheep (*19*). Therefore, it may take high doses of calystegines to produce histologic lesions. This is similar to the situation found with castanospermine, an inhibitor of α- and β-glucosidases. This compound only produced lesions in rats after feeding doses of 100mg/kg body weight for 20 days (*20*).

Table III. **Concentration of Calystegines Giving 50% Inhibition of Rat Digestive Glycosidase Activities.**

Substrate	$IC_{50}{}^a$ (μM) of Calystegines						
	A_3	A_5	B_1	B_2	B_3	B_4	C_1
Maltose	NI^b	NI	NI	640	NI	NI	190
Sucrose	NI	NI	NI	500	NI	NI	160
Palatinose	NI	NI	NI	270	NI	NI	230
Trehalose	12	NI	260	9	92	9.8	740
Cellobiose	1000	NI	25	80	NI	380	6.6
Lactose	110	NI	2.6	7.8	NI	110	0.38

^{a}Concentration giving 50% inhibition.
^{b}No inhibition (less than 50% inhibition at 1mM).
SOURCE: Adapted with permission from reference 17. Copyright 1996 Elsevier.

Table IV. Effects of Calystegines on Mammalian Liver Lysosomal Glycosidases.

Enzyme	$IC_{50}{}^a$ and Ki (μM) of Calystegines			
	A_3	B_1	B_2	C_1
β-Glucosidase				
Bovine	NI^b	360 (Ki, 150)	NI	60 (Ki, 15)
Human	NI	50 (Ki, 10)	NI	3 (Ki, 1.5)
Rat	90 (Ki, 24)	4.6 (Ki, 1.9)	15 (Ki, 8.9)	3.6 (Ki, 1)
α-Galactosidase				
Bovine	NI	NI	270 (Ki, 72)	NI
Human	410	NI	140 (Ki, 185)	NI
Rat	NI	NI	21 (Ki, 4.8)	NI
β-Xylosidase				
Human	400	50 (Ki, 19)	100 (Ki, 60)	5 (Ki, 0.13)
Rat	850	120	NI	NI

^{a}Concentration giving 50% inhibition.
^{b}No inhibition (less than 50% inhibition at 1mM).
SOURCE: Adapted with permission from reference 11. Copyright 1997 Oxford University Press.

Potato peel as a by-product of various food manufacturing processes is often fed to animals (*21*). However, perhaps this practice should be carefully monitored as the tuber periderm contains the highest concentration of the calystegines. From the

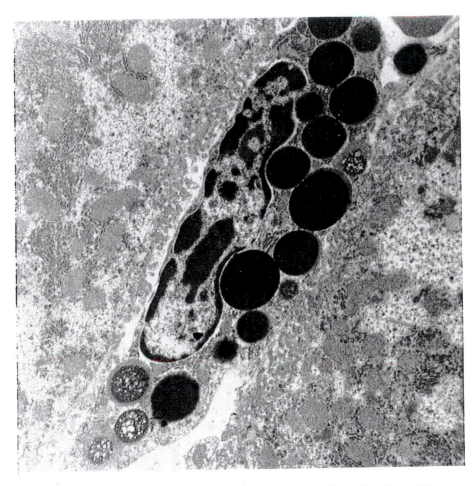

Figure 1. Electron photomicrograph of a mouse liver Kupffer cell (7000X mag) showing vacuolation caused by calystegine A_3 delivered by ALZET osmotic pump at 10mg/kg for 28 days.

evidence presented, the most likely effects of the alkaloids would be gastrointestinal disturbance through inhibition of digestive glycosidases, fatigue and possible vacuolation of tissues. However, ruminants in particular possess a vast array of micro-organisms in their gastrointestinal tracts that aid their digestion. Therefore, it was considered that this microbial population might possibly modify or break down the calystegines consumed and so reduce the potential risk posed by consumption of these alkaloids. This theory was investigated using the *in vitro* gas production technique (*22*). Strained rumen fluid was obtained from two Whether sheep being fed an *ad libitum* hay diet and this was used to measure the rate of digestion of freeze dried, powdered perennial ryegrass to which had been added either a mixture of 1mg of each of calystegines A_3, B_1 and B_2 or swainsonine. The cultures were incubated at 39°C for 7 days and the gas pressure was measured every 6 hours but no differences were observed between the treated and control cultures that had no additions. At the end of the experiment the cultures were filtered and the alkaloids analysed by GC-MS using trimethylsilyl-derivatisation (*17*) after cation exchange chromatography. Swainsonine and calystegine A_3 were found to be completely recoverable from the cultures and so it can be assumed that *in vivo* calystegine A_3 could have been absorbed by these sheep. However, calystegines B_1 and B_2 were completely metabolised by the rumen microbes during the trial. Several compounds appeared to be produced from these calystegines, including calystegine A_5, but the major product could not be identified and this will be isolated for structural determination. Ultimately, it may prove possible to identify specific rumen micro-organisms which can rapidly deactivate certain of the calystegines and so reduce the potential risk of poisoning.

Currently, it is not clear if feeding of potato material to livestock has any detrimental effect on animal performance that can be associated specifically with the calystegines. However, now that methods have been developed to routinely analyse for these alkaloids it will be possible to rapidly investigate whether they are implicated in toxicity cases in the future. Traditionally, toxicity associated with potatoes has been attributed to the steroidal glycoalkaloids. However, in humans the symptoms reported have included nausea, vomiting, diarrhoea, severe abdominal pain and occasional neurological disturbances (*4*). No evidence has ever been presented that can link these symptoms to the glycoalkaloids. The potato glycoalkaloids α-chaconine and α-solanine are teratogenic in laboratory animals but they are not glycosidase inhibitors, whereas the gastrointestinal disturbances associated with acute toxicoses could certainly be attributable to inhibition of the digestive glycosidases caused by the calystegines. However, the incidence of acute poisoning from potatoes is relatively low so it would appear that calystegines are consumed on a daily basis in common foods with no adverse effects. Perhaps only the ingestion of excessive amounts of calystegines would cause adverse effects or they may only be detrimental to those individuals with a genetic predisposition to low levels of β-glucosidase or α-galactosidase who might suffer more noticeably from inhibition of their residual enzymes. However, it should be remembered that the calystegines form one class of a group of glycosidase-inhibiting alkaloids known to have a wide range of potential therapeutic uses, such as in the treatment of cancer and viral diseases and so consuming certain of the calystegines may actually be beneficial in the human diet. However, it is suggested that concentrations of these alkaloids in common foods are monitored, particularly when it is considered that certain processed potato products can sometimes contain higher levels of the calystegines than those commonly found in the fresh tubers. Also, consumption of non-traditional plant parts of the Solanaceae and Convolvulaceae should be avoided until the effects of the calystegines on humans are better understood.

Acknowledgments.

New work described was partly funded by the U.K. Ministry of Agriculture, Fisheries and Food and the Biotechnology and Biological Sciences Research Council.

Literature Cited.

1. Hellenäs, K.-E.; Branzell, C.; Johnsson, H.; Slanina, P. *J. Sci. Food Agric.* **1995**, *68*, 249-255.
2. Nash, R.J.; Rothschild, M.; Porter, E.A.; Watson, A.A.; Waigh, R.D.; Waterman, P.G. *Phytochemistry* **1993**, *34*, 1281-1283.
3. Tepfer, D. A.; Goldmann, A.; Pamboukdjian, N.; Maille, M.; Lepingle, A.; Chevalier, D.; Denaire, J.; Rosenberg, C. *J. Bacteriol.* **1988**, *170*, 1153-1161.
4. Molyneux, R. J.; Nash, R. J.; Asano, N. In *Alkaloids: Chemical and Biological Perspectives*; Pelletier, S. W., Ed.; Elsevier Science Ltd., Oxford, 1996, Vol. 11, pp 303-343.
5. Asano,N.; Kato, A.; Miyauchi, M.; Kizu, H.; Tomimori,T.; Matsui, K.; Nash, R.J.; Molyneux, R.J. *Eur. J. Biochem.* **1997**, *248*, 296-303.
6. Asano, N.; Oseki, K.; Tomioka, E.; Kizu, H.; Matsui, K. *Carbohydr. Res.* **1994**, *259*, 243-255.
7. Nash, R. J.; Watson, A. A.; Asano, N. In *Alkaloids: Chemical and Biological Perspectives*; Pelletier, S. W., Ed.; Elsevier Science Ltd., Oxford, 1996, Vol. 11, pp 345-376.
8. Molyneux, R.J.; James, L.F.; Ralphs, M.H.; Pfister, J.A.; Panter, K.P.; Nash, R.J. In *Plant-associated toxins - Agricultural, Phytochemical and Ecological Aspects*, Colegate, S.M.;. Dorling, P.R., Eds. CAB International, Wallingford, U.K., 1994, pp 107-112.
9. Dorling, P. R. In *Lysosomes in Biology*; Dingle, J. T.; Dean. R. T., Eds. Elsvier, Amsterdam, 1984, pp 347-379.
10. Asano, N.; Kato, A.; Oseki, K.; Kizu, H.; Matsui, K. *Eur. J. Biochem.* **1995**, *229*, 369-376.
11. Asano, N.; Kato, A.; Matsui, K.; Watson, A.A.; Nash, R.J.; Molyneux, R.J.; Hackett, L.; Topping, J.; Winchester, B. *Glycobiology.* **1997**, *7*, 1085-1088.
12. Menzies, J.S.; Bridges, C.H.; Bailey, E.M. Jr. *Southwest Vet.* **1979**, 32, 45-49.
13. Pienaar, J.G.; Kellerman, T.S.; Basson, P.A.; Jenkins, W.L.; Vahrmeijer, *Onderstepoort J. Vet. Res.* **1976**, *43*, 67-74.
14. de Balogh, K. K. I. M.; Dimande, A. P.; van der Lugt, J. J.; Molyneux, R. J.; Naude, T. W.; Welman, W. G. In *Toxic Plants and Other Natural Toxicants*;

Garland, T.; Barr, A. C., Eds., CAB International, Wallingford, 1998, pp 428-434.

15. Griffiths, R.C.; Watson, A.A.; Kizu, H.; Asano, N.; Sharp, H.; Jones, M.G.; Wormald, M.; Fleet, G.W.J.; Nash, R.J. *Tetrahedron Lett.* **1996**, *37*, 3207-3208.

16. Asano, N.; Kato, A.; Kizu, H.; Matsui, K.; Griffiths, R.C.; Jones, M.G.; Watson, A.A.; Nash, R.J. *Carbohydr. Res.* **1997**, *304*, 173-178.

17. Asano, N.; Kato, A.; Kizu, H.; Matsui, K.; Watson, A. A.; Nash, R. J. *Carbohydr. Res.* **1996**, *293*, 195-204.

18. Chotai, K.; Jennings, C.; Winchester, B.; Dorling, P. *J. Cell Biochem.* **1983**, *21*, 107-117.

19. Stegelmeier, B.L.; James, L.F.; Panter, K.E.; Gardner, D.R.; Ralphs, M.H.; Pfister, J.A. *J. Anim. Sci.* **1998**, *76*, 1140-1144.

20. Stegelmeier, B.L.; Molyneux, R.J.; Elbein, A.D.; James, L.F. *Vet. Pathol.* **1995**, *32*, 289-298.

21. Zhao, J.; Camire, M.E.; Bushway, R.J.; Bushway, A.A. *J. Agric. Food Chem.* **1994**, *42*, 2570-2573.

22. Theodorou, M. K.; Williams, B. A.; Dhanoa, M. S.; McAllan, A. B.; France J. *Animal Feed Science and Technology* **1994**, *48*, 185-197.

Chapter 10

New Alkaloids from *Phalaris* Spp.: A Cause for Concern?

Neil Anderton[1], Peter A. Cockrum[1], Steven M. Colegate[1,3], John A. Edgar[1], and Kirsty Flower[2]

[1]Plant Toxins Unit, CSIRO Division of Animal Health, Private Bag 24, Geelong, Victoria 3220, Australia
[2]Kybybolite Research Centre, P.O. Box 2, Kybybolite, South Australia 5262, Australia

Phalaris spp. grasses are useful pasture components but have been associated with neurological and sudden death intoxication syndromes. Despite agronomic development of *Phalaris* spp. to produce "low-toxicity" cultivars, outbreaks of intoxication have continued to occur. These outbreaks could result from a combination of poorly understood environmental or animal factors exacerbating the effect of the low concentration of known toxic alkaloids. Alternatively, previously unrecognised alkaloids could have intrinsic toxicity. Recent investigations have revealed the presence of the cardioactive N-methyltyramine in *P. aquatica* cultivars, and alkaloids of unknown toxicity i.e., the oxindoles coerulescine and horsfiline, and the furanobisindole phalarine, in *P. coerulescens* cultivars. The structures of coerulescine and phalarine were determined using NMR spectroscopy and mass spectrometry.

The grass *Phalaris aquatica* (syn. *P. tuberosa*) has been introduced into Australia from its native Mediterranean habitat. Selected for its drought tolerance, the winter growing perennial also withstands waterlogging and is especially adaptable to heavier soils in areas of 400-650 mm annual rainfall. However, use of the grass as a pasture component has been complicated by the seed-shedding characteristics of the early imports and an associated toxicity.

The seed-shedding problems were able to be addressed once a cultivar was isolated that displayed a high seed retention. Development of this cultivar allowed the simple and economic harvesting of seed for commercial distribution (*1*, *2*). Thus, *P. aquatica* has become firmly established in the southern, temperate areas of Australia.

[3]Corresponding author.

Toxicity was first associated with *P. aquatica* by McDonald (*3*) who reported a muscular incoordination or "staggers" syndrome in sheep and cattle grazing *Phalaris* pastures. Cattle and sheep seemed to be more susceptible to fresh new growth of the plant whilst horses seemed unaffected by the *Phalaris* toxins despite extended grazing on pastures that caused adverse effects in cattle and sheep. Subsequently considerable effort has been directed at developing "non-toxic" cultivars of *Phalaris* (*1, 2, 4*)

Clinical Effects of *Phalaris* Intoxication

Early field reports of *Phalaris* intoxication described a sudden death, cardiac effect and a neurological staggers syndrome in sheep and cattle (*3, 5*). An initial differentiation of *Phalaris* spp. intoxication into peracute, acute and chronic syndromes (*6, 7*) referred to the sudden death, cardiac effect; the rapid, reversible onset of the staggers syndrome; and the production of irreversible neurological deficits respectively. Since this early differentiation of *Phalaris* spp. intoxication syndromes, doubt has been cast upon the existence of an acute neurological syndrome which is truly separate from the chronic staggers syndrome (*8*). In addition, the sudden death syndrome has been refined to include the cardiac effect described by Gallagher *et al.* (*6*) and a polioencephalomalacia effect (*9*). Tissue and cellular anoxia due to transiently high nitrate and cyanide content of the plant have also been implicated (*9*). Thus Bourke has suggested the following differentiation of *Phalaris* spp. intoxication syndromes (*10*) :

1. Neurological Staggers
    Sheep will show neurological signs including head and body tremors; incoordination; muscle asynergy and limb paresis; shaking and nodding of the head; twitching of the ears, lips and tail; hopping or bounding movements; splayed digits and disturbed equilibrium which can result in falling to lateral or sternal recumbency and difficulty in regaining a standing position; and knuckling over at the fetlock joints and standing or walking on their front knees. Cattle are affected differently in that they show difficulty in chewing and swallowing resulting in a failure to thrive. They can also show protrusion of the tongue with increased drooling. Cattle are only mildly affected in terms of incoordination and limb paresis but do become hyperexcitable and will crash against fences and gates when disturbed.

    1.1. *Immediate onset* :   neurological signs are observed within days of access to *Phalaris* spp. pasture and may persist for several weeks. Recovery can be rapid if the animals are removed from the pasture (*6*).

    1.2. *Delayed onset* :   long-lasting neurological signs may be observed several months after the animals have been removed from the pasture (*11,12*).

2. Sudden Death

Deaths occur within 12 – 16 hours of allowing stock access to *Phalaris* spp. pasture. Frequently there are no preemptive clinical signs and the animals are simply found dead when next checked after being put onto the pasture.

2.1. *Cardiac effect* : when animals have been observed, signs included an arrhythmic tachychardia leading to ventricular fibrillation and cardiac arrest, with no preceding neurological signs. Post-mortem examination revealed severe congestion of the liver, kidney and spleen, and epicardial hemorrhages characteristic of acute, sudden onset heart failure (*6, 7*).

2.2. *Polioencephalomalacia* : examination of field evidence where sheep were found dead within 12 – 16 hours of being allowed access to *Phalaris* spp. pasture indicated an involvement of polioencephalomalacia, a degenerative lesion of the brain, in the etiology of the disease (*9*). Only sheep have been shown to be affected and surviving animals show signs characteristic of polioencephalomalacia including apparent blindness, depression, head and body tremors, aimless wandering, opisthotonus and either cerebral convulsions or coma. Microscopic changes consistent with polioencephalomalacia, or its early stages in dead sheep, are observed in affected animals.

Causative Factors of *Phalaris* spp. Intoxication

Environmental factors that seem to influence the outbreak of poisoning episodes include (*13*):
- high soil nitrogen content, frosts, high ambient temperatures, shading of plants (cloudy conditions)
- young plants are more toxic than mature plants
- pastures are likely to be more hazardous in the early morning than later in the day
- freshly growing plant is more toxic than hay made from the pasture
- hungry animals being lightly stocked on newly emerging *Phalaris* pasture at the break of the season after a period of drought or low rainfall

Culvenor *et al.*, (*14*) isolated three dimethyltryptamine alkaloids (**1 – 3**) from *P. tuberosa* (now known as *P. aquatica*) which they suggested might be the chemical causes of the poisoning episodes. Indeed, some of the observed environmental influences could be attributable to affecting the amount of alkaloid biosynthesized or retained by the plant.

The involvement of these dimethylaminoethylindole alkaloids in *Phalaris* spp. intoxication was first investigated by Gallagher *et al.*, who administered (intravenous, subcutaneous and oral) pure alkaloids to sheep and reported the observation of clinical signs commensurate with the proposed peracute and acute poisoning syndromes (*6, 7*).

However, further field and experimental observations led to the conclusion that the indolylethylamines could not alone be responsible for the cardiac presentation of the sudden death syndrome (*4, 8*). Indeed, Oram (*4*) concluded that, although administration of the pure alkaloids could undoubtedly induce toxic reactions, the level of tryptamine alkaloids in the plant was not a good indication of the toxicity of the herbage, declaring the possible presence of unidentified toxins which may or may not act synergistically with the tryptamines. To complicate the issue further, electrocardiographic investigation of the effect of bufotenine (5-hydroxy-N,N-dimethyltryptamine, **2**) in sheep revealed a marked cardiac effect. Following intravenous injection of bufotenine ($0.125 - 0.5$ mg kg^{-1}) to conscious sheep, a transient bradycardia was observed $5 - 18$ seconds after administration of the alkaloid, followed immediately by ventricular arrhythmia and tachycardia. The results indicated that bufotenine was more cardioactive than the methylated derivative (**3**) (Stewart, G. A.; Culvenor, C. C. J.; Anderton, N.; Dyke, T. M., unpublished data).

Other alkaloids that have been isolated from *Phalaris* spp. include N-methyltryptamine (**4**) and 5-methoxy-N-methyltryptamine (**5**), hordenine (**6**), the indolylmethylamines (gramines, **7 - 10**) and the tetrahydro-β-carbolines (**11, 12**). The metho cations of the dimethyltryptamine (**1**) and its 5-methoxy derivative (**3**) and the tetrahydo-β-carbolines (**11, 12**) have also been isolated as natural products and separated from their respective bases by chromatography on neutral polystyrene resin (*15*).

The amounts of these alkaloids extracted from the *Phalaris* spp. can vary but typical yields (% of dry weight) in the plants are 0.04% of dimethyltryptamines, 0.004% tetrahydro-β-carbolines, 0.009% hordenine, and up to 0.3% gramine. Oram (*4*) has reported that levels of total tryptamine alkaloid can rise to almost 0.3% (dry matter) in autumn (April) but that poisoning cases occurred in winter (June) when the total tryptamine content was up to five times lower.

Bourke (*10*) has determined that the approximate minimum oral doses of the four classes of alkaloids i.e., the tryptamines, β-carbolines, tyramines and gramines, required to produce significant neurological signs are 40 mg kg^{-1}, 240 mg kg^{-1}, 480 mg kg^{-1}, and 720 mg kg^{-1} respectively. Based on these estimates, and the average daily intake for sheep of 1047 g (dry weight) of *Phalaris* sp. (*23*) Bourke has suggested that it is most unlikely that the tyramines or gramines could be ingested in sufficient quantity to elicit nervous signs. Indeed, the tryptamines would require several days of exposure whilst the β-carbolines would require several months in order to cause nervous signs. These conclusions are based upon the administration of pure alkaloids with no consideration given to synergistic effects of other alkaloids, or the potentiating effects of environmental factors or animal nutrition status.

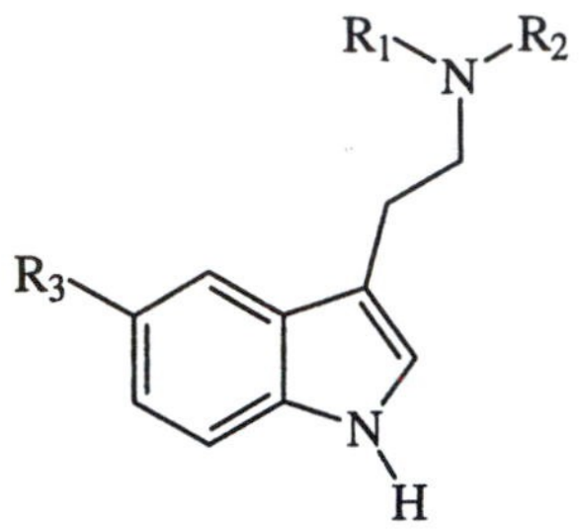

1 - 5

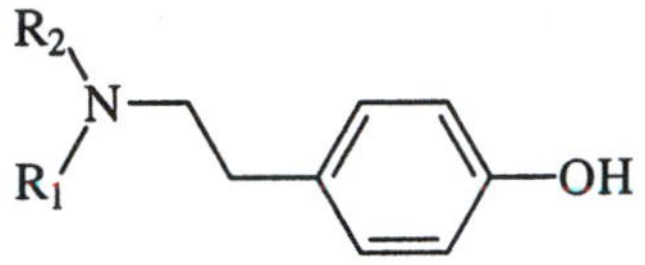

6

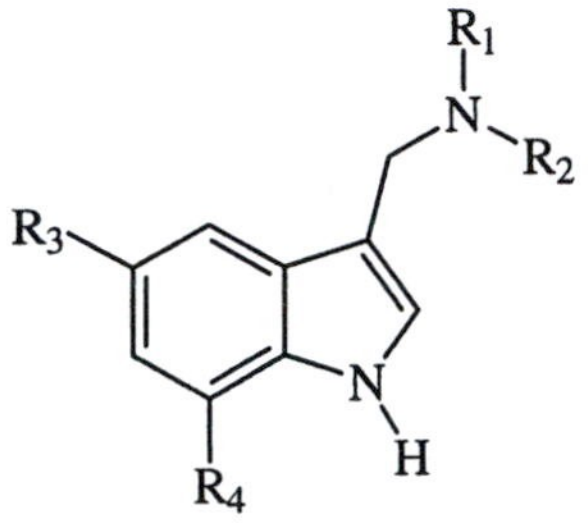

7 - 10

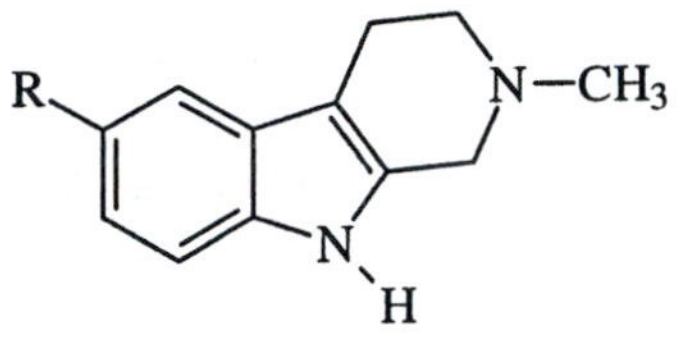

11 - 12

Structure	R	R₁	R₂	R₃	R₄	Reference
1	-	Me	Me	H	-	*14*
2	-	Me	Me	OH	-	*14*
3	-	Me	Me	OMe	-	*14*
4	-	H	Me	H	-	*19*
5	-	H	Me	OMe	-	*16*
6	-	Me	Me	-	-	*16, 17*
7	-	Me	Me	H	H	*14, 17*
8	-	Me	Me	H	OMe	*22*
9	-	Me	Me	OMe	H	*22*
10	-	Me	Me	OMe	OMe	*22*
11	H	-	-	-	-	*20, 21*
12	OMe	-	-	-	-	*17, 18, 20*

The Search for Other Toxins

Despite the fact that *Phalaris* has been the subject of agronomic development in order to produce low-alkaloid, and presumably low-toxicity cultivars, there have been many instances of poisoning that have occurred on the "low-toxicity" varieties (*24*). This, in addition to the conclusions of Oram (*4*) and Bourke *et al.*, (*8, 9*) that toxic episodes do not correspond to high levels of tryptamine alkaloids, that the tryptamine alkaloids are not responsible for the cardiac presentation of the sudden death syndrome, and that unknown toxins may be responsible for, or contribute to the poisoning syndromes (including the polioencephalomalacia-related sudden death), has stimulated the search for other *Phalaris* spp. toxins.

N-methyltyramine, a Potential Cardiac Toxin.   Using a rat atrial muscle bioassay, extracts of *Phalaris aquatica* cv Sirolan were screened for cardioactive components (*25*). The crude extract was fractionated using reverse phase and weak cation exchange chromatography. Test samples of extract fractions were applied to an organ bath in which the atrial preparation was suspended. Active samples caused a contraction of the atrial tissue.

The only active compound isolated was identified, using nuclear magnetic resonance (NMR) spectroscopy and fast atom bombardment mass spectrometry (FABMS), as N-methyltyramine (**13**). The spectroscopic properties of the isolated compound were identical to those of synthetic N-methyltyramine prepared by the method of Mangino *et al.* (*26*).

13

Studies with blocking agents confirmed the natural product as an indirectly-acting sympathomimetic amine, stimulating the release of norepinephrine at the nerve synapse. This finding was consistent with literature reports of the activity of tyramine and N-methyltyramine (*27*). Significant cardiac effects were observed in sheep which received oral doses of $300 - 400$ mg kg^{-1} of N-methyltyramine.

The action of N-methyltyramine is amplified by the presence of monoamine oxidase inhibitors (*28*) which prevent the degradation of amines thereby increasing their bioavailablity. This has important implications for synergistic effects on the action of N-methyltyramine since the co-occurring tryptamines are reported to be monoamine oxidase inhibitors (*7*) and animals on a low nutritional plane, as a result of low protein diets or starvation during transport, will have lower levels of monoamine oxidases (*27*). This was investigated by treating sheep with the monoamine oxidase inhibitor tranylcypromine before oral administration of N-methyltyramine (25). In this case, significant tachychardia and elevation of blood pressure was induced with a dose of 20 mg kg^{-1}. No sudden deaths were recorded with this experimental treatment but care was taken not to excite or distress the animals thereby possibly avoiding the precipitation of cardiac arrest.

Recent Poisoning Outbreaks. Several instances of *Phalaris*-related sheep sudden deaths were investigated in May 1997 (mid to late autumn) (CSIRO, Plant Toxins Unit, unpublished). Sheep were found dead after being put onto freshly sprouting *Phalaris* pasture 12 – 16 hours prior i.e., overnight. Grass samples were collected at the time of the poisonings and, in one case (Darrawill North, Victoria, Australia), the brains of affected sheep were removed for histological examination. The *Phalaris aquatica* cultivars involved were Australian and Sirosa.

Histological examination of the brain tissue (P. Hooper, personal communication; 29) detected evidence indicative of the early stages of polioencephalomalacia. It has been suggested that the rapid development of polioencephalomalacia is a result of the presence of a thiamine (vitamin B$_1$) or pyridoxine (vitamin B$_6$) antagonist in the plant (9). Such a toxin could conceivably lead to brain degeneration characteristic of polioencephalomalacia by affecting the integrity of the blood vessels supplying the brain and causing a fatal reduction in oxygen and nutrient supply (29).

Analysis of the alkaloid content of the *Phalaris* samples by high performance liquid chromatography (HPLC) (Anderton, N.; Cockrum, P. A.; Colegate, S. M.; Edgar, J. A.; Flower, K., *Phytochemical Analysis* **1998**, in press) indicated the presence of N-methyltyramine (**13**, 0.012% w/w dry weight), gramine (**7**, 0.001% w/w dry weight), dimethyltryptamine (**1**, 0.012% w/w dry weight) and 5-methoxydimethyltryptamine (**3**, 0.001% w/w dry weight) (Figure 1).

Generously assuming the sheep ate 2 kg (dry weight) of the *Phalaris* overnight, this would still only amount to an ingestion of about 240 mg of N-methyltyramine (**13**) which approximates to 5 mg kg^{-1}, considerably less than the 300 – 400 mg kg^{-1} needed to induce significant heart effects (25). However, conclusions are difficult to reach since little is known about the synergistic effects of the significant quantity of dimethyltryptamine (**1**) that was also present in the plants.

***Phalaris*-associated Sudden Death of Horses** McDonald (3) has noted that horses seem unaffected by *Phalaris aquatica* pasture that was detrimental to cattle and sheep. It is the experience of the authors that in the Geelong area of Victoria it is common to see horses agisted for many years on *P. aquatica*-dominant pasture without any ill-effect. However, Bourke (New South Wales Agriculture Department, personal communication) has noted sudden death of horses associated with *Phalaris* spp. ingestion.

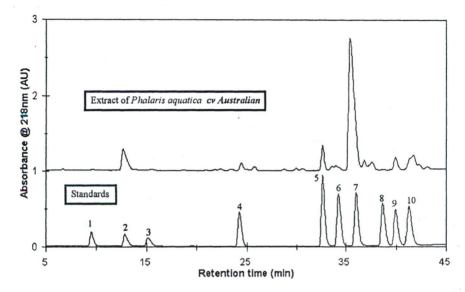

Figure 1. HPLC of *Phalaris* alkaloid standards and an alkaloidal extract of *P. aquatica* cv Australian collected at the time and site of poisoning outbreak. 1 : tyramine; 2 : N-methyltyramine; 3 : hordenine; 4 : bufotenine; 5 : gramine; 6 : tryptamine; 7 : dimethyltryptamine; 8 : 5-methoxytryptamine; 9 : 5-methoxydimethyltryptamine; 10 : 2-N-methyl-1,2,3,4-tetrahydro-β-carboline.

Over a period of 5 years from 1988 to 1993, 8 horses of various ages died suddenly on a hobby farm property about 60 km to the north west of Melbourne, Australia. In the last case, a horse was put onto the same pasture as three horses that died previously. After about 22 days, the horse died within 10 minutes of collapsing following spontaneous wild galloping. There was no obvious gross or microscopic pathology and tests for strychnine, lead, arsenic, barbiturate, cyanide and organophosphates were negative. Notable in the pasture was *P. coerulescens* (Dr Rex Oram, CSIRO Plant Industry, Canberra, personal communication). The period of grazing on the pasture ranged from 3 weeks to 8 months with most deaths occurring in autumn and winter with the regrowth of *Phalaris* after the summer. In May 1998, another two horses died suddenly in the same area and on *P. coerulescens* dominant pasture.

Samples of *P. coerulescens* were collected within a few days of the deaths from the properties where the horse deaths occurred. The plant was macerated and extracted into dilute hydrochloric acid within two hours of collection. Isolation, identification and quantitation of the alkaloids in the plant involved cation exchange chromatography, thin layer chromatography (TLC), gas chromatography-mass spectrometry (GC/MS), HPLC and NMR spectroscopy (Anderton, N.; Cockrum, P. A.; Colegate, S. M.; Edgar, J. A.; Flower, K., *Phytochemical Analysis* **1998,** in press).

Previous work on the phytochemical analysis of *P. coerulescens*, that was being investigated by the South Australian Research and Development Institute for possible agronomic development to meet pasture needs on poor soils in low rainfall areas, revealed the presence of the tetrahydro-β-carbolines (**11, 12**) and the oxindoles, coerulescine (**14**) and horsfiline (**15**), as the major alkaloid components (*30*). The combined yield of oxindoles and tetrahydro-β-carbolines isolated from the imported *P. coerulescens* accessions was estimated at 260 mg kg^{-1} (dry weight of plant).

14   **15**

Coerulescine (**14**) was a newly described alkaloid, whereas the 5-methoxy derivative (horsfiline, **15**) had previously only been isolated from the south-east Asian medicinal plant *Horsfieldia superba* (Myristicaceae) (*31*). Detection of these alkaloids in the *P. coerulescens* extracts was initially achieved by GC/MS. The TLC visualising agent response was different to the usual *Phalaris* alkaloids in that they did not colorize with acidified anisaldehyde reagent, however they were visualized with Dragendorff's reagent.

Structural elucidation of coerulescine (**14**) began with the determination of a molecular formula. A molecular mass of 202 was confirmed by chemical ionization MS and a high resolution mass measurement of the molecular ion peak indicated a molecular formula $C_{12}H_{14}N_2O$ for coerulescine (**14**). The fragmentation pattern of the molecular ion was remarkably similar to that expected of a tetrahydro-β-carboline, and high resolution mass measurements on some fragment ions confirmed the association of the oxygen atom with the indole portion of the molecule. The alkaloid was readily acetylated (acetic anhydride/pyridine) or methylated (MeI/K_2CO_3/acetone) to yield the mono acetyl (M$^+$ m/e 244) or mono methyl (M$^+$ m/e 216) derivative respectively. Extensive NMR spin mapping (^{1}H, ^{13}C, ^{1}H-^{1}H COSY, ^{1}H-^{13}C DEPT, ^{1}H-^{13}C HMQC and ^{1}H-^{13}C HMBC) allowed the structure determination of coerulescine although the relative stereochemistry around the spiro carbon atom remains undefined. Tentative identification of the 5-methoxy derivative of coerulescine (**15**, horsfiline) was based upon comparison of its mass spectral data with those of coerulescine reported above and those reported for horsfiline in the literature (*31*).

Therefore, because of the presence of tetrahydro-β-carbolines, the unknown toxicity of oxindoles and the presence of unidentified tryptamine-like alkaloids, it seems that the *P. coerulescens* accessions analyzed may have the potential to be toxic to grazing animals. This work with the *P. coerulescens* cultivars also highlighted the presence of alkaloids that reacted to the TLC visualisation reagent in the same way as the tryptamine, tyramine and tetrahydro-β-carboline alkaloids. The more non-polar of these, obtained in approximately 0.003% yield, was optically active, $[\alpha]^{20}_D$ -92° (0.0075, CH$_3$OH). Despite not being amenable to gas chromatography the compound could readily be detected using liquid chromatography/mass spectrometry (LC/MS). A molecular mass of 404 was indicated by direct insertion probe, electron impact ionization mass spectrometry (DIP-EIMS) and by electrospray ionization mass spectrometry (ESIMS) or atmospheric pressure chemical ionization mass spectrometry (APCIMS) (M$^+$ + H, 405). High resolution DIP-EI mass measurements suggested a molecular formula $C_{24}H_{28}N_4O_2$. ESIMS of the acetylation (acetic anhydride/pyridine) or methylation (methyl iodide/potassium carbonate/acetone) products of the compound indicated formation of a monoacetyl (M$^+$ + H, 447) or a monomethyl (M$^+$ + H, 419) derivative respectively.

The proton decoupled ^{13}C NMR spectrum of the parent alkaloid consisted of 22 sharp resonance signals and one low intensity, broadened resonance signal (δ 44.3ppm) which was subsequently assigned to a dimethylamine group. The total number of carbons detected thereby supported the proposed molecular formula based upon the high resolution mass measurement. A DEPT135 NMR experiment further defined the resonance signals as 6 methine, 4 methylene, 4 methyl (once again, the low intensity resonance signal for the dimethylamine group was very broad) and 10 quaternary carbons. Extensive NMR investigation (^{1}H-^{1}H COSY, ^{1}H-^{13}C DEPT, ^{1}H-^{13}C HMQC and ^{1}H-^{13}C HMBC) of the parent alkaloid and its monoacetyl derivative clearly established the presence of the hexahydro-β-carboline (A) and oxygenated methoxy-gramine (B) partial structures.

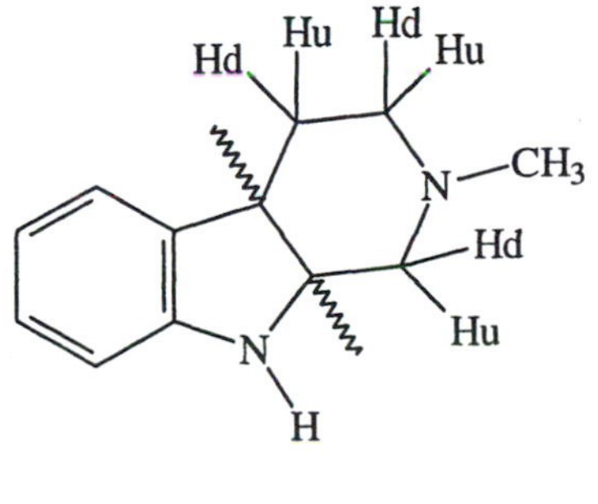

Partial Structure A

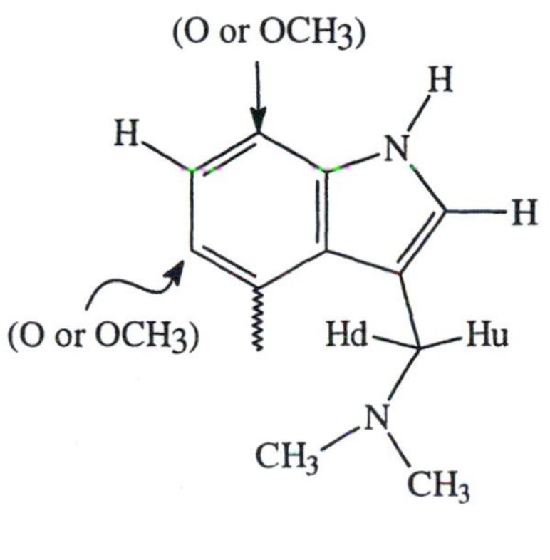

Partial Structure B

These structures were supported when the degree of fragmentation in the ESIMS of the parent alkaloid and its derivatives was enhanced by increasing the cone voltage of the source. Product ions indicating losses of -NMe$_2$, -CH$_2$NMe$_2$, -OMe, -NMe, –(CH$_2$)$_2$NMe and -(CH$_2$)$_2$ NMeCH$_2$ were observed. APCI MS/MS on the M$^+$+ H ion for the parent alkaloid (m/z 405) confirmed the loss of N(CH$_3$)$_2$ to yield m/z 360, and subsequent APCI MS/MS on m/z 360 confirmed the losses of CH$_3$ (m/z 345) and (CH$_2$)NCH$_3$ (m/z 317). In addition to showing fragmentations similar to the ESIMS, the DIP-EI mass spectrum of this compound included ions at m/z 186 and 143 which are particularly supportive of the partial structure A in that the EI mass spectrum of 2-N-methyl-1,2,3,4-tetrahydro-β-carboline (11) shows a molecular ion peak at m/z 186 and a base peak at m/z 143 corresponding to loss of –CH$_2$NCH$_3$.

Long range carbon-hydrogen correlations and 2-dimensional nuclear Overhauser effect spectroscopy (NOESY) established the linking of partial structures A and B to form the furanobisindole (16) that was subsequently given the trivial name phalarine (Anderton, N.; Cockrum, P. A.; Colegate, S. M.; Edgar, J. A.; Flower, K.; Gardner, D.; Willing, R. I., *Phytochemistry* **1999**, in press).

Whether or not the oxindoles (14, 15) or the furanobisindole (16) are toxic to

16

livestock remains to be determined. However, it may be appropriate to note that these compounds are the major alkaloids isolated from *P. coerulescens* collected at the time and site of the equine sudden deaths.

Comparison (GC/MS, TLC) of the alkaloid extracts of the imported *P. coerulescens* from the South Australian agronomic research with those established plants associated with the equine sudden deaths, consistently revealed that the putative methoxycoerulescine (horsfiline, 15), and its tetrahydro-β-carboline analogue (12), were predominant in the established variety (Figure 2).

Further investigation (TLC, HPLC) of the horse death-associated *P. coerulescens* revealed the significant presence of the furanobisindole, phalarine (16) and another, as yet unidentified major component of similar retention and color reaction to phalarine but which did not correspond to any of the tyramine, tryptamine, gramine or tetrahydo-β-carboline standards used for comparison.

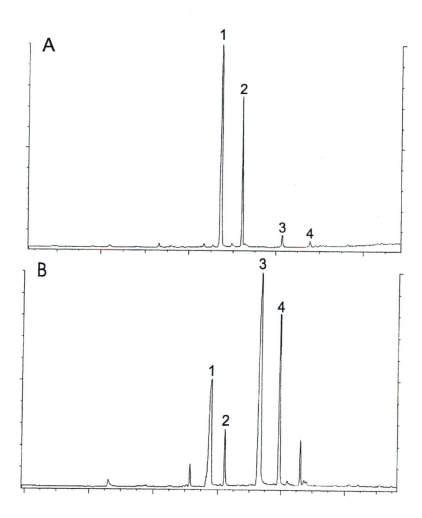

Figure 2. Gas chromatography (MS detection) of *P. coerulescens* alkaloid extracts. **A** : Grass imported from Portugal for agronomic development. **B** : Established grass collected at site of horse poisonings. 1 : coerulescine (**14**), 2 : 2-N-methyl-1,2,3,4-tetrahydro-β-carboline (**11**), 3 : horsfiline (**15**, methoxycoerulescine), 4 : 6-methoxy-2-N-methyl-1,2,3,4-tetrahydro-β-carboline (**12**).

Conclusions and Further Work

The *Phalaris aquatica* samples examined from the sites of recent instances of sheep poisoning showed the presence of N-methyltyramine (**13**) and dimethyltryptamine (**1**) at a combined level of approximately 0.024% (w/w, dry weight of plant). Taken in isolation, this alkaloid content would seem inadequate to illicit fatal neurological or cardiac effects. However, it is possible that synergistic effects of alkaloids, or the potentiating influence of environmental or animal nutritional factors may play a role in the toxicology. Nevertheless, the pathology strongly supported the involvement, or co-occurrence of polioencephalomalacia in the sudden death of these sheep less than 16 hours after being allowed access to the *P. aquatica*-dominant pasture. Further work is necessary to determine whether thiamine or pyridoxine antagonists are produced by the plant.

The *P. coerulescens* samples collected from the site with a history of sudden equine death, do not contain the tryptamine and tyramine alkaloids that have been associated with the immediate onset staggers and the cardiac presentation of the sudden death syndromes of *Phalaris aquatica* cultivar intoxication in sheep. However, the samples contain appreciable quantities of the tetrahydro-β-carboline alkaloids (**11, 12**), the newly described oxindoles (**14, 15**) and the furanobisindole (**16**), in addition to one further unidentified major alkaloid component and several minor unidentified alkaloids (Colegate, S. M.; Edgar, J. A.;. Gardner, D. R., unpublished). Whilst the tetrahydo-β-carbolines have been suggested as causative factors in the delayed onset of irreversible, neurological clinical signs (*10*), further work to establish the toxicity of these new alkaloids, especially with respect to sudden death of horses, will be necessary from animal health and plant agronomic aspects.

Acknowledgments

The graphics compilation assistance of Frank Filippi and Ted Stephens (CSIRO, Animal Health, Geelong) is gratefully acknowledged, as are the useful discussions on the clinical differentiation of *Phalaris* spp. intoxication between Dr Chris Bourke (New South Wales Agriculture Department, Orange, NSW) and SMC.

Literature Cited

1. Oram, R. N.; Schroeder, H. E., *Aust. J. Exptl. Agric.* **1992**, *32*, 261.
2. Oram, R. N., *Aust. J. Exptl. Agric.* **1996**, *36*, 913.
3. McDonald, I. W., *Aust. Vet. J.* **1942**, *18*, 182.
4. Oram, R. N., *Proceedings of the XI International Grasslands Congress;* Qld., Australia, 1970; pp. 785-788.
5. Moore, R. M.; Arnold, G. W.; Hutchings, R. J.; Chapman, H. W., *Aust. J. Sci.* **1961**, *24*, 88.
6. Gallagher, C. H.; Koch, J. H.; Moore, R. M.; Steel, J. D., *Nature* **1964**, *204*, 542.
7. Gallagher, C. H.; Koch, J. H.; Hoffman, H., *Aust. Vet. J.* **1966**, *42*, 279.
8. Bourke, C. A. ; Carrigan, M. J.; Dixon, R. J., *Aust. Vet. J.* **1988**, *65*, 218.

9.   Bourke, C. A.; Carrigan, M. J., *Aust. Vet. J.* **1992**, *69*, 165.
10. Bourke, C. A., In *Plant-associated toxins: Agricultural, Phytochemical and Ecological Aspects;* Colegate, S. M.; Dorling, P. R., Eds; CAB International: Wallingford, 1994; Chapter 93, pp. 523-528.
11. Nicholson, S. S.; Olcott, B. M.; Usenik, E. A.; Casey, H. W.; Brown, C. C.; Urbatsch, L. E.; Turnquist, S. E.; Moore, S. C., *JAVMA* **1989**, *195*, 345.
12. Bourke, C. A., Carrigan, M. J.; Seaman, J. T.; Evers, J. V., *Aust. Vet. J.* **1987**, *64*, 31.
13. Everist, S. L., *Poisonous Plants of Australia, 2nd Edition*; Angus&Robertson; Melbourne, 1981, pp. 340 –344.
14. Culvenor, C. C. J.; Dal Bon, R.; Smith, L. W., *Aust. J. Chem.* **1964**, *17*, 1301.
15. Frahn, J. L.; Illman, R. J., *J. Chromatography* **1973**, *87*, 187.
16. Wilkinson, S., *J. Chem. Soc.* **1958**, 2079.
17. Audette, R. C. S.; Bolan, J.; Vijayanagar, H. M.; Bilous, R.; Clark, K., *J. Chrom.* **1969**, *43*, 295.
18. Shannon, P. V. R.; Leyshon, W. M., *J. Chem. Soc.* **1971**, C, 2837.
19. Williams, M.; Barnes, R. F.; Cassady, J. M., *Crop Science* **1971**, *11*, 213.
20. Frahn, J. L.; O'Keefe, D. F., *Aust. J. Chem.* **1971**, 24, 2189.
21. Gander, J. E.; Marum, P.; Marten, G. C.; Hovin, A. W., *Phytochemistry* **1979**, *15*, 737.
22. Mulvena, D. P.; Picker, K.; Ridley, D. D.; Slaytor, M., *Phytochemistry* **1983**, *22*, 2885.
23. Hogan, J. P.; Weston, R. H.; Lindsay, J. R., *Aust. J. Agric. Res.* **1969**, *20*, 925.
24. Kennedy, D. J.; Cregan, P. D.; Glastonbury, J. R. W.; Golland, D. T.; Day, D.G., *Aust. Vet. J.* **1986**, *63*, 88.
25. Anderton, N.; Cockrum, P. A.; Walker, D. W.; Edgar, J. A., In *Plant-associated toxins: Agricultural, Phytochemical and Ecological Aspects;* Colegate, S. M.; Dorling, P. R., Eds; CAB International: Wallingford, 1994; Chapter 49, pp. 269-274.
26. Mangino, M. M.; Libbey, L. M.; Scanlan, R. A., *IARC Sci. Pubs* **1982**, 41, 57.
27. Camp, B. J., *Amer. J. Vet. Res.* **1970**, *31*, 755.
28. Tiller, J. W. G.; Maguire, K. P.; Davies, B. M.; Dowling, J. T.; Tung, L. H.; Rand, M. J., *Human Psycopharmacology* **1990**, *5*, 313.
29. Bourke, C. A., *Proceedings of the 9th Australian Agronomy Conference*; Wagga Wagga, New South Wales, 1998; pp. 326 – 329.
30. Anderton, N.; Cockrum, P. A.; Colegate, S. M.; Edgar, J. A.; Flower, K.; Vit, I.; Willing, R. I., *Phytochemistry* **1998**, *48*, 437.
31. Jossang, A.; Jossang, P.; Hadi, H. A.; Sévenet, T.; Bodo, B., *J. Org. Chem.* **1991**, *56*, 6527.

Chapter 11

Natural Toxins from Poisonous Plants Affecting Reproductive Function in Livestock

K. E. Panter[1], D. R. Gardner[1], L. F. James[1], B. L. Stegelmeier[1],
and R. J. Molyneux[2]

[1]Agricultural Research Service, Poisonous Plant Research Laboratory,
U.S. Department of Agriculture, Logan, UT 84341
[2]Agricultural Research Service, Western Regional Research Center,
U.S. Department of Agriculture, Albany, CA 94710

Certain poisonous plants and toxins therefrom cause various and often detrimental effects on reproductive function in livestock. Recent research efforts at the Poisonous Plant Research Laboratory have focused on three groups of plants with specific classes of toxins affecting different aspects of reproduction. Certain *Astragalus* and *Oxytropis* species called locoweeds contain the indolizidine alkaloid swainsonine and affect most aspects of reproduction from ovarian and testicular function to embryonic death and abortion. Ponderosa pine and related species contain labdane resin acids that cause abortion or premature parturition in late term pregnant cattle. These labdane resin acids include isocupressic acid (ICA) and two ICA derivatives, succinyl and acetyl ICA, as the active compounds. *Lupinus* spp., *Conium maculatum* and *Nicotiana* spp. contain neurotoxic and teratogenic alkaloids. Lupines contain quinolizidine and piperidine alkaloids that are responsible for induced fetal cleft palate and multiple congenital contracture (MCC) malformations in cattle. *Conium maculatum* and *N. glauca* contain piperidine alkaloids which have caused cleft palate and MCC in pigs, cattle, sheep and goats. A goat model, where *N. glauca* plant material or anabasine-rich extracts therefrom are gavaged during specific stages of pregnancy, has been established to study the mechanism of action of the induced cleft palate and MCC in cattle, and has recently been characterized to study the etiology of cleft palate formation and *in utero* repair in humans.

Locoweeds

Locoweeds [certain species of the *Oxytropis* and *Astragalus* genera containing the indolizidine alkaloid swainsonine (*1*)] reduce reproductive performance in livestock.

Most aspects of reproduction are affected, including mating behavior and libido in males, behavioral estrus and conception in females, fetal growth and development and neonatal/maternal behavior. While extensive research has been done to characterize and describe the histological changes, we have just scratched the surface of understanding the magnitude of the physiological problems, the mechanism of action of reproductive dysfunction and management strategies needed to prevent losses.

Swainsonine

Once animals begin to graze locoweed, measurable increases in serum swainsonine with concomitant decreases in α-mannosidase activity (Figure 1) occur within one to two days. While these measurable changes are diagnostic, the rapid clearance of swainsonine from serum ($t_{1/2} \approx 20$ hrs) and accompanying recovery of α-mannosidase activity ($t_{1/2} \approx 65$ hours) limits serum analysis of these parameters as a reliable test for locoweed exposure after animals have not ingested locoweed for several days (2). Currently, diagnosis of locoweed poisoning relies on history of locoweed ingestion, behavioral changes, loss of condition and, in terminal cases, histological evidence of neurovisceral vacuolation. Histological lesions induced by locoweed and purified swainsonine (2-4) have been compared and found to be the same. Lesions appear to develop in a threshold-like fashion since the severity of lesions does not increase at higher locoweed doses (2,4). Animal tissues that accumulate high swainsonine concentrations (such as liver and kidney) develop lesions more rapidly, and at lower dosages of locoweed, than other organ systems (such as blood and muscle). Even though α-mannosidase activity recovers quickly, tissue repair and return to normal organ function occur more slowly (2).

Effects on Female Reproduction. Locoweeds affect almost every aspect of reproduction in the female such as estrus behavior, estrous cycle length, ovarian function, conception, embryonic and fetal viability, and maternal/infant bonding (Panter, K. E., unpublished data; 3,5).

Locoweed fed to cattle and sheep at various times and dosages temporarily altered ovarian function, increased estrous cycle length, altered breeding behavior and reduced conception rates. Recent feeding trials with locoweeds (*A. mollissimus, A. lentiginosus* and *O. sericea*) in cycling ewes demonstrated that after 20 days of locoweed feeding at 10-15% of their diet, estrus was delayed and shortened, conception rates decreased and the number of viable embryos collected from superovulated ewes was reduced (Panter, unpublished data).

While only a few abnormal morula-stage embryos were collected from ewes fed locoweed for 30 days (Panter, unpublished data), recent *in vitro* data demonstrate that swainsonine added to culture media at different concentrations (up to 6.4 µg/mL) did not directly interfere with oocyte maturation (IVM), *in vitro* fertilization (IVF), or embryo growth and development when cultured *in vitro* (IVC) (*6*). Pregnancy rates were not different from controls when swainsonine-cultured (IVM/IVF/IVC) bovine embryos were transferred to recipient cows. This research suggests that the effects of locoweed on early embryo viability and development may not be from direct effects of swainsonine but rather are secondary, and result from the effects of locoweed (swainsonine) on maternal aspects of reproduction such as the pituitary/hypothalamic axis where glycoprotein gonadotropins are produced and released. Thus, embryonic effects from locoweed *in vivo* may result from secondary maternal pathways affected by locoweed.

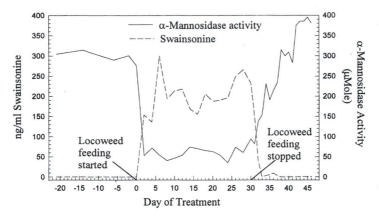

Figure 1. Average increase in bovine serum swainsonine with associated decrease in α-mannosidase activity in 4 cows fed locoweed (*Oxytropis sericea*). Figure adapted from *2*.

Mature cycling cows fed *Oxytropis sericea* at 20% of their diet for 30 days showed moderate signs of toxicity. The estrous cycle length increased during locoweed feeding (Figure 2) and conception was delayed (Panter, unpublished data). After feeding stopped, normal estrous cycle length returned relatively soon and cows bred normally, although conception was delayed in some cows (repeat breeders) for up to three estrous cycles. In another study cycling heifers fed *O. sericea* equivalent to three dosages of swainsonine (0.25, 0.75 and 2.25 mg/kg/day) showed ovarian dysfunction in a dose-dependent pattern. Heifers receiving the highest dose for 45 days had enlarged ovaries by day 20 of locoweed feeding. Observation of these ovaries by ultrasound suggested that both the luteal phase (observation of corpus luteum) and follicular phase (observation of a follicular cyst) were prolonged and persisted throughout the feeding period (Figure 3; Panter, unpublished data). Within 30 days after locoweed feeding had stopped, ovaries appeared normal via ultrasound, and 15 days later when heifers were necropsied the ovaries were similar in appearance to those of controls. Further studies are needed to fully characterize and define the effects of locoweed on ovarian function in cattle and sheep.

While gross and microscopic lesions in the dam may begin to resolve quickly after locoweed ingestion ceases, effects on the fetus may be prolonged and severe enough to result in abortion, small and weak offspring at birth, or reduced maternal/infant bonding and impaired nursing ability of the neonate (5, 7). Locoweed ingestion by pregnant ewes during gestation days 100-130 disrupted normal maternal infant bonding compared to control ewe-lamb pairs (5). Lambs from mothers ingesting locoweed failed to suckle within 2 hours after birth, were slower to stand, and were less vigorous than control lambs. Thus, maternal ingestion of locoweed disrupted the learning ability of their neonatal lambs (5). Swainsonine is also excreted in the milk and can result in further intoxication of nursing offspring or exacerbate intoxication when offspring begin to graze locoweeds and continue to nurse their locoweed-grazing mothers (8,9).

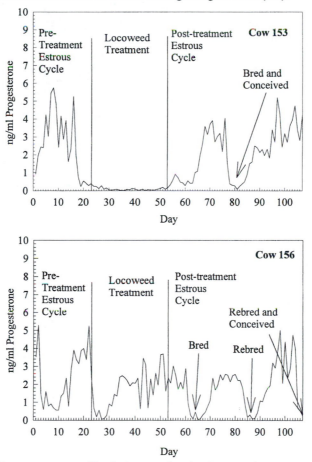

Figure 2. Progesterone profiles in 2 cows showing increased estrous cycle lengths.

Effects on Male Reproduction. The effects of locoweed ingestion are equally detrimental to male reproductive function. Panter *et al.* (*10*) reported transient

158

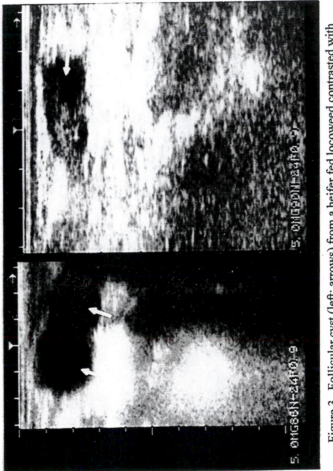

Figure 3. Follicular cyst (left; arrows) from a heifer fed locoweed contrasted with a normal ovary (right; arrow) from a heifer in the follicular phase of the estrous cycle.

degenerative changes in the seminiferous, epididymal, and vas deferens epithelia after feeding locoweed to yearling rams for 70 days. Clinically, there were changes in behavior, reduced libido, and loss of body condition. Grossly, there were no observed changes in testicular circumference or tissue appearance. Histologically, there was foamy, cytoplasmic vacuolation in the epithelium of the seminiferous tubules, epididymis and vas deferens and reduced spermatozoa production. Semen contained significantly more abnormal spermatozoa, including retained perinuclear cytoplasmic droplets, detached tails, bent tails and marked decreases in motility (*10*). These changes in spermatozoa were transient and by 70 days after locoweed feeding ended, the rams appeared clinically normal.

Recent feeding trials in two-year-old rams demonstrated that changes in breeding behavior occurred within 30 days after locoweed feeding started. Exposure of these rams to ewes in estrus often precipitated uncontrolled muscular tremors, proprioceptive deficits, and anxiousness (Panter, unpublished data). Even though these rams had no microscopic changes in semen quality, motility or metabolism at this time, there was a reduction in conception rates when these rams were bred to control ewes (Panter, unpublished data; *11*). After 60 days on locoweed, there were severe and clinically overt neurological deficits in the rams and microscopic changes in spermatozoa morphology and motility but metabolic activity of spermatozoa was still not significantly reduced (*11*). Spermatogenesis and spermatozoa defects resolved by 60 to 90 days after locoweed feeding stopped, but five of seven rams continued to lose weight (wasting) and neurologic deficits intensified until they were euthanized. Histological evaluation of tissues will be reported elsewhere.

Ortiz *et al.* (*12*) reported several delayed effects after feeding *Oxytropis sericea* to breeding age ram lambs for 35 days. Subsequently, 35 days after locoweed feeding had stopped, they found reduced sperm motility and decreased scrotal circumference in all treated rams. There was also reduced testosterone responses to gonadotrophin releasing hormone (GnRH) challenge, suggesting that locoweed affected testicular function in these rams. This delayed effect is expected as the normal cycle of spermatogenesis in sheep is about 60 to 70 days and Panter *et al.* (*10*) demonstrated that increased spermatozoa abnormalities peaked after continuous feeding of locoweed for 70 days.

While research and field observations have demonstrated that locoweed affects almost every aspect of reproduction in livestock, several questions still need to be answered. How much locoweed can livestock eat, and over what time period, before reproduction declines? What functions are affected first? When are reproductive effects irreversible? What are the modes of action? How long does it take for reproductive function to return to normal once locoweed ingestion stops? Answers to these questions will aid management decisions to improve reproductive performance and allow better utilization of locoweed-infested ranges.

Ponderosa Pine and Related Species

Ponderosa pine needles (PN) induce abortion in cattle when eaten during the last trimester of gestation (*13,14*). Isocupressic acid (ICA), a labdane resin acid, and two

ICA derivatives were identified as the abortifacient toxins (*15,16*; Figure 4). Oral and intravenous (iv) administration of ICA induced abortions in a dose-dependent manner with the higher doses inducing abortion in a shorter period of time. Acetyl and succinyl ICA derivatives, naturally present in PN, were abortifacient when administered orally but not iv. Both have been shown to be hydrolyzed in the rumen to ICA (*17*). Isocupressic acid metabolites include imbricataloic acid, agathic acid, dihydroagathic acid, and tetrahydroagathic acid (Figure 5). Pine needles are not abortifacient in goats and sheep nor was ICA abortifacient in goats when administered orally or iv. Isocupressic acid metabolites, similar to those detected in cow serum, were detected in goat serum (Gardner, D. R., unpublished data).

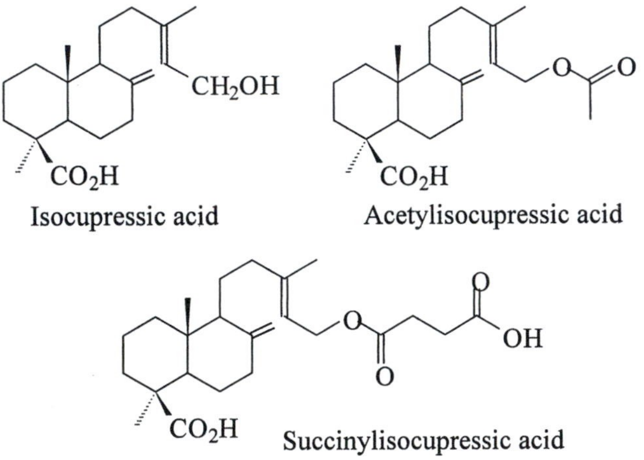

Figure 4. Abortifacient compounds identified in Ponderosa pine needles.

Serum samples taken from cows 15 minutes after iv infusion with ICA revealed that ICA metabolism or distribution and excretion are rapid since only residual amounts of ICA were detected (*17*). Metabolism studies using homogenates of bovine liver determined that the ICA was rapidly metabolized to agathic acid and dihydroagathic acid with a $t_{\frac{1}{2}}$ of 15 minutes (Gardner, unpublished data). Similar metabolism occurred in goat, sheep, pig, guinea pig, and rat liver homogenates although the guinea pig and rat livers were less efficient. Metabolism occurred in the liver homogenate supernatant and not in the microsomal fraction.

Twenty-three other tree and shrub species from throughout the western and southern states were analyzed for ICA (*18*). Significant levels (>0.5% dry weight) were detected in *Pinus jefferyi* (Jeffrey pine), *P. contorta* (lodgepole pine), *Juniperus scopulorum* (Rocky Mountain juniper) and *J. communis* (common juniper) and from *Cupressus macrocarpa* (Monterey cypress) from New Zealand and Australia. Abortions were induced when lodgepole pine and common juniper containing 0.7% and 2.5% ICA, respectively, were experimentally fed to pregnant cows, inducing abortions in 9 and 3.5 days, respectively (*18*). This research confirmed field reports of lodgepole pine needle abortion in British Columbia, Canada (France, B., personal communication). Monterey cypress is known to cause abortions in cattle in New Zealand and Southern Australia and contained ICA levels of 0.89% to 1.24%. Similar labdane resin acids are present in broom snakeweed, but the abortifacient and toxic components have not been identified.

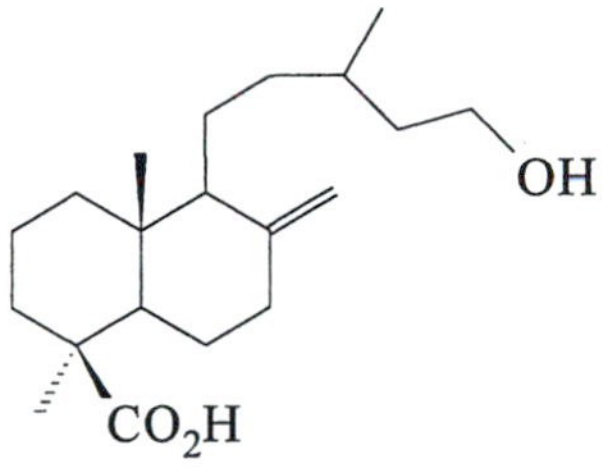

Imbricataloic acid

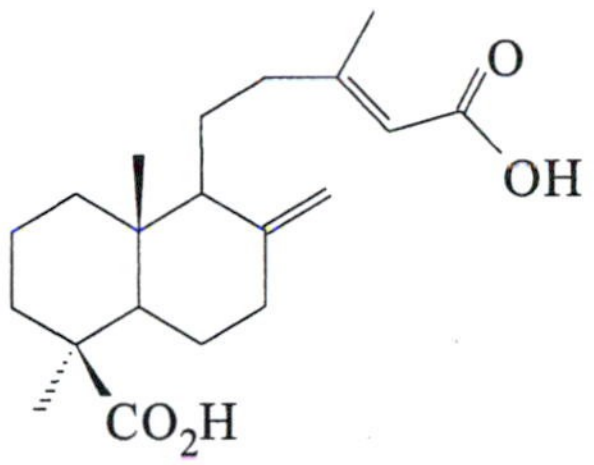

Agathic acid

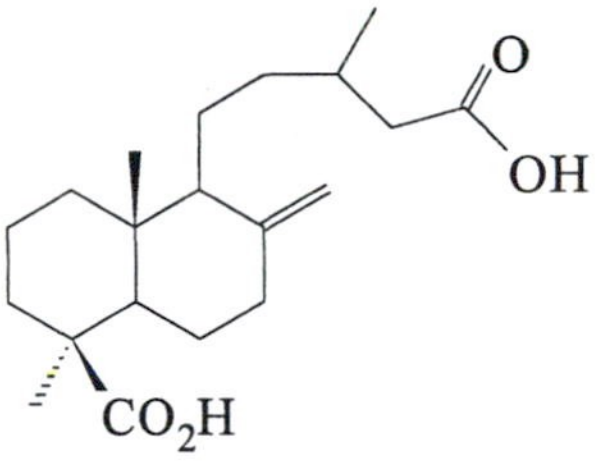

Dihydroagathic acid

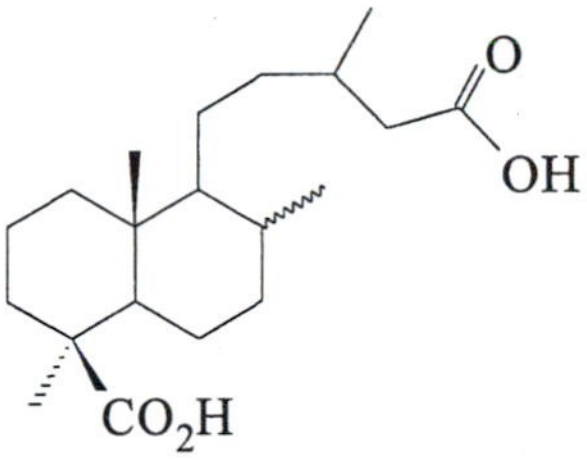

Tetrahydroagathic acid

Figure 5. Serum metabolites of the Ponderosa pine needle toxin ICA.

Occasional toxicoses from PN have been reported in field cases but are rare and have only occurred in pregnant cattle. No toxicity other than abortion in cattle has been demonstrated from ICA or ICA derivatives. However, the abietane-type resin acids in PN were shown to be toxic but not abortifacient at high doses when administered orally to cattle, goats, and hamsters, causing nephrosis, edema of the central nervous system, myonecrosis, and gastroenteritis (*19*). While we believe these resin acids may contribute to the occasional toxicoses reported in the field, we do not believe they contribute to the abortions. Most cow losses in the field are associated with difficult parturition or post abortion toxemia due to retained fetal membranes.

Cattle readily graze PN, especially during the winter months in the western U.S. Pine needle consumption increases during cold weather, with increased snow depth, and when other forage is reduced (*20*). Cattle are easily averted to green PN using an emetic (lithium chloride) paired with PN consumption, but aversions extinguish if cattle ingest dry needles intermingled with dormant grasses.

Currently, recommendations to remove pregnant cattle in the last trimester of pregnancy from Ponderosa pine-infested pastures or fencing around pine trees are the only preventive measures to ensure no losses from grazing pine needles. Current and future research centers around metabolism of the abortifacient labdanes, mechanism of action of the induced parturition, and treatment of premature calves and retained fetal membranes to ultimately reduce losses for livestock producers.

Lupinus spp., *Conium maculatum*, and *Nicotiana glauca*

Lupinus spp. (lupines) contain quinolizidine and piperidine alkaloids, whilst *Conium maculatum* (poison-hemlock) and *Nicotiana glauca* (wild tree tobacco) contain piperidine alkaloids. Piperidine and quinolizidine alkaloids are widely distributed in nature and most possess a certain level of toxicity. Some are teratogenic, depending on structural characteristics which are only partially understood.

Lupines have caused large losses to the sheep and cattle industries in the past and they continue to cause significant losses to the cattle industry in the western U.S. Large calf losses from congenital birth defects have been recorded in Oregon, Idaho, Montana, and Washington from 1992-1997. While huge death losses were reported in sheep in the early 1900's, the teratogenic effects (crooked calf disease; Figure 6) are responsible for most recent losses associated with lupine. In 1992, 56% of the calves from a single herd of cows either died or were destroyed (*21*). In the spring of 1997 over 4000 calves (>35% of the calf crop) in Adams County, Washington were destroyed due to lupine-induced crooked calf disease. Similar but less severe losses were reported in Montana, Oregon, and British Columbia.

Poison-hemlock (*Conium maculatum*) has historic significance as the "tea" used for execution in ancient Greece and the decoction used to execute Socrates (*22*). Toxicoses in livestock frequently occur and field incidences of teratogenic effects in cattle and pigs have been reported (*23,24*). The teratogenic effects are the same as those induced in cattle by lupines, *i.e.*, cleft palate and multiple congenital skeletal contractures (MCC).

While *Nicotiana glauca* has not been associated with field cases of teratogenesis in livestock, there have been reported cases of overt poisoning (*25*). The domesticated relative, *N. tabacum,* caused epidemic proportion outbreaks of malformations in newborn pigs in the late 1960's in Kentucky after tobacco stalks were fed to pregnant sows. Experimentally, *N. glauca* is teratogenic in cattle, pigs, sheep, and goats and was used to establish that anabasine was the teratogenic alkaloid in *N. tabacum*. Using *N. glauca* as the anabasine source to induce teratogenesis, the goat has been established as a model to study the mechanism of action of teratogenic piperidine alkaloids and the induction of cleft palate for both animal and human studies. These malformations are of the same type as those in lupine-induced crooked calf disease.

Lupine Toxicity. Stockmen have long recognized the toxicity of lupines, especially in late summer and fall when the pods and seeds are present. The clinical signs of poisoning begin with nervousness, depression, grinding of the teeth, frothing around the mouth, relaxation of the nictitating membrane of the eye, frequent urination and defecation and lethargy. These progress to muscular weakness and fasciculations, ataxia, collapse, sternal recumbency leading to lateral recumbency, respiratory failure, and death. Signs may appear as early as one hour after ingestion and progressively get worse

Figure 6. Lupine-induced crooked calf, demonstrating the contracture defects in the front legs.

over the course of 24 to 48 hours even if further ingestion does not occur. Generally, if death does not occur within this time frame, the animal recovers completely.

More than 150 quinolizidine alkaloids have been structurally identified from genera of the Leguminosae family, including *Lupinus, Laburnum, Cytisus, Thermopsis* and *Sophora (26)*. Quinolizidine alkaloids occur naturally as N-oxides as well as free bases, but very little research has been done on the toxicity or teratogenicity of the N-oxides. However, the structurally similar pyrrolizidine N-oxides have been shown to be reduced to the corresponding free bases in the rumen, and it seems likely that quinolizidine alkaloids could undergo a similar conversion (Molyneux, R. J., personal communication).

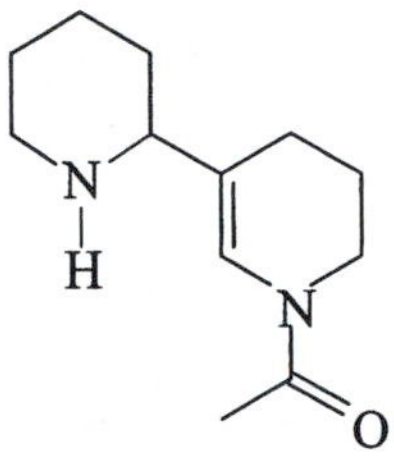

Ammodendrine 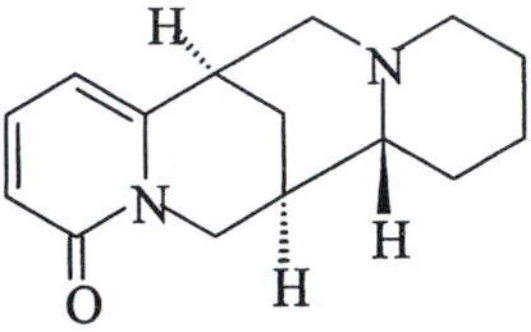   Anagyrine

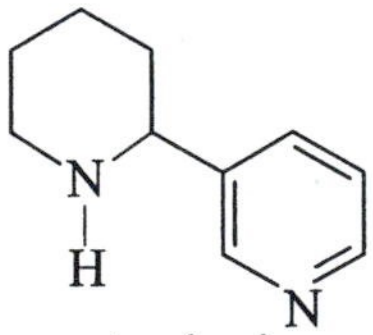

Anabasine 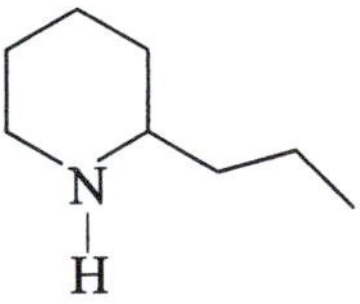   Coniine 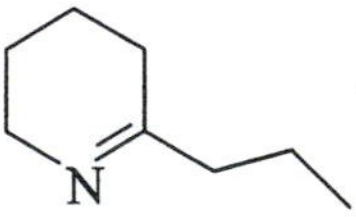   γ-Coniceine

Figure 7. Teratogenic alkaloids from *Lupinus* spp., *Conium maculatum* and *Nicotiana* spp. that cause crooked calf disease and associated birth defects.

Eighteen western American lupine species have been shown to contain the teratogen anagyrine (Figure 7) with 14 of these containing teratogenic levels (27). Lupine alkaloids are produced by leaf chloroplasts and are translocated via the phloem and stored in epidermal cells and in seeds (28). Little is known about individual alkaloid toxicity, however, 14 alkaloids isolated from *Lupinus albus*, *L. mutabilis*, and *Anagyris foetida* were analyzed for their affinity to nicotinic and/or muscarinic acetylcholine receptors (26). Of the 14 compounds tested, the α-pyridones (*N*-methyl cytisine and cytisine) showed the highest affinities at the nicotinic receptor (IC_{50} of 0.05 and 0.14μM for nicotinic *vs* 417 and 400 μM for muscarinic receptors, respectively), while several quinolizidine alkaloid types including the teratogen anagyrine (IC_{50} 132 μM at muscarinic receptor *vs* 2096 μM at the nicotinic receptor) were more active at the muscarinic receptor. If one compares binding affinities of the teratogen anagyrine for nicotinic versus muscarinic receptors, there is 16 times greater binding affinity to muscarinic receptors. Information about the maternal or fetal mechanism of toxicity or teratogenicity of anagyrine can perhaps be implied from this comparison.

Piperidine and quinolizidine alkaloid content and profile vary between lupine species and in individual plants depending on environmental conditions, season of the year and stage of plant growth (29). Typically, alkaloid content is highest during early growth stages, decreases through the flower stage, and increases in the seeds and pods. This knowledge has been used in management strategies to reduce losses. Site differences in alkaloid levels have been described and are substantial (30). Total alkaloid content decreases as elevation increases and was shown to be six times higher in plants at 2700m vs plants collected at 3500m. This phenomenon persisted even when seedlings from the highest and lowest elevations were grown under identical greenhouse conditions, suggesting genetic differences as plants adapted to elevation. For many lupines, the time and degree of seeding varies from year to year. Some species of lupines are readily grazed by livestock and are acceptable forage under certain range conditions. Most deaths have occurred under conditions in which animals consume large amounts of pods or toxic plants in a brief period. This may happen when livestock are driven through an area of heavy lupine growth, unloaded into such an area, trailed through an area where the grass is covered by snow but the lupine is not, or when animals are forced to eat the plants due to over-grazing. A recent report described the death of 10 yearling stocker calves after grazing Lupinus argenteus containing predominantly the piperidine alkaloids ammodendrine and N-methyl ammodendrine (Figures 7, 8; Panter, unpublished data). Most poisonings occur in the late summer and fall because seed pods are present and lupine remains green after other forage has matured or dried. Most calf losses occur because of teratogenic effects resulting from their mothers grazing lupine plants or pods during susceptible stages of pregnancy.

Conium maculatum and *Nicotiana* spp. Toxicity. The clinical signs of poisoning from ingestion of *Conium maculatum* and *N. glauca* are similar in all animal species thus far tested and appear to be the same as those caused by lupine in cattle. They include early signs of nervousness, occlusion of the eyes by the nictitating membrane (most pronounced in pigs and occasionally seen in cows, sheep and goats) and progressing quickly through a pattern of nervous system stimulation with peripheral and local effects which include frequent urination and defecation, dilated pupils, trembling, incoordination, and excessive salivation. The stimulation soon progresses to depression resulting in relaxation, recumbency, and eventually death from respiratory paralysis if the dosage is high enough. Cattle, pigs, goats, and elk have demonstrated a preference for *Conium maculatum* plant once they have acquired a taste for it (31-34).

Coniine, γ-coniceine, and N-methyl coniine (Figures 7, 8) are the principal alkaloids in *Conium maculatum* with relative concentration depending on the stage of plant growth. Gamma-coniceine, a metabolic precursor of coniine and N-methyl coniine, is at highest concentration in early plant growth producing coniine and N-methyl coniine as the plant matures (35,36). Coniine, γ-coniceine and N-methyl coniine have been shown to be toxic and teratogenic (37,38). Structural differences impart significant differences in toxicity (γ-coniceine>coniine>N-methyl coniine; Table I), but teratogenic potency is unknown although we believe it is related to toxicity (38).

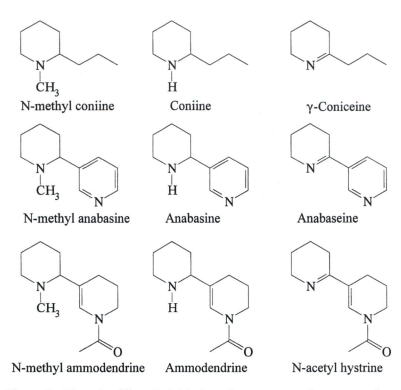

N-methyl coniine   Coniine   γ-Coniceine

N-methyl anabasine   Anabasine   Anabaseine

N-methyl ammodendrine   Ammodendrine   N-acetyl hystrine

Figure 8. Nine piperidine alkaloids from *Lupinus* spp., *Conium maculatum* and *Nicotiana* spp. showing structural similarities that enhance toxic characteristics (toxicity data shown in Table I).

Table I. Relative Toxicity of Nine Piperidine Alkaloids from *Lupinus formosus*, *Conium maculatum*, and *Nicotiana* spp. with Similar Structural Characteristics

Plant	Alkaloids and Toxicity (LD_{50}, mg/kg) iv mouse		
Lupinus formosus	N-methyl ammodendrine 110.7	ammodendrine 134.4	N-acetyl hystrine 29.7
Conium maculatum	N-methyl coniine 20.5	coniine 11.4	γ-coniceine 2.5
Nicotiana glauca	N-methyl anabasine 12.5	anabasine 1.6	anabaseine 1.1

Field cases of the toxic effects of poison-hemlock have been reported in cattle (*31,39,40*), swine (*23*), horses (*41,42*), sheep (*42*), goats (*33*), elk (*34*), turkeys (*43*), quail and chickens (*44*), wild geese (Converse, K., personal communication) and humans (*22,45*). The teratogenic effects have been experimentally induced in cattle (*42*), swine (*46,47*), goats (*48*) and sheep (*49*).

Teratogenicity. The teratogenic effects of *Lupinus* spp., *Conium maculatum,* and *Nicotiana* spp. are similar, and the mechanism of action is believed to be similar in susceptible livestock species (*50*). Current research at the Poisonous Plant Research Laboratory includes defining the specific periods of gestation susceptible to teratogens, differences in susceptibility among livestock species, alkaloid structure-activity relationships, mechanisms of action, and management strategies to reduce losses. A goat model using milled *Nicotiana glauca,* and extracts therefrom, has been established to study the mechanism of action of the cleft palate and MCC. Subsequently, this model has recently been characterized to study the mechanism of cleft palate formation and to develop surgical techniques for *in utero* repair in humans (*51*).

Susceptible Periods of Gestation. The periods of gestation when the fetus is susceptible to these plant teratogens have been partially defined in cattle, sheep, goats, and swine (Table II; *21,46,47,52-56*). The severity and type of the malformations also depend on the alkaloid dosage ingested, the stage of pregnancy when the plants are eaten, and the length of time ingestion takes place. In swine, only cleft palate occurred when *Conium maculatum* was fed during days 30-41 of gestation (*46*). Skeletal defects, predominantly the forelimbs, spine, and neck without cleft palate, were induced when pregnant sows were fed *Conium maculatum* during gestation days 40-53 (*47*). When feeding included days 50-63, rear limbs were affected also. When the feeding period included days 30-60, all combinations of the defects described occurred. In sheep and goats, the teratogenic insult period is similar to pigs and includes days 30-60 (*48,54*). In goats, a narrow period for cleft palate induction only was defined to include days 35-41 (*55*). The critical gestational period for exposure in cattle is 40-70 days with susceptible periods extending to 100 days (*21,57*). The cleft palate induction period in cattle was recently defined as gestation days 40-50 (*56*).

Table II. Susceptible Periods of Gestation for Alkaloid-Induced Cleft Palate and Multiple Congenital Contractures (MCC) in Cattle, Sheep, Goats, and Swine

Defect	Cattle	Sheep	Goats	Swine
Cleft palate	40-50	35-41	35-41	30-41
MCC[a]	40-70 40-100	30-60	30-60	40-53 50-63 30-60

[a]MCC–multiple congenital contractures–include arthrogryposis, scoliosis, kyphosis and torticollis. Rib cage anomalies and asymmetry of the head also occur.

Livestock Species Differences. The syndrome known as crooked calf disease (Figure 6) associated with lupine ingestion was first reported in the late 1950's and included various skeletal contracture-type birth defects and occasionally cleft palate (*52,53,58-61*). Epidemiologic evidence and chemical comparison of teratogenic and non-teratogenic lupines has determined that the quinolizidine alkaloid anagyrine is the teratogen (*62*; Figure 7). A second teratogen, ammodendrine, was found in *Lupinus formosus* and induced the same type of skeletal birth defects (*63,64*; Figure 7). Further research determined that the anagyrine-containing lupines only caused birth defects in cattle and did not affect sheep or goats. No breed predilection or genetic susceptibility to the lupine-induced condition has been determined in cattle. The piperidine-containing lupine *L. formosus* caused birth defects experimentally in cattle and goats (*50,63*). This led to speculation about possible metabolism or absorption differences between cattle and small ruminants. Keeler and Panter (*63*) hypothesized that the cow might metabolize the quinolizidine alkaloid anagyrine to a complex piperidine, meeting the structural characteristics determined for the simple teratogenic piperidine alkaloids in poison-hemlock (*37*). This was supported by feeding trials with other piperidine alkaloid-containing plants, extracts, and pure compounds. Even though comparative studies supported the hypothesis that the cow may convert the quinolizidine alkaloid anagyrine to a complex piperidine by ruminal metabolism, recent evidence reporting the absorption and elimination patterns of many of the quinolizidine alkaloids, including anagyrine, in cattle, sheep, and goats does not support this theory (*65*). Further research on this is currently ongoing at the Poisonous Plant Research Laboratory in Logan, Utah.

Structure-Activity Relationship. Keeler and Balls (*37*) fed commercially available structural analogs of coniine to pregnant cows to compare structural relationships to teratogenic effects. Results suggested that the piperidine alkaloids must meet certain structural criteria to be teratogenic. Based on these data, Keeler and Balls (*37*) speculated that the piperidine or 1, 2-dehydropiperidine alkaloids with a side chain of at least three carbon atoms in length adjacent to the nitrogen atom might be considered potential teratogens. Note that the piperidine alkaloids in Figure 8 meet these criteria. Additionally, the 1, 2-dehydropiperidine alkaloids (γ-coniceine, anabaseine, N-acetyl hystrine) are more toxic than either the piperidine or N-methyl piperidine analogues (Table I; *38*).

While all the alkaloids in Table I are believed to have teratogenic activity, only coniine and anabasine have been experimentally tested in their purified form (*24,37*). Coniine, a simple piperidine from poison-hemlock, and anabasine, a simple piperidine from tree tobacco (*Nicotiana glauca*), induced the same defects in cattle, sheep, pigs and goats (*24,37,46-49,54,66,67*). In other experiments, plant material containing predominantly γ-coniceine and ammodendrine were teratogenic in cattle and goats (*32, 37,56,63*).

Mechanism of Action. The proposed mechanism of action for *Lupinus, Conium maculatum*, and *N. glauca*-induced contracture defects and cleft palate involves a chemically induced reduction in fetal movement much as one would expect with a sedative, neuromuscular blocking agent, or anesthetic (*48*). This mechanism of action was supported by experiments using radio ultrasound where a direct relationship between reduced fetal activity and severity of contracture-type skeletal defects and cleft palate in sheep and goats was recorded. Further research suggests that this inhibition of fetal movement must be over a protracted period of time during specific stages of gestation. For example, fresh *Conium maculatum* plant was fed to pregnant sheep and goats during gestation days 30-60 and fetal movement monitored over a 12 hour period at 45, 50, and 60 days gestation (*48*). *Conium maculatum* plant inhibited fetal movement for 5 to 9 hours after gavage, but by 12 hours fetal movement was similar to that of controls. The lambs and kids had no cleft palates and only slight to moderate carpal flexure (buck knees), which spontaneously resolved a few weeks after birth. On the other hand, *Conium maculatum* seed (with higher teratogen concentration) or *N. glauca* plant inhibited fetal movement during the entire 12-hour period between dosages (two times daily) over the treatment period of 30 to 60 days. Severe limb, spine, and neck defects and cleft palate occurred.

Further ultrasonographic studies showed that strong fetal movement becomes evident in the untreated goat at about day 35 gestation and that these first movements are extension-type of the fetal head and neck. The heads of fetuses under the influence of anabasine through days 35-41 of gestation remained tightly flexed against the sternum and no movement was seen. Subsequently, the newborn goats had cleft palate but no other defects. Panter and Keeler (*55*) suggested that these cleft palates were caused by mechanical interference by the tongue between palate shelves during programmed palate closure time (day 38 in goats; between days 40 to 50 in cows).

In addition to the ultrasonographic studies, which provide direct evidence of reduced fetal movement, the nature of some of the defects in calves from cows gavaged with *L. formosus* and in goats gavaged with *Conium maculatum* seed or *N. glauca* plant offers other evidence of the importance of lack of fetal movement in normal development. These defects included depressions in the rib cage, legs, or spinal column suggesting a mechanical impact from pressure of a sibling or the head turned back on the rib cage. Based on ultrasonographic studies, the action of teratogens appears to be directly on the fetus (inhibited fetal movement) rather than via maternal toxicity. Fetal movement inhibition persists between doses for a much greater duration than do signs of overt toxicity in the dam. Furthermore, manual manipulation of the fetus at the time of ultrasound examination during feeding trials revealed that there was adequate space in the uterus for normal body movement, yet the fetus remained totally immobile.

Even though research at the Poisonous Plant Research Lab has been limited to the three genera mentioned above, there are others that contain piperidine and quinolizidine alkaloids structurally similar to what would be expected to be toxic and teratogenic. These include species of the genera *Genista, Prosopis, Lobelia, Cytisus, Sophora, Pinus, Punica, Duboisia, Sedum, Withania, Carica, Hydrangea, Dichroa, Cassia, Ammondendron, Liparia, Colidium* and others (*54*). Many plant species or varieties from these genera may be included in animal and human diets, however, toxicity and teratogenicity are a matter of dose, rate of ingestion, and alkaloid level and composition in the plant.

Prevention and Treatment. Prevention of poisoning and birth defects induced by *Lupinus, Conium maculatum,* and *Nicotiana* spp. can be accomplished by using a combination of management techniques: 1) coordinating grazing times to avoid the most toxic stage of plant growth such as early growth and seed pod stage for lupine and early growth and green seed stage for *Conium maculatum;* 2) changing time of breeding, either advancing, delaying or changing from spring to fall calving, thereby avoiding exposure to the teratogens at the most susceptible period of gestation; 3) reducing plant population through herbicide treatment; 4) managing grazing to maximize grass coverage; and 5) intermittent grazing, allowing short duration grazing of lupine pastures with frequent rotation when cows are first observed grazing lupine plants. The risk is reduced when lupine is in flower or post-seed stage and when poison-hemlock has matured.

Conclusion

In this review we have briefly discussed the effects of some poisonous plants and toxins characterized therefrom on reproductive performance in livestock. While the effects described vary depending on the plant species or livestock species involved, the critical issue remains that natural toxins from poisonous plants have powerful and often detrimental effects on biological systems. These effects on reproductive function may be subtle like those described for locoweeds before overt toxicosis becomes evident; they may be obvious and dramatic as described for pine needle abortion in cattle; or the observed effects may be delayed yet shocking as is the case when offspring are born with severe skeletal defects or cleft palate many months after the true insult period. These effects are significant and continue to cause large economic losses for livestock producers. Poisonous plant research provides new information and tools to better manage livestock grazing systems to reduce losses and enhance product quality. Additional spin-off benefits from this research include animal models, new techniques, novel compounds and management strategies that will enhance animal and human health.

References

1. Molyneux, R. J.; James, L. F. *Science* **1982**, *216*, 190.
2. Stegelmeier, B. L.; James, L. F.; Panter, K. E.; Molyneux, R. J. *Amer. J. Vet. Res.* **1995**, *56*, 149.
3. James, L. F.; Panter, K. E.; Broquist, H. P.; Hartley, W. J. *Vet. Hum. Tox.* **1991**, *33*, 217.
4. Stegelmeier, B. L.; Molyneux, R. J.; Elbein, A. D.; James, L. F. *Vet. Pathol.* **1995**, *32*, 289.
5. Pfister, J. A.; Astorga, J. B.; Panter, K. E.; Molyneux, R. J. *Appl. Anim. Behav. Sci.* **1993**, *36*, 159.
6. Wang, S.; Holyoak, G. R.; Panter, K. E.; Liu, G.; Bunch, T. J.; Evans, R. C.; Bunch T.D. *Biology of Reproduction* **1998**, *58(1)*, 65.

7. Panter, K. E.; Bunch, T. D.; James, L. F.; Sisson, D. V. *Amer. J. Vet. Res.* **1987**, 686.
8. James, L. F.; Hartley, W. J. *Amer. J. Vet. Res.* **1977**, *38*, 1263.
9. Ralphs, M. H.; Graham, D.; James, L. F.; Panter, K. E. *Rangelands* **1994**, *16*, 35.
10. Panter, K. E.; James, L. F.; Hartley, H. J. *Vet. Human Toxicol.* **1989**, *31*, 42.
11. Wang, S.; Holyoak, G. R.; Panter, K. E.; Liu, G.; Evans, R. C.; Bunch, T. D. *Proc. Soc. Exp. Biol. Med.* **1998**, *217*, 197.
12. Ortiz, A. R.; Hallford, D. M.; Galyean, M. L.; Schneider, F. A.; Kridli, R. T. *J. Anim. Sci.* **1997**, *75*, 3229.
13. James, L. F.; Short, R. E.; Panter, K. E.; Molyneux, R. J.; Stuart, L. D.; Bellows R. A. *Cornell Vet.* **1989**, *79*, 53.
14. James, L. F.; Molyneux, R. J.; Panter, K. E.; Gardner, D. R.; Stegelmeier, B. L. *Cornell Vet.* **1994**, *84(1)*, 33.
15. Gardner, D. R.; Molyneux, R. J.; James, L. F.; Panter, K. E.; Stegelmeier, B. L. *J. Agric. Food Chem.* **1994**, *42(3)*, 756.
16. Gardner, D. R.; Panter, K. E.; Molyneux, R. J.; James, L. F.; Stegelmeier, B. L. *J. Agric. Food Chem.* **1996**, *44(10)*, 3257.
17. Gardner, D. R.; Panter, K. E.; Molyneux, R. J.; James, L. F.; Stegelmeier, B. L.; Pfister, J. A. *J. Nat. Tox.* **1997**, *6*, 1.
18. Gardner, D. R.; Panter, K. E.; James, L. F.; Stegelmeier, B. L. *Vet. Hum. Toxicol.* **1998**, *40(5)*, 260.
19. Stegelmeier, B. L.; Gardner, D. R.; James, L. F.; Panter, K. E.; Molyneux, R. J. *Vet. Pathol.* **1996**, *33*, 22.
20. Pfister, J. A.; Adams, D. C. *J. Range Manage.* **1993**, *46*, 394.
21. Panter, K. E.; Gardner, D. R.; Gay, C. C.; James, L. F.; Mills, R.; Gay, J. M.; Baldwin, T. J. *J. Range Manage.* **1997**, *50*, 587.
22. Daugherty, C. G. *J. Med. Biography* **1995**, *3*, 178.
23. Edmonds, L. D.; Selby, L. A.; Case, A. A. *J. Am. Vet. Med. Assoc.* **1972**, *160*, 1319.
24. Keeler, R. F.; Balls, L. D.; Panter, K.E. *Cornell Vet.* **1981**, *71*, 47.
25. Plumlee, K. H.; Holstege, D. M.; Blanchard, P. C.; Fiser, K. M.; Galey, F. D. *J. Vet. Diagn. Invest.* **1993**, *5*, 498.
26. Schmeller, T.; Sauerwein, M.; Sporer, F.; Wink, M.; Muller, W. E. *J. Nat. Prod.* **1994**, *57*, 1316.
27. Davis, A. M.; Stout, D. M. *J. Range Manage.* **1986**, *39*, 29.
28. Wink, M.; Meibner, C.; Witte, L. *Phytochemistry* **1995**, *38*, 139.
29. Wink, M.; Carey, D. B. *Biochem. Systematics and Ecology* **1994**, *22*, 663.
30. Carey, D. B.; Wink, M. *J. Chem. Ecol.* **1994**, *20*, 849.
31. Penney, H. C. *The Veterinary Record* **1953**, *65*, 669.
32. Panter, K. E. *Toxicity and Teratogenicity of Conium maculatum in Swine and Hamsters*; Ph.D. thesis, University of Illinois, Urbana., **1983**; 106 p.
33. Copithorne, B. *The Veterinary Record* **1937**, *49*, 1018.
34. Jessup, D. A.; Boermans, H. J.; Kock, N. D. *J. Am. Vet. Med. Assoc.* **1986**, *189*, 1173.
35. Cromwell, B. T. *Biochem J.* **1956**, *64*, 259.

36. Leete, E.; Olson, J. O. *J. Am. Chem. Soc.* **1972**, *94(15)*, 5472.
37. Keeler, R. F.; Balls, L. D. *Clin. Toxicol.* **1978**, *12*, 49.
38. Panter, K. E.; Gardner, D. R.; Shea, R. E.; Molyneux, R. J.; James, L. F. In *Toxic Plants and Other Natural Toxicants*; Garland, T.; Barr, A. C., Eds.; CAB International: Wallingford, UK, **1998**; pp. 345-350.
39. Kubik, I. M.; Rejholec, J.; Zachoval, J. *Veterinarstvi* **1980**, *30*, 157.
40. Galey, F. D.; Holstege, D. M.; Fisher, E. G. *J. Vet. Diagn. Invest.* **1992**, *4*, 60.
41. MacDonald, H. *Vet. Rec.* **1937**, *49*, 1211.
42. Keeler, R. F.; Balls, L. D.; Shupe, J. L.; Crowe, M. W. *Cornell Vet.* **1980**, *70*, 19.
43. Frank, A. A.; Reed, W. M. *Avian Diseases* **1987**, *31(2)*, 386.
44. Frank, A. A.; Reed, W. M. *Avian Diseases* **1990**, *34(2)*, 433.
45. Frank, B. S.; Michelson, W. B.; Panter, K. E.; Gardner, D. R. *West. J. Med.* **1995**, *163*, 573.
46. Panter, K. E.; Keeler, R. F.; Buck, W. B. *Am. J. Vet. Res.* **1985**, *46*, 2064.
47. Panter, K. E.; Keeler, R. F.; Buck, W. B. *Am. J. Vet. Res.* **1985**, *46*, 1368.
48. Panter, K. E.; Bunch, T. D.; Keeler, R. F.; Sisson, D. V.; Callan, R. J. *Clin. Toxicol.* **1990**, *28*, 69.
49. Panter, K. E.; Bunch, T. D.; Keeler, R. F.; Sisson, D. V. *Clin. Toxicol.* **1988**, *26*, 175.
50. Panter, K. E.; Gardner, D. R.; Molyneux, R. J. *J. Nat. Toxins* **1994**, *3(2)*, 83.
51. Weinzweig, J.; Panter, K. E.; Pantaloni, M.; Spangenberger, A.; Harper, J. S.; Edstrom, L. E.; Gardner, D. R.; Wierenga, T. *Plastic and Reconstructive Surgery* **1999**, *104(2)*, 419.
52. Shupe, J. L.; Binns, W.; James, L. F.; Keeler, R. F. *J. Am. Vet. Med. Assoc.* **1967**, *151*, 198.
53. Shupe, J. L.; James, L. F.; Binns, W. *J. Am. Vet. Med. Assoc.* **1967**, *151*, 191.
54. Keeler, R. F.; Crowe, M. W. *Cornell Vet.* **1984**, *74*, 50.
55. Panter, K. E.; Keeler, R. F. *J. Nat. Toxins* **1992**, *1*, 25.
56. Panter, K. E.; Gardner, D. R.; Molyneux, R. J. *J. Nat. Toxins* **1998**, *7*, 131.
57. Shupe, J. L.; Binns, W.; James, L. F.; Keeler, R. F. *Aust. J. Agric. Res.* **1968**, *19*, 335.
58. Palotay, J. L. *Western Veterinarian* **1959**, *6*, 16.
59. Wagnon, K. A. *J. Range Manage.* **1960**, *13*, 89.
60. Binns, W.; James, L. F. Proceedings Western Sec. Amer. Soc. Anim. Prod. **1961**, *12, LXVI,* 1.
61. Shupe, J. L.; James, L. F.; Binns, W.; Keeler, R. F. *The Cleft Palate J.* **1968**, *1*, 346.
62. Keeler, R. F. *Teratology* **1973**, *7*, 31.
63. Keeler, R. F.; Panter, K. E. *Teratology* **1989**, *40*, 423.
64. Keeler, R. F.; Panter, K. E. In *Poisonous Plants, Proceedings of the Third International Symposium,* James, L. F.; Keeler, R. F.; Bailey, E. M.; Cheeke, P. R.; Hegarty, M. P., Eds.; Iowa State Press: Ames, IA, **1992**; pp. 239-244.
65. Gardner, D. R.; Panter, K. E. *J. Nat. Toxins* **1993**, *2(1)*, 1.
66. Keeler, R. F.; Crowe, M. W.; Lambert, E. A. *Teratology* **1984**, *30*, 61.
67. Keeler, R. F. *Clin. Toxicol.* **1979**, *15*, 417.

Chapter 12

Mechanistic Investigation of *Veratrum* Alkaloid-Induced Mammalian Teratogenesis

William Gaffield[1], John P. Incardona[2], Raj P. Kapur[3], and Henk Roelink[4]

[1]Western Regional Research Center, Agricultural Research Service,
U.S. Department of Agriculture, Albany, CA 94710
Departments of [2]Pediatrics, [3]Pathology, and [4]Biological Structure,
University of Washington, Seattle, WA 98195

For many years, the teratogenic *Veratrum* alkaloids have offered vast potential to serve as molecular probes for investigation of several mammalian developmental processes involving craniofacial, limb, and foregut morphogenesis. Research on cyclopamine-treated chick embryo neural tube and somites has revealed disruption of dorsoventral patterning that occurs due to inhibition of Sonic hedgehog signaling. Because cyclopamine-induced Sonic hedgehog signal blockage is not rescued upon the addition of exogenous cholesterol, the teratogenic properties of cyclopamine are possibly derived from a direct interaction of the alkaloid with elements in the signal transduction cascade. Recent studies in organ systems other than the neural tube have shown that cyclopamine inhibition of Sonic hedgehog signaling promotes pancreatic development and inhibits hair follicle morphogenesis.

The Cyclops in ancient Greek mythology exhibited a single, central eye (*1*); in our modern era cyclopia represents the most severe manifestation of holoprosencephaly, a malformation sequence that is defined by impaired cleavage of the embryonic forebrain (*2*). Historical accounts of cyclopia likely were inspired by the craniofacial appearance of individuals afflicted with the syndrome. Holoprosencephaly occurs in approximately 1 of 16000 live human births although the overall frequency of incidence is appreciably greater because 99% of affected embryos abort spontaneously (*3*). Thus, holoprosencephaly is considerably more common in early embryogenesis, occurring in 1 of 250 abortions (*4*). Various degrees of facial dysmorphism are commonly associated with holoprosencephaly in humans. For example, a spectrum of facial types that represent holoprosencephaly in order of decreasing severity includes (Figure 1); (a) cyclopia with single median eye and absence of proboscis, (b) single median eye with varying degrees of doubling, (c) formation of an abnormal proboscis, (d) ethmocephaly (closely-spaced eyes,i.e., ocular hypotelorism, with an abnormal proboscis), (e) cebocephaly (a misshapen nasal chamber and a single-nostril nose), and (f) cleft lip with ocular hypotelorism (*5*). Humans suffering from the mildest form of the syndrome have anomalies as minor as a single upper central incisor (*6*). Developmental delay occurs in most afflicted individuals, although the extent of delay

174

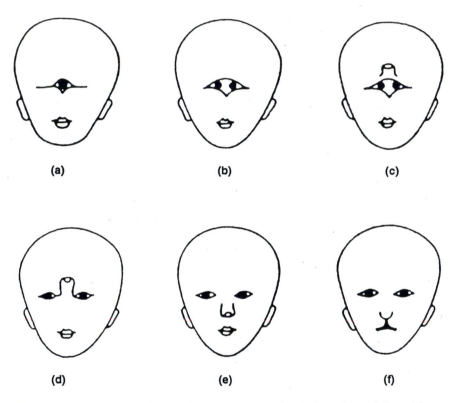

Figure 1. Holoprosencephaly in humans. See text for designations. Adapted from reference 5.

varies in correlation with the severity of the brain malformation. Although infants with severe cyclopia generally do not live longer than one week, those with less severe facial phenotypes may survive beyond a year. However, while the morphological abnormalities comprising the holoprosencephaly manifold have been well-characterized, knowledge of the biochemical mechanisms responsible for inducing this series of malformations has remained obscure (6).

Causative Factors of Holoprosencephaly

Maternal diabetes has been shown epidemiologically to increase the risk of holoprosencephaly in infants of mothers 200-fold over that of the general population (7). Environmental agents that have been reported to induce holoprosencephaly in animal models include ethanol (8) and retinoic acid (9). However, the *Veratrum* alkaloids clearly are most prominent of the chemical agents that induce a wide array of the malformations incorporated in the holoprosencephaly spectrum, including cyclopia (10,11). Ingestion of these plants, or steroidal alkaloids derived from them, by sheep early in their gestation was conclusively shown to result in the birth of cyclopic lambs bearing a single median eye (12). Keeler's extensive research efforts culminated in the isolation and characterization of three structurally-related jerveratrum alkaloids from *Veratrum californicum*; jervine (1) (13), 11-deoxojervine (cyclopamine)(2) (14), and 3-glucosyl-11-deoxojervine (cycloposine) (3) (15), all of which induced severe craniofacial malformations in the offspring of pregnant ewes when administered on the 14th day of gestation (16). Administration of cyclopamine induced cyclopia also in rabbits (17) and chick embryos (18), with the latter becoming recently a favored experimental animal model for investigation of molecular biological mechanisms underlying the induction of holoprosencephaly (19,20).

Hedgehog-mediated Morphogenesis in Vertebrates

Early embryogenesis is characterized by several morphogenetic events, collectively known as gastrulation, which lead to the formation of three primary germ layers. The origin of all mammalian tissues can be traced to one of these three layers: endoderm (gut); mesoderm (muscle, bone, and connective tissue); and, ectoderm (epidermis and neural tissue) (21). Members of the Hedgehog (Hh) family of secreted proteins act as intercellular signals sent between germ layers, and function in a multitude of developmental processes ranging from neuronal specification to bone morphogenesis (22). A very striking aspect of the *Sonic hedgehog* gene is its key role in developmental patterning of the mammalian head and brain as revealed in mouse embryos lacking functional copies of *Sonic hedgehog* that displayed severe holoprosencephaly, including cyclopia (23). Furthermore, loss-of-function mutation at the human *Sonic hedgehog* locus was shown to be associated with several forms of holoprosencephaly, *e.g.*, closely-spaced eyes, single-nostril nose, midline cleft lip, and a single central upper incisor (24). Because administration of either cyclopamine or jervine to gastrulation-stage mammalian embryos induced cyclopia at high incidence, the *Veratrum* alkaloids were prime candidates to serve as potential inhibitors or disrupters of the Sonic hedgehog-mediated patterning of the neural tube (25,26).

Sonic hedgehog is expressed in several organizing centers such as the notochord and prechordal mesoderm. A view of the role of Sonic hedgehog in vertebral development is shown diagrammatically (Figure 2) (27): Sonic hedgehog presented directly by the notochord serves as an inducing cue to the adjacent neural tube and in response to this signal the central midline cells of the neural tube form the floor plate (a); subsequently, the induced floor plate assumes production of Sonic hedgehog as in (b); this signaling protein then induces neuroblasts to differentiate into motor neurons on either side of the floor plate as in (c) and (b). Sonic hedgehog released by

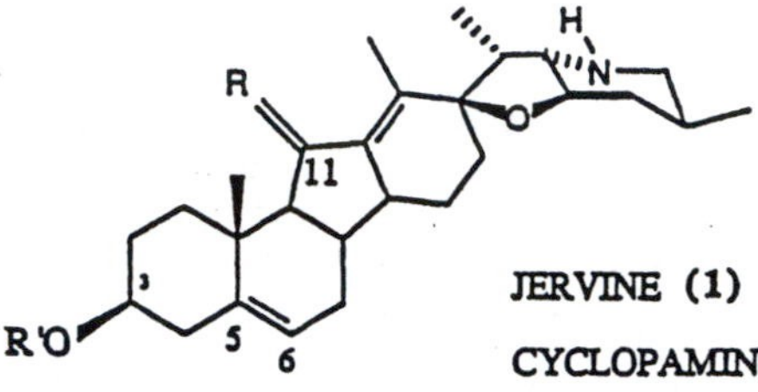

	R'	R
JERVINE (1)	H	O
CYCLOPAMINE (2)	H	H₂
CYCLOPOSINE (3)	D-Glc	H₂

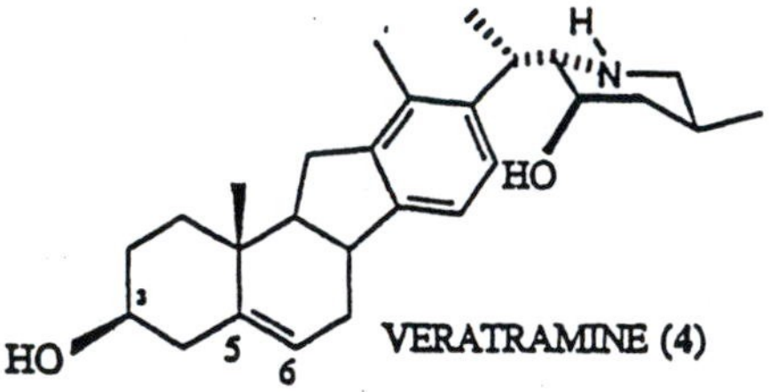

VERATRAMINE (4)

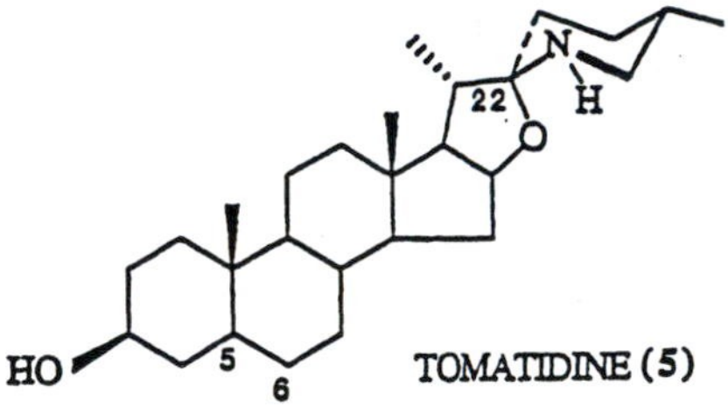

TOMATIDINE (5)

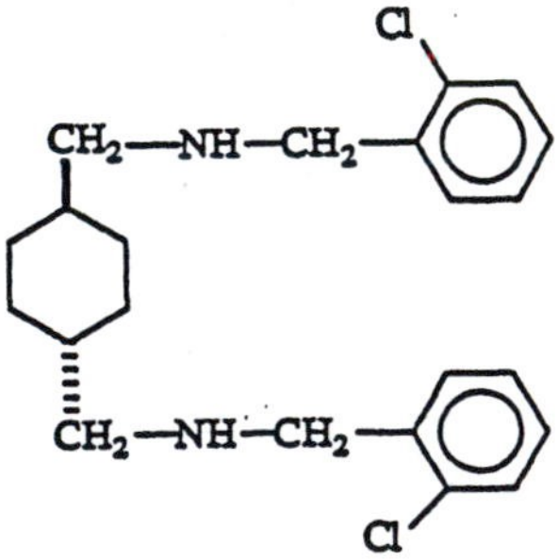

AY9944 (6)

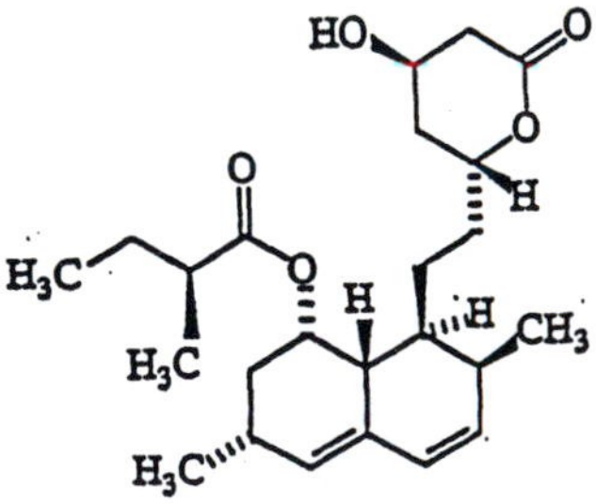

LOVASTATIN (7)

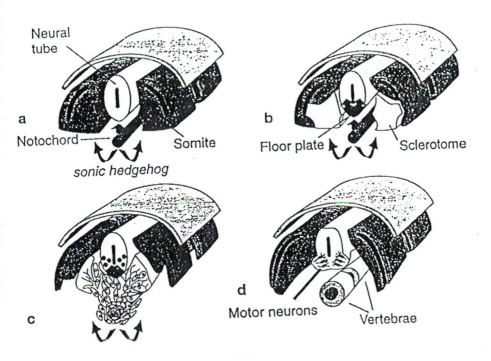

Figure 2. Sonic hedgehog morphogenesis in vertebrates. Abstracted from reference 27.

the notochord into its surroundings induces the sclerotome cells to detach from the somites as in (b), migrate around the notochord as in (c), and form the vertebral bodies as shown in (d). A range of facial malformations associated with holoprosencephaly was observed in chick embryos after treatment with cyclopamine (19,20). An undivided monoventricular telencephon occurred with other severe facial anomalies that were consistent with interruption of inductive events between prechordal mesoderm and the anterior neural plate. In contrast, chick embryos treated with veratramine (4), an acid aromatization product of cyclopamine that does not induce cyclopia in sheep (28), failed to develop holoprosencephalic malformations (20). To overcome the inherent variability of in ovo treatments, neural plate tissue was dissected from the embryo to provide an explant assay (19, 20).

Cyclopamine Disrupts Shh-dependent Neural Tube and Somites Patterning

Examination of a suite of Sonic hedgehog-dependent cell types, in the neural tube and somites of chick embryos with cyclopamine-induced malformations, has shown that essentially all aspects of Sonic hedgehog signaling in these tissues are interrupted by treatment with the alkaloid (20). A failure of dorsoventral patterning in the thoracic neural tube was demonstrated in several ways (20). For example, Sonic hedgehog was always detected in the notochord of cyclopamine-treated embryos, because its initial expression is dependent on factors other than itself; however, Sonic hedgehog floor plate cells were absent in trunk regions of treated embryos, compared to normal floor plate generation of Sonic hedgehog. A downstream transcription factor induced by Sonic hedgehog, HNF-3β, was also absent from the floor plate of treated (Figure 3, column 3) compared to untreated (Figure 3, column 1) embryos. This observation is consistent with a reduction of HNF-3β expression observed in the ventral (front-side) neural tube of cyclopamine-treated hamster embryos (29). Ventral progenitor cells (Nkx 2.2+) were often absent in trunk regions of treated embryos and motor neurons (isl 1/2+) either were absent or reduced to a few midline cells (Figure 3, column 4) (20). Sonic hedgehog suppresses the development of Lim 1/2+ interneuron cells within the ventral neural tube (30) and the repression of Pax 6+ and Pax 7+ cells ventrally is an early event in normal Sonic hedgehog signaling (23). Pax 6 and Pax 7 are important transcription factors involved in dorsoventral patterning that are encoded by the Pax gene family. In cyclopamine-treated embryos, Lim 1/2+ interneurons appeared in the normal dorsal location but also appeared aberrantly in the ventral neural tube, including the ventral midline. Pax 6+ cells were expressed throughout the ventral neural tube in trunk regions of cyclopamine-treated embryos. Although complete failure of repression of the dorsal marker Pax 7 was not attained in the neural tube, the Pax 7 domain often extended ventrally in treated embryos. In somites, Sonic hedgehog signaling mediates the repression of dorsal cell types such as dermatome which express Pax 7 (31). However, in cyclopamine-treated embryos, Pax 7+ cells were present throughout the somites indicating development of dermatome at the expense of sclerotome required to form the vertebrae. Hence, instead of a solid block of tissue with dorsal Pax 7+ cells as observed in control embryos, the somites of cyclopamine-treated embryos were hollow in cross-section with near-circumferential Pax 7 expression (20). Thus, cyclopamine-induced malformations in chick embryos are correlated with interruption of Sonic hedgehog-mediated dorsoventral patterning of the neural tube and somites (20). Cell types normally induced in the ventral neural tube by Sonic hedgehog are either absent or appear aberrantly at the ventral midline after cyclopamine treatment, while dorsal cell types normally repressed by Sonic hedgehog

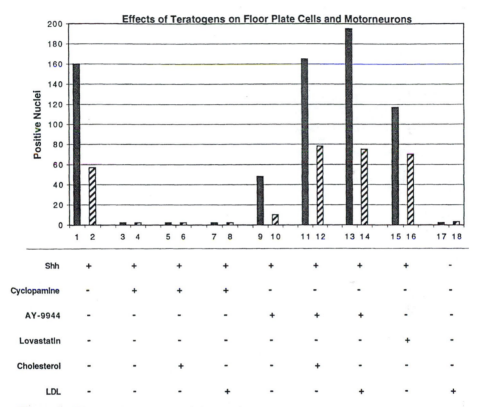

Figure 3. Comparison of the effects of teratogens and cholesterol synthesis inhibitors on the induction of floor plate cells (HNF-3β+) and motorneurons (isl 1/2+) by recombinant Shh in chick embryo explants. Ordinate values are number of positive nuclei counted for the appropriate antigen. Derived from reference 20.

appear ventrally. Somites in cyclopamine-treated embryos display Pax 7 expression throughout, indicating failure of sclerotome induction.

Cyclopamine Inhibits Shh Signal Transduction

The Sonic hedgehog protein is synthesized as an inactive precursor (45 kDa) which must undergo autoproteolytic cleavage into two similar sized fragments; a 20 kDa amino-terminal fragment that carries the developmental signal, and a 25 kDa carboxyl-terminal fragment that serves both as the endopeptidase for cleavage and as a lipid transferase (32,33). Hedgehog autoprocessing proceeds through a two-step sequence (34); in the first reaction, the thiol group of a cysteine residue serves as a nucleophile to attack the carbonyl group of the preceding amino acid residue, forming a thioester linkage in place of the peptide bond that results in a nitrogen to sulfur shift. In the second reaction, the activated thioester intermediate is subjected to nucleophilic attack by the 3-β-hydroxyl group of a cholesterol molecule, resulting in cleavage of the thioester with release of the carboxyl-terminal Sonic hedgehog fragment and formation of an ester linkage between cholesterol and the carboxyl terminus of the amino-terminal Sonic hedgehog fragment (35). Modification of the amino-terminal Sonic hedgehog protein with cholesterol causes association of this signal-carrying protein with the cell membrane. The possible effects of *Veratrum* teratogens upon the processing of Sonic hedgehog protein were evaluated; both the autoproteolytic cleavage of the precursor and the addition of a lipophile (cholesterol) to the amino-terminal fragment (19,20). Although high concentrations of cyclopamine inhibited the post-translational modification of Sonic hedgehog by cholesterol, maximal inhibition of Sonic hedgehog signaling was attained at a concentration 200-fold lower than that significantly affecting Sonic hedgehog processing (20). Cyclopamine reduced floor plate induction by 95% at a concentration of 50nM and completely blocked Pax 7 repression and induction of floor plate and motor neurons at a concentration of 120nM (20). Measurable inhibition of Sonic hedgehog response was observed at cyclopamine concentrations as low as 10-20nM while the non-teratogenic alkaloid tomatidine (5) partially blocks signaling only at concentrations of 12μM. Thus, both research groups that have investigated cyclopamine-induced effects on Sonic hedgehog response have concluded that cyclopamine exerts its teratogenic effects on developing embryos by inhibiting signal transduction (19,20).

Cyclopamine's Blocking Effect is not due to Inhibition of Cholesterol Metabolism

Sterols are required in vertebrates to serve as major membrane components and as precursors for bile acid and steroid hormone synthesis (36). Unexpectedly, cholesterol is required also for mammalian morphogenesis (35). The association of holoprosencephaly with reduced cholesterol levels in mammalian embryos suggests that cyclopamine might induce holoprosencephaly by interfering with cholesterol synthesis or uptake. The Smith-Lemli-Opitz (SLO) syndrome is an inborn disorder of sterol metabolism that expresses characteristic congenital malformations and dysmorphias (37,38). The SLO phenotype comprises anomalies such as microcephaly, cleft palate, failure to thrive, and mental retardation. A deficiency of the ultimate step of cholesterol biosynthesis has long been known to cause SLO syndrome because a decrease of plasma cholesterol accompanied by the accumulation of its precursor, 7-dehydrocholesterol, is always observed. Recently, mutations of the human sterol Δ^7 - reductase gene have been shown to cause SLO syndrome (39,40). Cholesterol is synthesized from mevalonate via a series of intermediates beginning with HMG-CoA reductase catalyzed synthesis of mevalonic acid (Figure 4). Lovastatin, which is

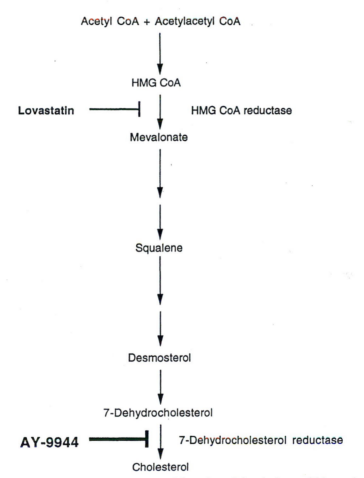

Figure 4. Schematic representation of a portion of the cholesterol biosynthesis pathway.

non-teratogenic, inhibits cholesterol synthesis at this early stage without the accumulation of precursor sterols (*41*). From mevalonic acid, the cholesterol biosynthesis proceeds via a series of isoprenoid intermediates to squalene, which undergoes cyclization to the first sterol precursor. Several demethylations and reductions eventually produce 7-dehydrocholesterol. Rearrangement of the double bond to the 5,6-position present in cholesterol is catalyzed by 7-dehydrocholesterol reductase. AY-9944 (*6*), a compound that induces holoprosencephalic malformations in rat embryos (*42*), inhibits the terminal step of cholesterol biosynthesis leading to an accumulation of various sterol intermediates (*43*). To elaborate potential mechanisms by which Sonic hedgehog signaling might be disrupted, the effect of three potential inhibitors, each of which might impair cholesterol biosynthesis, was examined in chick embryo explants. The compounds selected were cyclopamine and two inhibitors of cholesterol synthesis, AY-9944 which induces holoprosencephaly and lovastatin (*7*) which is non-teratogenic (*20*).

From these experiments, a model was derived that describes the difference between the mechanism of terata induction by cyclopamine on one hand and AY-9944 on the other (*20*). The Sonic hedgehog receptor is a complex of two proteins; Patched (Ptc) which possesses multiple membrane-spanning domains that form a non-covalent complex with Smoothened (Smo), an atypical member of the G-protein-coupled receptor family (*44,45*). The Sonic hedgehog, Patched, and Smoothened receptor complex ultimately affects the transcription of Sonic hedgehog-responsive genes through an incompletely understood signal cascade. Research on *Drosophila* homologs has led to formulation of a model where Patched binds to Hedgehog and in response to this binding, Patched, which normally represses Smoothened, releases this inhibition allowing Smoothened to activate the transcription of downstream target genes via the cubitus interruptis (C_i) transcription factor (*22*). C_i is a member of the Gli family of zinc finger transcription factors, of which mouse analogs are Gli 1, Gli 2 and Gli 3.

Sonic hedgehog-responsive cells that do not have an exogenous supply of cholesterol are shown in Figure 5 and Sonic hedgehog-responsive cells that have been provided an exogenous source of cholesterol, such as yolk, LDL, or free cholesterol, are represented in Figure 6. The cholesterol biosynthetic pathway is active in chick neural plate explants, the target of AY-9944 which is 7-dehydrocholesterol reductase is present, and treatment of the explant with AY-9944 therefore interrupts the response to Sonic hedgehog (Figure 5)(Figure 3, columns 9 and 10), presumably through the action of a sterol precursor or metabolite. The response to Sonic hedgehog is blocked also by cyclopamine under these conditions (Figure 5)(Figure 3, columns 3 and 4). However, with exogenous cholesterol available, cholesterol synthesis is down-regulated, the target of AY-9944 is either absent or present only at low levels, and treatment of the explant with AY-9944 no longer interrupts Sonic hedgehog signaling (Figure 6)(Figure 3, columns 11 and 12) . Conversely, the potential target of cyclopamine must still be available even in the presence of an exogenous source of cholesterol because the alkaloid continues to inhibit Sonic hedgehog signaling (Figure 6)(Figure 3, columns 5 and 6). Thus, the Sonic hedgehog blocking effects of cyclopamine are not secondary to alteration of cholesterol metabolism (*20*). The non-teratogenic cholesterol inhibitor lovastatin does not produce an accumulation of aberrant sterols nor adversely affect Sonic hedgehog signaling either in the presence or absence of exogenous cholesterol (Figures 5,6)(Figure 3, columns 15 and 16). The failure to rescue Sonic hedgehog signaling upon addition of exogenous cholesterol in cyclopamine-treated explants suggests that cyclopamine-induced teratogenesis results from a more direct interaction with certain elements in the Sonic hedgehog signal transduction cascade (*20*). Processes that might potentially be adversely affected by

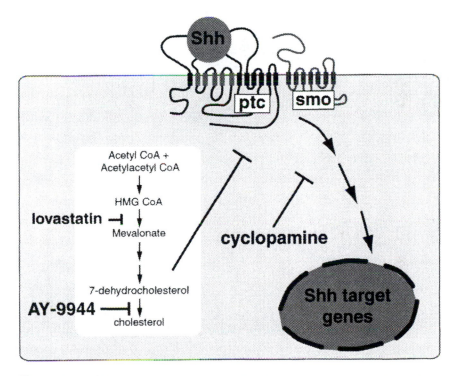

Figure 5. Model describing inhibition of Shh signal transduction by AY-9944 and cyclopamine in Shh-responsive cells with no exogenous supply of cholesterol.

(Adapted with permission from reference 20. Copyright 1998 Company of Biologists Ltd.)

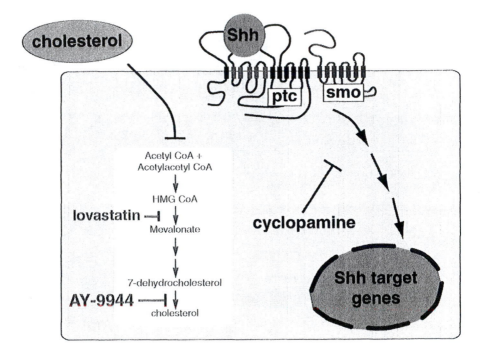

Figure 6. Model describing inhibition of Shh signal transduction by cyclopamine in Shh-responsive cells that have an exogenous supply of cholesterol.

(Adapted with permission from reference 20. Copyright 1998 Company of Biologists Ltd.)

aberrant membrane sterols or serve as possible targets of *Veratrum* alkaloids include; presentation of the Patched/Smoothened complex on the cell surface, internalization of receptor/ligand complexes, or conformational changes induced upon ligand binding that are required for Smoothened activity (*20*).

Cyclopamine Inhibition of Other Organ Development

In research involving a mammalian organ other than the neural tube, whisker pad explants obtained from mice that were treated with cyclopamine resulted in striking inhibition of hair follicle morphogenesis (*46*). An essential role for Sonic hedgehog has been demonstrated during hair follicle morphogenesis, where it is required for normal advancement beyond the hair germ stage of development. The hair follicle is a source of stem cells and site of origin for certain epithelial skin tumors (*47*). The commonest form of skin cancer, basal cell carcinoma, is apparently caused by mutations in genes involved in the Sonic hedgehog signaling pathway (*48*). Unregulated activation of the same pathway also underlies many primitive neuroectodermal tumors in infants.

Finally, Sonic hedgehog has been shown to be required for the growth and differentiation of the oesophagus, trachea, and lung in mice amid suggestions that mutations in Sonic hedgehog and its signaling components may be involved in human foregut defects including oesophageal atresia/stenosis (*49*). The association of holoprosencephaly with foregut anomalies has been observed in some humans, suggesting the possibility that common signals operate both in the forebrain and foregut during development. Experiments conducted in sheep had shown that, in addition to the early induction of craniofacial malformations, administration of *Veratrum* to pregnant ewes later in gestation on days 31-33 induced tracheal stenosis (*50*). Related research has demonstrated that exposure to cyclopamine promotes pancreatic development in embryonic chicks (*51*). Apparently, cyclopamine inhibition of Sonic hedgehog permits the expansion of the endodermal region of the foregut where Sonic hedgehog signaling does not occur, resulting in pancreatic differentiation in a larger region of the foregut endoderm.

Literature Cited

1.   Mandelbaum, A. *The Odyssey of Homer, Book IX;* University of California Press: Berkeley, CA, 1990; pp. 171-192.
2.   DeMyer, W.; Zeman, W.; Palmer, C.G. *Pediatrics* **1964**, *34*, 256.
3.   Roach, E.; DeMyer, W.; Conneally, P.M.; Palmer, C.; Merritt, A.D. *Birth Defects* **1975**,*11*, 294.
4.   Matsunaga, E.; Shiota, K. *Teratology* **1977**, *16*, 261.
5.   Nishimura, H.; Okamoto, N. *Sequential Atlas of Human Congenital Malformations: Observation of Embryos, Fetuses, and Newborns*; University Park Press: Baltimore, MD,1976, pp. 34-36.
6.   Siebert, J.R.; Cohen, Jr., M.M.; Sulik, K.K.; Shaw, C.-M.; Lemire, R.J.; *Holoprosencephaly: An Overview and Atlas of Cases*, Wiley-Liss: New York, 1990.
7.   Barr, Jr., M.; Hanson, J.W.; Currey, K.; Sharp, S.; Toriello, H.; Schmickel, R.D.; Wilson, G.N. *J. Pediatr.* **1983**, *102, 565.
8.   Siebert, J.R.; Astley, S.J.; Clarren, S.K. *Teratology* **1991**, *44*, 29.
9.   Sulik, K.K.; Dehart, D.B.; Rogers, J.M.; Chernoff, N. *Teratology* **1995**, *51*, 398.
10.   Keeler, R.F. In *Isopentenoids in Plants:Biochemistry and Function*; Nes, W.D.; Fuller, G.; Tsai, L.-S. Eds.; Dekker: New York, N Y, 1984, pp. 531-562.

186

11. Keeler, R.F. In *Alkaloids: Chemical and Biological Perspectives*; Pelletier, S.W., Ed.; Wiley- Interscience: New York, NY,1986, Vol. 4; pp. 389-425.
12. Binns, W.; James, L.F.; Shupe, J.L.; Everett, G. *Am. J.Vet. Res.* **1963**, *24*, 1164.
13. Keeler, R.F.; Binns, W. *Can. J. Biochem.* **1966**, *44*, 819.
14. Keeler, R.F. *Phytochemistry* **1969**, *8*, 223.
15. Keeler, R.F. *Steroids* **1969**,*13*, 579.
16. Keeler, R.F.; Binns, W. *Teratology* **1968**, *1*, 5.
17. Keeler, R.F. *Lipids* **1978**, *13*, 708.
18. Bryden, M.M.; Perry, C.; Keeler, R.F. *Teratology* **1973**, *8*, 19.
19. Cooper, M.K.; Porter, J.A.; Young, K.E.; Beachy, P.A. *Science* **1998**, *280*, 1603.
20. Incardona, J.P.; Gaffield, W.; Kapur, R.P.; Roelink, H. *Development* **1998**, *125*, 3553.
21. Kimelman, D.; Griffin, K.J.P. *Cell* **1998**, *94*, 419.
22. Hammerschmidt, M; Brook, A.; McMahon, A.P. *Trends Genet.* **1997**,*13*,14.
23. Chiang, C.; Litingtung, Y.; Lee, E.; Young, K.E.; Corden, J.L.; Westphal, H.; Beachy, P.A. *Nature* **1996**, *383*, 407.
24. Roessler, E.; Belloni, E.; Gaudenz, K.; Jay, P.; Berta, P.; Scherer, S.W.; Tsui, L.-C.; Muenke, M. *Nature Genet.* **1996**, *14*, 357.
25. Gaffield, W.; Keeler, R.F. *J. Nat. Toxins* **1996**, *5*, 25.
26. Gaffield, W.; Keeler, R.F. *J. Toxicol., Toxin Revs.* **1996**, *15*, 303.
27. Mueller, W. *Developmental Biology*; Springer-Verlag: New York, NY, 1996, pp.180-183.
28. Keeler, R.F.; Binns, W. *Proc. Soc. Exp. Biol .Med.* **1966**, *123*, 921.
29. Coventry, S.; Kapur, R.P.; Siebert, J.R. *Pediatr. Dev. Path.* **1998**, *1*, 29.
30. Tsuchida, T.; Ensini, M.; Morton, S.B.; Baldassare, M.; Edlund, T.; Jessell, T.M.; Pfaff, S.L.*Cell* **1994**, *79* , 957.
31. Fan, C.-M.; Porter, J.A.; Chiang, C.; Chang, D.T.; Beachy, P.A.; Tessier-Lavigne, M.*Cell* **1995**, *81*, 457.
32. Lee, J.J.; Ekker, S.C.; von Kessler, D.P.; Porter, J.A.; Sun, B.I.; Beachy, P.A.*Science* **1994**, *266*, 1528.
33. Bumcrot, D.A.; Takada, R.; McMahon, A.P. *Mol.Cell. Biol* **1995**, *15*, 2294.
34. Porter, J.A.; Ekker, S.C.; Park, W.-J.; von Kessler, D.P.; Young, K.E.; Chen, C.-H.; Ma,Y.; Woods, A.S.; Cotter, R.J.; Koonin, E.V.; Beachy, P.A. *Cell* **1996**, *86*, 21.
35. Porter, J.A.; Young, K.E.; Beachy, P.A. *Science* **1996**, *274*, 255.
36. Gibbons, G.F.; Mitropoulos, K.A.; Myant, N.B. *Biochemistry of Cholesterol*; Elsevier: Amsterdam, 1982.
37. Tint, G.S.; Irons, M.; Elias, E.R.; Batta, A.K.; Frieden, R.; Chen T.S.; Salen, G. *N. Engl.. J. Med.* **1994**, *330*,107.
38. Salen, G.; Shefer, S.; Batta, A.K.; Tint, G.S.; Xu, G.; Honda, A.; Irons, M.; Elias, E.R. *J. Lipid Res.* **1996**, *37*, 1169.
39. Wassif, C.A.; Maslen, C.; Kachilele-Linjewile, S.; Lin, D.; Linck, L.M.; Conner, W.E.; Steiner, R.D.; Porter, F.D. *Am. J. Hum. Genet.* **1998**, *63*, 55.
40. Fitzky, B.U.; Witsch-Baumgartner, M.; Erdel, M.; Lee, J.N.; Paik, Y.-K.; Glossman, H.; Utermann, G.; Moebius, F.F. *Proc. Natl. Acad. Sci., USA* **1998**,*95*, 8181.
41. Alberts, A.W.; Chen, J.; Kuron, G.; Hunt, V.; Huff, J.; Hoffman, C.; Rothrock, J.; Lopez, M.; Joshua, H.; Harris, E., *et al. Proc. Natl. Acad. Sci., USA* **1980**,*77*, 3957.
42. Roux, C.; Aubry, M.M. *C.R. Seances Soc. Biol . Fil.* **1966**, *160*, 1353.
43. Wolf, C.; Chevy, F.; Pham, J.; Kolf Clauw, M.; Citadelle, D.; Mulliez, N.; Roux, C. *J. Lipid Res.* **1996**, *37*, 1325.

44.   Marigo, V.; Davey, R.A.; Zuo, Y.; Cunningham, J. M.; Tabin, C.J. *Nature* **1996**, *384,* 176.
45.   van den Heuvel, M.; Ingham, P.W. *Nature* **1996**, *382*, 547.
46.   Chiang, C.; Swan, R. Z.; Grachtchouk, M.; Bolinger, M.; Litingtung, Y.; Robertson, E.K.; Cooper, M.K.; Gaffield, W.; Westphal, H.; Beachy, P.A.; Dlugosz, A.A. *Dev. Biol.* **1999**, *205*, 1.
47.   Hansen, L.A.; Tennant, R.W. *Proc. Natl Acad. Sci., USA* **1994**, *91*, 7822.
48.   Oro, A.E.; Higgins, K.M.; Hu, Z.; Bonifas, J.M.; Epstein, Jr.; E.H.; Scott, M.P.*Science,* **1997**, *276,* 817.
49.   Litingtung, Y.; Lei, L.; Westphal, H.; Chiang, C. *Nature Genet.* **1998**, *20,* 58.
50.   Keeler, R.F.; Young, S.; Smart, R. *Teratology* **1985**, *31,* 83.
51.   Kim, S.K.; Melton, D.A. *Proc. Natl. Acad. Sci.,USA* **1998**, *95,* 13036.

Chapter 13

Analysis and Fractionation of Cigarette Tar Extracts: What Causes DNA Damage?

Koni Stone[1] and William A. Pryor[2]

[1]Department of Chemistry, California State University, Stanislaus, Turlock, CA 95382
[2]Biodynamics Institute, 711 Choppin Hall, Louisiana State University, Baton Rouge, LA 70803

Previously, we have shown that aqueous cigarette tar (ACT) extracts from mainstream or sidestream cigarette smoke contain a stable radical signal that binds to and damages DNA (deoxyribonucleic acid) in mammalian cells. We have also shown that aged solutions of catechol contain a similar radical signal and cause DNA damage. We have fractionated these ACT solutions and the fractions were analyzed by UV and EPR spectroscopy and GC-MS (Gas Chromatography-Mass Spectrometry). The fractions were also analyzed for DNA nicking activity and only the fractions containing phenolic species (as determined by mass spectroscopy) caused significant DNA damage in rat thymocytes. These DNA damaging fractions also produced hydrogen peroxide, hydroxyl radicals, and superoxide. The tar radicals associate with DNA and cause DNA damage, and this is how cigarette tar may be involved in the toxicity associated with cigarette smoking.

Cigarette smoking has been correlated with an increased risk for a number of diseases including cancer, heart disease and emphysema (1-7). It is estimated that 60% of all hospitalizations are due to smoking related illnesses and over 450,000 Americans die each year from these afflictions. Additionally, environmental tobacco smoke (ETS) causes the death of 3000 Americans per year and it has been classified as a class A carcinogen (8).

Several research groups have proposed that DNA damage caused by cigarette smoke is due to the free radicals that are known to be present in cigarette tar (9, 10). Cigarette tar contains high concentrations ($>10^{17}$ spins/gram) of stable radicals that can be directly observed by EPR. The most prevalent radical in cigarette tar is a mixture of semiquinones, quinones and hydroquinones (9,11).

Aqueous Cigarette Tar Extracts (ACT)

Tobacco smoke can be separated into two phases, a particulate phase (tar) and gas-phase smoke. Both gas-phase and particulate phase cigarette smoke contain oxidants that can damage biomolecules (12). The oxidants in gas-phase smoke are shorter lived and may not survive long enough to reach DNA in a cell nucleus. Cigarette tar contains a long lived radical that has been shown to bind to DNA. Cigarette tar can be separated from gas phase constituents by using a Cambridge filter, 99.9% of the particles greater than one micron are trapped on this glass fiber filter. These filters are then soaked in pH 7 phosphate buffers at 37°C in the dark to prepare ACT solutions. The semiquinone radical present in cigarette tar can be extracted into aqueous solutions and ACT solutions from either mainstream or sidestream smoke contain the tar radical. Environmental tobacco smoke, ETS, consists primarily of the smoke that comes off the burning end of the cigarette, thus we trapped sidestream smoke as a model of ETS. Lung tissue is continually bathed in an aqueous solution that solubilizes and transports the water-soluble components of the tar. Thus, we believe that aqueous cigarette tar (ACT) extracts closely model the chemical mixtures that a smoker's lung cells encounter.

ACT solutions bind and nick DNA. These ACT solutions nick plasmid DNA, producing nicks that are not easily repairable (*13,14*). The tar-radical induced DNA nicks may be repaired by mechanisms that often induce errors in the DNA code, thus causing mutations that may lead to carcinogenesis. We have shown that the radical present in these ACT solutions becomes associated with DNA in rat alveolar macrophages (RAM) (*15*). Viable RAM were incubated with ACT solutions, made from extracting either ETS or mainstream tar, and the DNA was then isolated via a modification of the alkaline elution method (*16*). Double stranded DNA was trapped on polycarbonate filters and these filters were then analyzed by EPR spectroscopy. DNA from RAM that had been incubated with ACT solutions contained the tar radical.

Using the FADU (Fluorescence Analysis of DNA Unwinding) assay (*17,18*) we showed that ACT solutions nick DNA in viable rat thymocytes and this nicking is concentration dependent. ACT solutions from either mainstream or sidestream smoke were incubated with viable rat thymocytes (*15,19*). Rat thymocytes were used as a model cell because they are easily harvested, they contain a large amount of DNA and the resultant cell population is very homogeneous. The nicking follows saturation kinetics, indicating that the tar radical binds to the DNA and then causes nicking. A comparison of the mainstream smoke and sidestream smoke extracts data is shown in Table I. Mainstream smoke is about four times more damaging than sidestream smoke.

Table 1. Comparison of tar extracts from mainstream and sidestream smoke and aged catechol solutions.

Parameter	Mainstream cigarette tar	Sidestream cigarette tar	Aged catechol solutions
Number of cigarettes, or cigarette equivalents (for aged catechol) needed to cause maximum DNA damage	0.24 cigarettes	1.0 cigarettes	0.34 cigarette equivalents
H_2O_2 production	25-60 µmol/mg	1-5 µmol/mg	2.5 - 4 µmol/mg
% Protection of DNA nicking			
by catalase[a]	65[b]	72[b]	16[b,c]
by superoxide dismutase	3[b]	1 ± 12	0[b]
by GSH (200 mM)	85 ± 7	87 ± 3	84 ± 4
by DTPA (20 mM)	28 ± 5	42 ± 8	not tested
by Desferoxamine (1.0 mM)	27 ± 9	not tested	29 ± 1

[a] Protection by boiled enzymes has been subtracted
[b] Average of two determinations
[c] Catechol is an effective inhibitor of catalase, thus no protection is observed
(Adapted from ref. 20)

ACT solutions produce oxidants. Hydroquinones and semiquinones react with molecular oxygen to produce superoxide (equations 1 and 2), superoxide can dismutate to form hydrogen peroxide (equation 3). Hydrogen peroxide can oxidize biomolecules, for example hydrogen peroxide can oxidize and inactivate the protein α-1-proteinase inhibitor (α 1PI). Oxidative inactivation of a1PI is thought to be responsible for the pathogenesis of emphysema, as summarized in a recent review (21).   Also, hydrogen peroxide can be reduced to the hydroxyl radical by metals ions such as ferrous iron (Fe^{+2}) as shown in equation 4.

$$QH_2 + O_2 \rightleftharpoons QH^{\cdot} + O_2^{\bullet -} + H^+ \tag{1}$$

$$QH^{\bullet} + O_2 \rightleftharpoons Q + O_2^{\bullet -} + H^+ \tag{2}$$

$$2O_2^{\bullet -} + 2H^+ \rightleftharpoons H_2O_2 + O_2 \tag{3}$$

$$H_2O_2 + Fe^{+2} \rightleftharpoons OH^{\bullet} + OH^- + Fe^{+3} \tag{4}$$

Superoxide and hydroxyl radicals are unstable and have short lifetimes. These radicals have both been observed in ACT by EPR spin trap methods (22). Hydroxyl radicals are known to cause nicks in DNA (23).

We have tested a number of inhibitors and these results are also summarized in Table I. Glutathione protects against DNA nicking, as it is known to form covalent adducts with quinones and hydroquinones (24). Also, this protection may be due to the well known ability of thiols to reduce/scavenge radicals (24). Superoxide dismutase doesn't protect against DNA damage because it produces hydrogen peroxide. The iron chelators, desferoxamine and DTPA do not provide protection against DNA damage because the iron is associated with the tar that is bound to the DNA. Thus, the chelators can not effectively interact with the metals. Based on these nicking and binding studies, we have proposed a model in which the tar radicals first bind to DNA and then produce hydrogen peroxide and hydroxyl radicals in close proximity to the DNA molecule; these hydroxyl radicals then nick the DNA (11,15,19).

Constituents of Cigarette tar. Cigarette tar contains over 5,000 different compounds including nicotine, nitrosamines, and polyphenols (25). Since many of these tar constituents are water soluble, ACT solutions are complex mixtures with hydroquinone and catechol as major components. Walters et al. fractionated cigarette tar and determined that the fraction containing catechol was tumorigenic in mice (26). The mutagenic activity of catechol is controversial and has been summarized by Hecht et al. (26). However, catechol has been determined to be carcinogenic in rats and mice and it is a cocarcinogen when incubated with benzo[a]pyrene (27,28).

Aged Catechol solutions. Aged alkaline solutions of catechol or hydroquinone autoxidize to form a radical that is similar to the radical observed in ACT solutions and these aged catechol solutions generate superoxide. This suggests that the EPR signal observed in the ACT solutions is due to oxidized forms of catechol and other polyhydroxyaromatic compounds (29). While fresh hydroquinone has been shown to cause DNA nicking, fresh solutions of catechol do not nick DNA (30,31,32). Thus, we used autoxidized solutions of catechol as a simplified model for ACT; these solutions contain a radical similar to the cigarette tar radical, yet any unreacted (unoxidized) material would have no DNA nicking activity. We have shown that aged solutions of catechol, like ACT, nick DNA; the nicking follows saturation kinetics, and the amount of nicking is dependant upon the concentration of radicals in the aged catechol solutions. (31)

Fractionation of ACT solutions. To further study the tar radical in ACT solutions, we used Sephadex chromatography to fractionate these solutions (32). The initial fractionation yielded 78 fractions that were analyzed by EPR and UV spectroscopy and assayed for H_2O_2 production. Based upon similar UV spectra, fractions from the Sephadex column were combined into eight major fractions. We have assayed these eight fractions for DNA nicking in rat thymocytes, and have analyzed these fractions by GC-MS and EPR spectroscopy. The fractions that contain the tar radical also produce superoxide, H_2O_2, and hydroxyl radicals; these are the only fractions that cause significant amounts of DNA nicking as measured by the FADU asssay in rat thymocytes. In fact, these fractions (V and VI) account for only 3 % of the total mass yet they cause 72% of the DNA damage. These data are summarized in Table 2 and they provide further support for our proposal that the cigarette tar semiquinone radical is critically involved in causing DNA damage.

Table II. Summary of Sephadex G-25 Fractionation of ACT solutions.

ACT fraction	Percent of mass	Percent of DNA damage[a]	H_2O_2 produced (μM)[b]	Radicals detected by EPR	Components Detected by GC-MS
I	36.8	0.9	1.8	none	
II	29.1	1.5	11.6	Unidentified	nicotine
III	11.5	7.2	23.5	none	nicotine
IV	15.3	5.8	21.5	none	H_2Q, catechol
V	2.0	40.4	144.0	•OH, o-Q$^{\bullet-}$, p-Q$^{\bullet-}$, O$_2^{\bullet-}$	H_2Q, catechol
VI	0.9	31.5	81.7	•OH, o-Q$^{\bullet-}$, O$_2^{\bullet-}$	catechol
VII	1.1	5.7	13.4	none	
VIII	3.2	7.1	15.6	none	

[a] Percent of total DNA damage as determined by the FADU assay. *(17)*
[b] Hydrogen peroxide concentrations were determined indirectly by measuring the uptake of Oxygen.
This table is based on data from reference 32.

Literature Cited

1.   Boring, C.C;, Squires, T.S.; Tong, T.; Heath, C.W. *J.Am.Med.Assoc.* **1993**, 270, 2541-2542.
2.   Boyle, P. *Lung Cancer* **1996**, 17, 1-60.
3.   Morabia, A.; Wynder, E.L. *Br.Med.J.* **1992**, 304, 541-543.
4.   Giovannucci, E.; Rimm, E.B.; Stampfer, M.J.; Colditz, G.A.; Ascherio, A.; Kearney, J.; Willett, W.C. J.Natl.Cancer Inst. **1994**, 86(3), 183-191.
5.   Giovannucci, E.; Colditz, G.A.; Stampfer, M.J.; Hunter, D.; Rosner, B.A.; Willett, W.C.; Speizer, F.E.. *J.Natl.Cancer Inst.* **1994**, 86(3), 192-199.
6.   Surgeon General Smoking and Health, U.S Department of Health, Education, and Welfare: Rockville, MD **1979**.
7.   Parkin, D.M.; Pisani, P.; Lopez, A.D.; Masuyer, E. *Int.J.Cancer* **1994**, 59, 494-504.

8. U.S.Environmental Protection Agency. Smoking and Tobacco Control Monograph 4, Bethesda, Maryland: National Institutes of Health, National Cancer Institute, **1993**, pp. 1-364.

9. Pryor, W.A.; Hales, B.J.; Premovic, P.I.; Church, D.F. *Science* **1983**, 220, 425-427.

10. Randerath, E.; Danna T.F.; Randerath, K. *Mutat. Res. 1992*, 268, 139-153.

11. Church, D.F.; Pryor, W.A. *Environ.Health Perspect. 1985*, 64, 111-126.

12. Pryor, W.A.; Prier, D.G.; Church, D.F.. *Environ.Health Perspect. 1983*, 47, 345-355.

13. Borish, E.T.; Cosgrove, J.P.; Church, D.F.; Deutsch, W.A.; Pryor, W.A. *Biochem.Biophys.Res.Commun. ,* **1985,** 133:780-786.

14. Borish, E.T.; Pryor, W.A.; Venugopal, S.; Deutsch, W.A.. *Carcinogenesis,* **1987**, 8,1517-1520.

15. Stone, K.; Bermúdez, E.; Pryor, W.A.. *Environ.Health Perspect.* **1994**, 102(Suppl 10), 173-178.

16. Kohn, K.W.; Erickson, L.C.; Ewig; R.A.G.; Friedman, C.A. *Biochemistry* **1976**, 15, 4629-4637,.

17. Birnboim, H.C. *Methods Enzymol.* **1990**, 186, 550-555.

18. McLean, J.R.; McWilliams, R.S.; Kaplan, J.G.; Birnboim, H.C. In: *Progress in Mutation Research*, Vol. 3, Bora, K.C. Ed. Elsevier/North-Holland Biomedical Press: Amsterdam, 1982; pp. 137-141.

19. Bermúdez, E.; Stone, K.; Carter, K.M.; Pryor, W.A. *Environ.Health Perspect.* **1994**, 102(Suppl 10):870-874.

20. Evans, M.D.; Pryor, W. A. *Am. J. Physiol. (Lung Cell. Mol. Physiol. 10)* 1994 , 266, L593-L611.

21. Zang, L.-Y.; Stone, K.; Pryor, W.A.. *Free Radic.Biol.Med.* **1995**, 19(2):161-167.

22. Pryor, W.A. *Br.J.Cancer* **1987**, 55, Suppl. VIII, 19-23.

23. Wynder, E.L.; Hoffmann, D. *Tobacco and Tobacco Smoke: Studies in Experimental Carcinogenesis*, Academic Press: New York,NY, 1967.

24. Walters, D.B.; Chamberlain, W.J.; Akin, F.J.; Snook, M.E,; Chortyk, O.T. *Anal.Chim.Acta* **1978**, 99, 143-150.

25. Hecht, S.S.; Carmella, S.G.; Mori, H.; Hoffmann, D. *J.Natl.Cancer Inst.* **1981**, 66,163-169.

26. Hirose, M.; Fukushima, S.; Tanaka, H.; Asakawa, E.; Takahashi, S.; Ito, N. *Carcinogenesis,* **1992**, 14, 525-529.

27. Sealy, R.C.; Felix, C.C.; Hyde, J.S.; Swartz, H.M. In: *Free radicals in biology*; Pryor, W.A.Ed; ;Academic Press Inc.: New York, NY, 1980, Vol 4; pp 209-259.

28. Leanderson, P.; Tagesson, C.. *Chem.Biol.Interact.* **1992,** 81,197-208.

29. Lewis, J.G.; Stewart, W.; Adams, D.O. *Cancer Res.* **1988**, 48, 4762-4765.

30. Walles, S.A.S.. *Cancer Lett.* **1992**, 63, 47-52.

31. Stone, K.; Bermùdez, E.; Zang, L.-Y.; Carter, K.M.; Queenan, K.E.; Pryor, W.A; *Arch.Biochem.Biophys.* **1995**, 319(1),196-203.

32. Pryor, W. A.;Stone, K.; Zang, L.Y.; Bermudez, E. *Chem. Res. Tox.* **1998,** 11, 441-448.

Chapter 14

Molecular Neurosurgery: Using Plant Toxins to Make Highly Selective Neural Lesions

R. G. Wiley

Departments of Neurology and Pharmacology, Vanderbilt University
and VAMC, Nashville, TN 37212–2637

Ribosome inactivating proteins found in many plants can be used to make targeted cytotoxins that selectively destroy specific types of cells. Ricin and volkensin have been used to make anatomically selective neural lesions. Immunotoxins consisting of monoclonal anti-neuronal antibodies armed with saporin (from *Saponaria officinalis*) can selectively destroy specific types of neurons such as the cholinergic neurons of the basal forebrain. This approach is valuable for animal modeling of neurodegenerative diseases such as Alzheimer's and Parkinson's disease. Recently, neuropeptides, such as substance P, have been armed with saporin and used to selectively destroy neurons expressing the appropriate neuropeptide receptors. This approach holds great promise for research in the neurobiology of pain and for treatment of chronic, intractable pain.

Experimental neural lesions have long played a key role in functional neuroanatomy research. Increasingly selective lesioning techniques have allowed more and more detailed analysis of neural structure-function relationships. Molecular neurosurgery is a recent advance in the lesion making art using targeting cytotoxins that seek out and destroy specific neurons that express a selected surface molecule. A number of molecular neurosurgery toxins have been developed that permit anatomically and cell type-selective lesions for research in functional neuroanatomy. The present chapter will review the development of the three classes of molecular neurosurgery strategies - suicide transport, immunolesioning and neuropeptide-toxins. The field has become too large to completely cover in the available space, therefore, selected examples of the experimental power of molecular neurosurgery will be presented. Several reviews have covered early (*1-3*) and selected aspects of the topic. (*4-6*) With the development of neuropeptide-toxins, the molecular neurosurgery approach promises to become clinically useful in addition to numerous experimental applications.

Suicide transport

Original strategy using ribosome inactivating proteins. The initial experimental problem that led to development of suicide transport was how to selectively destroy vagal baroreceptor afferent neurons. The approach that was envisioned consisted of applying a toxin to the baroreceptor axons low in the neck where they can be identified as a separate and unique nerve. For this approach to be successful, the toxin must be taken up by axons and transported back to perikarya and destroy just those cells, i.e. a toxic retrograde anatomical tracer. Initial experiments involved applying candidate toxins to the cervical vagus trunk with the intention of producing complete lesions of the nodose ganglion and the dorsal motor nucleus of the vagus. The first agents studied were low molecular weight drugs including doxorubicin that proved ineffective. The first effective agent was the plant-derived toxic lectin, ricin (from *Ricinus communis*). Ricin is a 65 kDa protein composed of a carbohydrate binding subunit (B chain) attached by a disulfide bond to the A chain which is an enzyme that irreversibly inactivates the large ribosomal subunit. Further experiments using vagal application of ricin and related toxic lectins, abrin (from *Abrus precatorius*), modeccin (from *Adenia digitata*) and volkensin (from *Adenia volkensii*) revealed a characteristic sequence of changes in vagal sensory and motor neurons beginning with profound chromatolysis and ending with dissolution of the poisoned neurons. (7) Experiments with the microtubule poison, vincristine (from *Vinca rosea*) confirmed that fast axonal transport delivered the toxic lectins to the perikarya. Experiments with ricin injection into various peripheral nerves have shown the general usefulness of intraneural ricin injections and the feasibility of concurrently protecting animals from systemic ricin poisoning using anti-ricin antiserum. Table I gives the most frequently used suicide transport agents. It is important to note that ricin and volkensin are highly toxic if injected parenterally or inhaled requiring careful handling and dosing. All of these agents should be handled using good lab practice including adding a dye such as Fast Green FCF to all solutions to visualize injections and spills. Concentrated sodium hypochlorite is used to inactivate waste toxin solutions and to clean glassware. Incineration is used for dry waste and carcasses, however, note that once injected into an animal, there is no danger of spread of toxin to other animals or lab personnel.

Table I. Suicide Transport Agents

Agent	Target	Safety
1. Ricin (*Ricinus communis*)	Peripheral neurons	Highly toxic
2. Volkensin (*Adenia volkensii*)	Most neurons, central or peripheral	Highly toxic
3. OX7-saporin	All rat neurons	Non-toxic

Peripheral nervous system - ricin. Ricin injection of peripheral nerves has been used to identify the cellular localization of neurotransmitter receptors. (*8,9*) The experimental strategy in all cases consisted of applying ricin to a peripheral nerve, allowing the motor neurons projecting through the nerve to completely disintegrate and then using autoradiographic techniques to visualize the loss of receptors where the missing motor neurons were located. These experiments provided evidence that substance P receptors but not muscarinic receptors were expressed by specific types of motor neurons. An elegant series of experiments used ricin to ablate medial rectus motor neurons and then analyze the functional and anatomic plastic consequences to remaining interneurons and oculomotor neurons. (*10,11*) These experiments showed reorganization within the oculomotor system after loss of medial rectus motor neurons resulting in new neural connections that do not normally exist. Additional uses of ricin lesioning of motor neurons have included analysis of glial responses, (*12-16*) changes in the blood-brain barrier in vicinity of dying neurons (*17,18*) and as a model for motor neuron degeneration. (*19,20*)

Peripheral nerve ricin injections have been shown to produce anatomically restricted lesions (*21-26*) which have been used for the purpose of studying sensory plasticity after loss of first order sensory neurons. These experiments include studies of anatomical plasticity (*24,27*) and studies of physiologic plasticity. (*28,29*) These experiments showed little or no plasticity at the level of the dorsal horn terminations of primary sensory neurons, but there was remarkable functional plasticity of the cortical somatosensory map that was similar to the effects of distal nerve transection suggesting that plasticity of the cortical map was not dependent on the presence or absence of primary sensory neurons. One group has raised questions about the anatomical selectivity after intraneural ricin injection, (*30,31*) but these experiments used the 120 kDa isoform rather than the more toxic 60 kDa form used in most other studies. This work pointed out the importance of using just enough ricin to lesion the target neurons. At higher doses, they reported spread of the lesion to adjacent neurons. Similar findings were reported when very high ricin doses were used along with systemic anti-ricin antiserum to protect animals against systemic toxicity. (*32*)

Not surprisingly, ablation of peripheral nerves with ricin has been used to study pain phenomena. Several investigators have studied autotomy behavior after ligating or transecting the sciatic nerve in rats (*33,34*) and found significant autotomy, but somewhat less than with ricin. Others have reported ablation of post-traumatic neuromas with intraneural ricin injection. (*35-37*) At this point, the effects of intraneural ricin on pain perception remains to be fully determined.

Central nervous system - volkensin. Ricin is not an effective suicide transport agent within rat CNS (*38*) probably because of penultimate sialic acid residues on many glycoproteins of adult CNS neurons (*39,40*) which prevents ricin binding. Abrin is also ineffective within the CNS, but two related toxic lectins, volkensin and modeccin, are effective suicide transport agents in rat CNS. (*7*) Volkensin has been used as a suicide transport agent in both the PNS and CNS. (*41-45*)

A typical use of volkensin is to selectively destroy the striatonigral projection neurons (*46-50*) and then study changes in striatal markers such as dopamine or glutamate receptors using radio-labeled ligands. A careful anatomic study established the selectivity of the lesion produced by nigral volkensin injection. (51) A number of other studies have successfully used volkensin for anatomical or neurochemical mapping. (41,52-61) However, volkensin is not effective on at least one pathway, the striatopallidal projection which is lesioned by the immunotoxin, OX7-saporin. (51)

Immunolesioning

The next development in this field was to turn to anti-neuronal immunotoxins consisting of a monoclonal antibody to a specific neural antigen armed to kill with a ribosome inactivating protein, saporin. Table II lists these agents.

Table II. Immunolesioning Agents

Immunotoxin	Target Antigen	Target neurons
1. OX7-saporin	Rat Thy 1	All
2. 192-saporin	p75^{NTR}	Cholinergic basal forebrain, Purkinje cells, postganglionic sympathetic, some primary sensory
3. Anti-DβH-saporin	Dopamine β-hydroxylase	noradrenergic, adrenergic
4. Anti-DAT-saporin	Dopamine transporter	dopaminergic in substantia nigra and ventral tegmental area

Saporin. Saporin is a ribosome-inactivating protein from *Saponaria officianlis*, the soapwort. (62) It has no binding subunit which means it cannot gain entry into a cell unless attached to a targeting vector such as a monoclonal antibody that will mediate endocytosis. Once inside the cell, saporin must escape the endosome into the cytoplasm where it irreversibly inhibits protein synthesis by enzymatically inactivating the large ribosomal subunit, just like ricin A chain. It is probable that only one molecule free in the cytoplasm is sufficient to kill a cell. In order to allow escape of saporin into the cytoplasm, a disulfide bond is used to couple to the targeting vector.

OX7-saporin. The first effective anti-neuronal immunotoxin was OX7-saporin composed of the MRC monoclonal antibody, OX7, which has high affinity for rat Thy 1. This construct was tested after initial attempts to make an anti-neuronal immunotoxin with ricin A chain were unsuccessful (1,63) and because comparison of OX7-saporin to OX7-ricin A chain showed the saporin conjugate to be more active *in vivo*. (64) Experiments with OX7 alone showed axonal transport by rat central and peripheral neurons. (65) OX7-saporin is an effective suicide transport agent in rat central and peripheral nervous system. (66) Interestingly, intraventricular injection of OX7-saporin selectively destroys cerebellar Purkinje neurons (67) probably due to the abundance on Thy 1 on Purkinje cells and the tendency of these cells to take up a variety of molecules from the cerebrospinal fluid. (68)

OX7-saporin has been used as a suicide transport agent in rat CNS in much the same way as volkensin with the advantages that it is less toxic and so far has been active on all pathways tested. Example applications have been to map dopamine receptors in striatum after destruction of striatopallidal neurons by injection of OX7-saporin into the ipsilateral globus pallidus. (*47,51*) Volkensin was ineffective in these experiments. Similarly, OX7-saporin produced more reliable lesions of corticothalamic projection neurons in experiments that showed localization of 5-HT1a and mu opiate receptors to these cells. (*69,70*) A novel application of OX7-saporin has been to selectively destroy the sensory innervation of the knee joint by intra-articular injection of the immunotoxin. (*71*)

192 IgG-saporin. Numerous authors have used 192-saporin to selectively destroy the cholinergic neurons of the rat basal forebrain. (*4-6*) This lesion mimics a key feature of Alzheimer's Disease and has revealed interesting information about the function of the cholinergic basal forebrain (CBF) consistent with the hypothesis that the CBF is involved in attentional processes. (*72-77*) Detailed analysis of this complex literature is beyond the scope of the present article. The major considerations about the use of 192-saporin to study the CBF has been the need to make very extensive lesions before a behavioral deficit can be detected and this usually requires intraventricular toxin injection. However, intraventricular injections produce concurrent destruction of cerebellar Purkinje neurons. Additional control experiments using OX7-saporin are needed to assess the contribution of Purkinje cells to the behavioral deficits seen with 192-saporin.

192-saporin targets the rat low affinity neurotrophin receptor, $p75^{NTR}$. A similar immunotoxin targeting the human low affinity neurotrophin receptor using the monoclonal antibody, ME20.4, recently has been reported effective in destroying the CBF in primates. (*78,79*) Experiments in primates should permit more detailed behavioral analysis of CBF function. Also, local injections of 192-saporin into target fields such as the cortex or hippocampus may prove useful in a variety of experiments. (*80-82*)

Anti-DβH-saporin. The immunotoxin, anti-DβH-saporin has been developed to selectively destroy noradrenergic and adrenergic neurons. (*83-85*) This agent is active against postganglionic sympathetic neurons after intravenous injection and against CNS noradrenergic and adrenergic neurons after intraventricular injection. Injection into the olfactory bulb will selectively destroy the noradrenergic neurons of the locus coeruleus that project to the injection site. (*86*) The first functional study using anti-DβH-saporin assessed opiate withdrawal after intraventricular injection. (*87*) No doubt this agent will find widespread use as a superior method for producing noradrenergic denervation.

Neuropeptide-toxin conjugates

Substance P-saporin. A conjugate of the neuropeptide, substance P, to saporin is active *in vitro* and *in vivo* to selectively destroy neurons that express the neurokinin-1 receptor. (*88*) Initial results with substance P-saporin injected into the lumbar spinal fluid of rats indicate selective destruction of neurokinin-1 receptor expressing neurons in the superficial dorsal horn which results in decreased hyperalgesia to hind paw capsaicin injection. (*89*)

Other peptide-toxin conjugates. The mu opiate peptide, dermorphin has been conjugated to saporin and found to destroy striatal neurons expressing the mu opiate receptor (Wiley and Lappi, unpublished). Other peptides have been reported to selectively target ricin A chain. (*90-95*) Table III lists these agents. Conjugates of gelonin to corticotrophin releasing factor have been reported to selectively destroy pituitary corticotropes. (*96-98*)

Table III. Neuropeptide-toxin conjugates

Agent	Target
1. Substance P-saporin	Neurokinin-1 receptor
2. Dermorphin-saporin	Mu opiate receptor
3. Oxytocin-ricin A chain	Oxytocin receptor
4. Atrial natriuretic peptide- ricin A chain	ANP receptors & clearance receptors
5. C-type natriuretic peptide- ricin A chain	CNP receptors
6. CRF-gelonin	CRF receptor (pituitary corticotropes)

Summary

Ribosome inactivating proteins from plants have proven useful in making selective neural lesions. Toxic lectins such as ricin and volkensin can produce anatomically restricted lesions by suicide transport. Immunotoxins consisting of monoclonal antibodies armed with saporin can produce highly selective lesions of specific types of neurons. Most recently, neuropeptide-saporin conjugates have been shown to selectively destroy neurons expressing the appropriate receptors for the neuropeptide. Numerous additional agents seem likely in the near future, some may prove therapeutically useful in treating pain.

References

1   Wiley, R. G. and Lappi, D. A. Suicide transport and immunolesioning. 1994. Austin, TX, R.G. Landes Co. Molecular Biology Intelligence Unit.
2   Wiley, R.G. *Trends.Neurosci.* **1992**, *15*, 285-290.
3   Contestabile, A.; Stirpe, F. *Eur.J.Neurosci.* **1993**, *5*, 1292-1301.
4   Wiley, R.G. *Ann.N.Y.Acad.Sci.* **1997**, *835*, 20-29.
5   Rossner, S. *Int.J.Dev.Neurosci.* **1997**, *15*, 835-850.

6   Schliebs, R.; Rossner, S.; Bigl, V. *Prog.Brain Res.* **1996**, *109* , 253-264.

7   Wiley, R.G.; Stirpe, F. *Brain Res.* **1988**, *438*, 145-154.

8   Helke, C.J.; Charlton, C.G.; Wiley, R.G. *Neuroscience* **1986**, *19*, 523-533.

9   Helke, C.J.; Charlton, C.G.; Wiley, R.G. *Brain Res.* **1985**, *328* , 190-195.

10   de la Cruz, R.R.; Pastor, A.M.; Delgado-Garcia, J.M. *Neuroscience* **1994**, *58*, 81-97.

11   de la Cruz, R.R.; Pastor, A.M.; Delgado-Garcia, J.M. *Neuroscience* **1994**, *58*, 59-79.

12   Graeber, M.B.; Streit, W.J.; Kreutzberg, G.W. *Acta Neuropathol.(Berl)* **1989**, *78*, 348-358.

13   Ling, E.A.; Kaur, C.; Wong, W.C. *Histol.Histopathol.* **1992**, *7* , 93-100.

14   Streit, W.J.; Kreutzberg, G.W. *J.Comp.Neurol.* **1988**, *268*, 248-263.

15   Sutin, J.; Griffith, R. *Exp.Neurol.* **1993**, *120*, 214-222.

16   Yamamoto, T.; Iwasaki, Y.; Konno, H.; Kudo, H. *Ann.Neurol.* **1986**, *20*, 267-271.

17   Bouldin, T.W.; Earnhardt, T.S.; Goines, N.D. *J.Neuropathol.Exp.Neurol.* **1991**, *50*, 719-728.

18   Bouldin, T.W.; Earnhardt, T.S.; Goines, N.D. *Neurotoxicology.* **1990**, *11*, 23-34.

19   Palladini, G.; Medolago-Albani, L.; Guerrisi, R.; Millefiorini, M.; Antonini, G.; Filippini, C.; Conforti, A.; Palatinsky, E. *Path.Biol.* **1986**, *34*, 1047-1053.

20   Yamamoto, T.; Iwasaki, Y.; Konno, H.; Kudo, H. *J.Neurol.Sci.* **1985**, *70*, 327-337.

21   Aldskogius, H.; Wiesenfeld-Hallin, Z.; Kristensson, K. *Brain Research* **1988**, *461*, 215-220.

2   Leong, S.K.; Tan, C.K. *J.Anat.* **1987**, *154*, 15-26.

23   Paul, I.; Devor, M. *J.Neurosci.Methods* **1987**, *22*, 103-111.

24   Pubols, L.M.; Foglesong, M.E. *J.Comp.Neurol.* **1988**, *275*, 271-281.

25   Yamamoto, T.; Iwasaki, Y.; Konno, H. *J.Neurosurg.* **1984**, *60*, 108-114.

26   Yamamoto, T.; Iwasaki, Y.; Konno, H. *Brain Research* **1983**, *274*, 325-328.

27   Pubols, L.M.; Bowen, D.C. *J.Comp.Neurol.* **1988**, *275*, 282-287.

28   Cusick, C.G.; Wall, J.T.; Whiting, J.H., Jr.; Wiley, R.G. *Brain Res.* **1990**, *537*, 355-358.

29   Wall, J.T.; Cusick, C.G.; Migani-Wall, S.A.; Wiley, R.G. *J.Comp.Neurol.* **1988**, *277*, 578-592.

30   Rivero-Melian, C.; Arvidsson, J. *Brain Res.* **1989**, *496*, 131-140.

31   Rivero-Melian, C.; Arvidsson, J. *Brain Res.* **1990**, *509*, 335-338.

32   Wiley, R.G.; Oeltmann, T.N. *J.Neurosci.Methods* **1989**, *27*, 203-209.

33   Blumenkopf, B. and Lipman, J. J. Studies in autotomy: Its pathophysiology and usefulness as a model of chronic pain. Pain 45, 203-209. 1991.

34   Wiesenfeld-Hallin, Z.; Nennesmo, I.; Kristensson, K. *Pain* **1987**, *30*, 93-102.

35   Brandner, M.D.; Buncke, H.J.; Campagna-Pinto, D. *J.Hand Surg.[Am].* **1989**, *14*, 710-714.

36   Cummings, J.F.; Fubini, S.L.; Todhunter, R.J. *Equine.Vet.J.* **1988**, *20*, 451-456.

37   Nennesmo, I.; Kristensson, K. *Acta Neuropathol.(Berl)* **1986**, *70*, 279-283.

38   Wiley, R.G.; Talman, W.T.; Reis, D.J. *Brain Res.* **1983**, *269*, 357-360.
39   Margolis, R.K.; Gomez, Z. *Brain Research* **1974**, *74*, 370-372.
40   Margolis, R.K.; Preti, C.; Lai, D.; Margolis, R.U. *Brain Research* **1976**, *112*, 363-369.
41   Contestabile, A.; Fasolo, A.; Virgili, M.; Migani, P.; Villani, L.; Stirpe, F. *Brain Res.* **1990**, *537*, 279-286.
42   Leanza, G.; Stanzani, S. *Neurosci.Lett.* **1998**, *244*, 89-92.
43   Nógrádi, A.; Vrbová, G. *Restor.Neurol.Neurosci.* **1997**, *11*, 37-45.
44   Nogradi, A.; Vrbova, G. *Neuroscience* **1992**, *50*, 975-986.
45   Nógrádi, A.; Vrbová, G. *Exp.Neurol.* **1994**, *129*, 130-141.
46   Harrison, M.B.; Roberts, R.C.; Wiley, R.G. *Brain Res.* **1993**, *630*, 169-177.
47   Harrison, M.B.; Wiley, R.G.; Wooten, G.F. *Brain Res.* **1992**, *590*, 305-310.
48   Harrison, M.B.; Wiley, R.G.; Wooten, G.F. *Brain Res.* **1992**, *596*, 330-336.
49   Harrison, M.B.; Wiley, R.G.; Wooten, G.F. *Brain Res.* **1990**, *528*, 317-322.
50   Tallaksen-Greene, S.J.; Wiley, R.G.; Albin, R.L. *Brain Res.* **1992**, *594*, 165-170.
51   Roberts, R.C.; Harrison, M.B.; Francis, S.M.; Wiley, R.G. *Exp.Neurol.* **1993**, *124*, 242-252.
52   Black, M.D.; Crossman, A.R. *Neurosci.Lett.* **1992**, *134*, 180-182.
53   Cevolani, D.; Strocchi, P.; Bentivoglio, M.; Stirpe, F. *Brain Research* **1995**, *689*, 163-171.
54   Chessell, I.P.; Francis, P.T.; Bowen, D.M. *Neurodegeneration* **1995**, *4*, 415-424.
55   Chessell, I.P.; Francis, P.T.; Pangalos, M.N.; Pearson, R.C.; Bowen, D.M. *Brain Res.* **1993**, *632*, 86-94.
56   Chessell, I.P.; Pearson, R.C.; Heath, P.R.; Bown, D.M.; Francis, P.T. *Exp.Neurol.* **1997**, *147*, 183-191.
57   Ferris, C.F.; Pilapil, C.G.; Hayden-Hixson, D.; Wiley, R.G.; Koh, E.T. *J.Neuroendocrinol.* **1992**, *4*, 193-205.
58   Francis, P.T.; Pangalos, M.N.; Pearson, R.C.; Middlemiss, D.N.; Stratmann, G.C.; Bowen, D.M. *J.Pharmacol.Exp.Ther.* **1992**, *261*, 1273-1281.
59   Glendenning, K.K.; Baker, B.N.; Hutson, K.A.; Masterton, R.B. *J.Comp.Neurol.* **1992**, *319*, 100-122.
60   Heath, P.R.; Chessell, I.P.; Sanders, M.W.; Francis, P.T.; Bowen, D.M.; Pearson, R.C. *Exp.Neurol.* **1997**, *147*, 192-203.
61   Pangalos, M.N.; Francis, P.T.; Pearson, R.C.A.; Middlemiss, D.N.; Bowen, D.M. *J.Neurosci.Meth.* **1991**, *40*, 17-29.
62   Lappi, D.A.; Esch, F.S.; Barbieri, L.; Stirpe, F.; Soria, M. *Biochem.Biophys.Res.Comm.* **1985**, *129*, 934-942.
63   DiStefano, P.S.; Schweitzer, J.B.; Taniuchi, M.; Johnson, E.M.J. *J.Cell Biol.* **1985**, *101*, 1107-1114.
64   Thorpe, P.E.; Brown, A.N.F.; Bremner, J.A.G.; Foxwell, B.M.J.; Stirpe, F. *J.Nat.Canc.Inst.* **1985**, *75*, 151-159.
65   LaRocca, C.D.; Wiley, R.G. *Brain Res.* **1988**, *449*, 381-385.
66   Wiley, R.G.; Stirpe, F.; Thorpe, P.; Oeltmann, T.N. *Brain Res.* **1989**, *505*, 44-54.

202

67 Davis, T.L.; Wiley, R.G. *Brain Res.* **1989**, *504*, 216-222.
68 Borges, L.F.; Elliot, P.J.; Gill, R.; Iversen, S.D.; Iversen, L.L. *Science* **1985**, *228*, 346-348.
69 Crino, P.B.; Vogt, B.A.; Volicer, L.; Wiley, R.G. *J.Pharmacol.Exp.Ther.* **1990**, *252*, 651-656.
70 Vogt, B.A.; Wiley, R.G.; Jensen, E.L. *Exp.Neurol.* **1995**, *135*, 83-92.
71 Salo, P.T.; Theriault, E.; Wiley, R.G. *J.Orthop.Res.* **1997**, *15*, 622-628.
72 Bushnell, P.J.; Chiba, A.A.; Oshiro, W.M. *Behav.Brain Res.* **1998**, *90*, 57-71.
73 Baxter, M.G.; Bucci, D.J.; Gorman, L.K.; Wiley, R.G.; Gallagher, M. *Behav.Neurosci.* **1995**, *109*, 714-722.
74 Baxter, M.G.; Bucci, D.J.; Sobel, T.J.; Williams, M.J.; Gorman, L.K.; Gallagher, M. *NeuroReport* **1996**, *7*, 1417-1420.
75 Bucci, D.J.; Chiba, A.A.; Gallagher, M. *Behav.Neurosci.* **1995**, *109*, 180-183.
76 Chiba, A.A.; Bucci, D.J.; Holland, P.C.; Gallagher, M. *J.Neurosci.* **1995**, *15*, 7315-7322.
77 McGaughy, J.; Kaiser, T.; Sarter, M. *Behav.Neurosci.* **1996**, *110*, 247-265.
78 Fine, A.; Hoyle, C.; Maclean, C.J.; Levatte, T.L.; Baker, H.F.; Ridley, R.M. *Neuroscience* **1997**, *81*, 331-343.
79 Mrzljak, L.; Levey, A.I.; Belcher, S.; Goldman-Rakic, P.S. *J.Comp.Neurol.* **1998**, *390*, 112-132.
80 Holley, L.A.; Wiley, R.G.; Lappi, D.A.; Sarter, M. *Brain Res.* **1994**, *663*, 277-286.
81 Sarter, M.; Holley, L.A.; Wiley, R.G. *Abs.Soc.Neurosci.* **1993**, *19*, abstract.
82 Sachdev, R.N.; Lu, S.M.; Wiley, R.G.; Ebner, F.F. *J.Neurophysiol.* **1998**, *79*, 3216-3228.
83 Picklo, M.J.; Wiley, R.G.; Lappi, D.A.; Robertson, D. *Brain Res.* **1994**, *666*, 195-200.
84 Picklo, M.J.; Wiley, R.G.; Lonce, S.; Lappi, D.A.; Robertson, D. *J.Pharmacol.Exp.Therap.* **1995**, *275*, 1003-1010.
85 Wrenn, C.C.; Picklo, M.J.; Lappi, D.A.; Robertson, D.; Wiley, R.G. *Brain Res.* **1996**, *740*, 175-184.
86 Blessing, W.W.; Lappi, D.A.; Wiley, R.G. *Neurosci.Lett.* **1998**, *243*, 85-88.
87 Rohde, D.S.; Basbaum, A.I. *J.Neurosci.* **1998**, *18*, 4393-4402.
88 Wiley, R.G.; Lappi, D.A. *Neurosci.Lett.* **1997**, *230*, 97-100.
89 Mantyh, P.W.; Rogers, S.D.; Honore, P.; Allen, B.J.; Ghilardi, J.R.; Li, J.; Daughters, R.S.; Lappi, D.A.; Wiley, R.G.; Simone, D.A. *Science* **1997**, *278*, 275-279.
90 Blackburn, R.E.; Samson, W.K.; Fulton, R.J.; Stricker, E.M.; Verbalis, J.G. *Am.J.Physiol.* **1995**, *269*, R245-R251
91 Blackburn, R.E.; Samson, W.K.; Fulton, R.J.; Stricker, E.M.; Verbalis, J.G. *Proc.Natl.Acad.Sci.U.S.A.* **1993**, *90*, 10380-10384.
92 Samson, W.K.; Alexander, B.D.; Skala, K.D.; Huang, F.L.; Fulton, R.J. *Can.J.Physiol.Pharmacol.* **1992**, *70*, 773-778.
93 Samson, W.K.; Alexander, B.D.; Skala, K.D.; Huang, F.L.; Fulton, R.J. *Ann.N.Y.Acad.Sci.* **1992**, *652*, 411-422.
94 Samson, W.K.; Huang, F.L.; Fulton, R.J. *Endocrinology* **1993**, *132*, 504-509.

95   Samson, W.K.; Huang, F.L.; Fulton, R.J. *J.Neuroendocrinol.* **1995**, *7*, 759-763.

96   Schwartz, J.; Penke, B.; Rivier, J.; Vale, W. *Endocrinology* **1987**, *121*, 1454-1460.

97   Schwartz, J.; Vale, W. *Endocrinology* **1988**, *122*, 1695-1700.

98   Schwartz, J.; Vale, W. *Methods Enzymol.* **1989**, *168*, 29-44.

Chapter 15

Plant-Associated Hepatogenous Photosensitization Diseases

Arne Flåøyen

National Veterinary Institute, P.O. Box 8156 Dep., N–0033 Oslo, Norway

Hepatogenous photosensitization of livestock is both economically important and an animal welfare problem in various parts of the world. The condition results when a toxin, whether produced by a higher plant, fungus or alga, causes liver damage or liver dysfunction resulting in retention of the photosensitizing agent phylloerythrin. Phylloerythrin is a metabolic product of chlorophyll produced by rumen microorganisms. Many hepatogenous photosensitization disorders are associated with ingestion of plants containing steroidal saponins. Typical of these diseases is the accumulation of insoluble Ca^{2+} salts of episarsasapogenin or epismilagenin glucuronides in liver cells and bile ducts. Sporidesmin produced by the fungus *Pithomyces chartarum* and the pentacyclic triterpene acids lantadene A and B from the plant *Lantana camara* cause severe problems in some countries. A cyclic hexapeptide produced by the fungus *Phomopsis leptostromiformis* and cyclic heptapeptides produced by the alga *Microcystis aeruginosa* are also known to cause hepatogenous photosensitization.

Hepatogenous photosensitization is of economic importance as well as an animal welfare problem in various parts of the world, particularly in sheep but also in cattle, goats, fallow-deer and horses. More than 500,000 head of sheep and goats have been photosensitized in severe outbreaks of geeldikkop (literally: yellow-thick-head) in South Africa (*69*), and in New Zealand the estimated annual production loss due to facial eczema can be as high as NZ$100 million (*52*). Occasionally in Norway in certain flocks, 30 to 50% of lambs grazing bog asphodel (*Narthecium ossifragum*) will suffer from alveld (literally: elf-fire) (*18*). In the USA hepatogenous photosensitization diseases occur sporadically in animals in various parts of the country.

Liver Damage causes Diseases of the Skin

Hepatogenous photosensitization occurs when a toxin, normally produced by a higher plant, a fungus or an alga, causes liver damage or liver dysfunction that results in retention of the photosensitizing agent, phylloerythrin (7,19,35). Phylloerythrin is a metabolic product of chlorophyll produced in sheep, goats and cattle by rumen microorganisms (63,65), and in horses probably by microorganisms in the large intestine. Unlike chlorophyll, which has a long phytyl hydrocarbon side chain preventing it from being absorbed from the digestive tract, phylloerythrin has lost the side chain and can be absorbed from the gut (Figure 1). Normally, phylloerythrin is conjugated by the liver and excreted into the bile but certain types of liver lesion or liver dysfunction lead to depression or cessation of hepatic elimination so that it enters the blood stream and reaches the skin where photochemical reactions with sunlight can occur.

An acute inflammatory response of the skin is induced when phylloerythrin reacts with sunlight. Information about the photodynamic action of phylloerythrin is not complete, and only the main principles can be outlined. Sunlight, in a wavelength range corresponding to the absorption spectrum of phylloerythrin (approximately 420 nm), excites the phylloerythrin molecules which generate oxidative reactions that require molecular oxygen to be available for the reaction to proceed. The excited phylloerythrin molecules interact and transfer their energy either directly to biomolecules or to oxygen thus causing formation of toxic singlet oxygen or oxygen radicals. These molecules cause pathological changes in the skin resulting in clinical signs (34).

Clinical Signs.

In general, only unpigmented animals or animals with unpigmented areas of the skin become photosensitized. The most salient clinical signs in sheep are restlessness, shade seeking, head shaking, scratching of the face and ears with the hind feet, and rubbing of the irritated skin against solid objects. The skin change develops rapidly, and includes edema and reddening. The eyelids, muzzle, and lips become swollen and turgid (Figure 2). However the most obvious signs in serious cases are the thickened edematous heavily drooping ears. Secondary bacterial infection of the skin is a normal finding and in severe cases jaundice, due to accumulation of bilirubin, is typical. Hepatogenous photosensitization is a painful condition and affected animals stop eating or drinking. A large proportion die from dehydration or secondary infections unless kept out of sunlight and given other appropriate treatment. In animals surviving the acute phase of the disease, the affected skin dries and cracks, the ears curl and a part, or all, of the ear may be lost from necrosis. Similar lesions can be seen in species other than sheep. In cattle there are typically lesions on unpigmented areas where hair is thin or absent such as on the udder, teats and escutcheon.

Photosensitization Agents

Plants containing Steroidal Saponins

In several parts of the world, hepatogenous photosensitization of sheep is associated with grazing of plants containing steroidal saponins (18). A common feature in all photosensitization diseases associated with ingestion of steroidal saponins is the appearance in the liver and in the bile ducts of biliary crystals, which (in the cases investigated to date) have been identified as insoluble salts of episarsasapogenin β-D-glucuronide and or epismilagenin

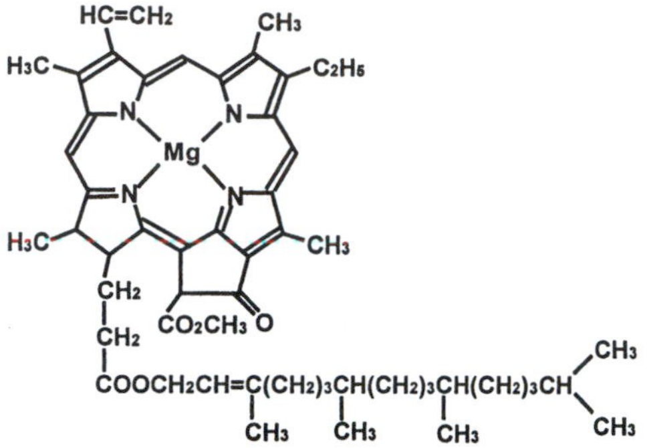

206

Figure 1. Chemical structure of chlorophyll (left) and phylloerythrin (right).

Figure 2. Lamb showing typical clinical signs of photosensitization.

β-D-glucuronide (*18*). Typically all these diseases occur sporadically and are difficult to reproduce by dosing experiments, whilst the saponin containing plants associated with them do not always seem to be toxic to the grazing animal. Several workers have suggested that saponins may cause the diseases (*2,15,36,42*), while other workers have questioned whether saponins, or sapogenins, are the sole cause (*20,22*).

Steroidal saponins are spirastones bearing one or more sugar chains, usually one at the C-3 carbon and one at C-26 (Figure 3) (*59*). The classical definition is based on their surface activity, for many saponins have detergent properties and give stable foams in water (*30*). Saponins have antifungal activity and the primary mode of action towards fungi is similar to that of polyene antibiotics, involving the formation of complexes with membrane sterols (*59*). The aglycon or non-saccharide portion of the saponin is called the *genin* or *sapogenin*. The steroidal skeleton of steroidal saponins, containing 27 carbon atoms, is built up of six isoprene units and is derived from *squalene* (*30*).

Agave lecheguilla is a conspicuous, long-lived perennial of the lily family with a thick fibrous toothed crown baring a cluster of thick, fleshy basal leaves and tall flower stalks. The plant grows abundantly over a large area of southwestern Texas, where it is reported to cause photosensitization of sheep (*42*). Smilagenin is the only sapogenin known to be present in the plant (*73*).

Tribulus terrestris is a prostrate creeping plant belonging to the Zygophyllaceae. It has a semi-perennial underground stem and root system and has been reported to cause photosensitization of sheep in South Africa, United States, Argentina, Australia and Iran (*4,8,26,35,44,71*). It also causes photosensitization of goats (*25,35*). The greatest losses occur in South Africa, where the disease is called geeldikkop (yellow-thick-head) (*35*).

T. terrestris contains several steroidal sapogenins includeing diosgenin, yamogenin, epismilagenin, tigogenin, neotigogenin, gitogenin and neogitogenin (*48*). Biliary crystals from sheep photosensitized after ingesting *T. terrestris* have been found to consist of both episarsasapogenin β-D-glucuronide and epismilagenin β-D-glucuronide (*47,48*).

Narthecium ossifragum is a loosely to densely clonal, perennial herb of the lily family. It can be up to 40 cm tall. The plant occurs on oligotrophic, mesotrophic, and eutrophic peat deposits in northern and western Europe (*70*). Photosensitization of sheep grazing the plant has been reported from Norway (*15,17*), The Faroe Islands (*21*) and the British Isles (*24*). Normally, only lambs between 2 and 6 months old become photosensitized, and more cases are seen in cold and rainy summers than in hot and dry summers.

N. ossifragum has been found to contain 4 sapogenins, of which sarsasapogenin and smilagenin are the most important (*49*). Studies on samples from the gastrointestinal tract of a sheep fed *N. ossifragum* have revealed 3 distinct regions of metabolic activity (Figure 4) (*23*). In the fore stomachs the ingested materials were hydrolyzed to the parent sapogenins before being oxidized at C-3 and reduced to give *epi* analogues of the ingested sapogenins. The sapogenins were probably absorbed in the small intestine, transported via the portal vein to the liver, where the sapogenins were conjugated and excreted into the bile as episarsasapogenin β-D-glucuronide and epismilagenin β-D-glucuronide. Finally, in the large intestine, the *epi*-sapogenin conjugates were hydrolyzed to free *epi*-sapogenins and excreted with the feces.

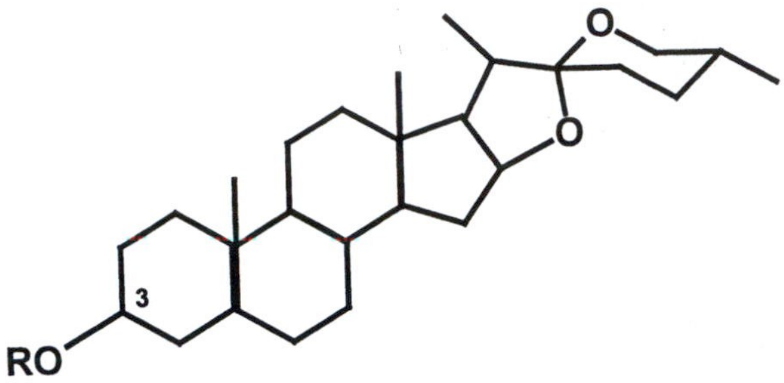

208

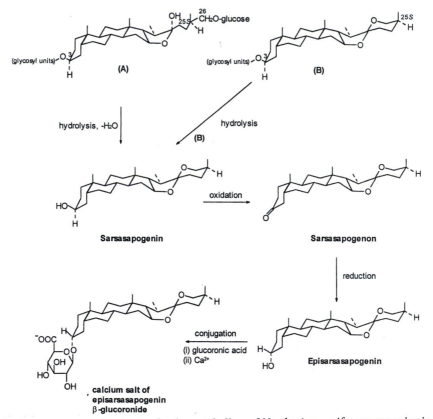

Figure 3. Typical Structure of a steroidal saponin (R = sugars) and a steroidal sapogenin (R = H).

Figure 4. Proposed pathways for the metabolism of *Narthecium ossifragum* saponins in sheep, illustrated for (25*S*)-saponins and sapogenins. [A] and [B] = plant saponins: furastonol and 5β-sarsasapogenin and 5β-furastonol types, respectively.

Brachiaria decumbens belongs to the grass family Panicoideae and is the major species used to improve pasture in many parts of the tropics because of its aggressive growth habit, efficient nitrogen utilisation, ability to withstand grazing, drought resistance, and relative freedom from pests and diseases. It has been reported to cause photosensitization of sheep and/or cattle in Australia, Malaysia, Indonesia, Papua New Guinea, Nigeria, and Brazil (*1,3,27,39,40,58*).

Yamogenin and diosgenin are the sapogenins reported to be present in *B. decumbens* (*68*), whereas episarsasapogenin and epismilagenin have been isolated from the ruminal content of sheep suffering from *B. decumbens* intoxication (*37*).

Panicum spp. are annual or perennial grasses widely distributed from the tropics to the warm temperate regions. Many of them are important pasture grasses or cereals and some are aggressive weeds (*10*).

Panicum dichotomiflorum is reported to have caused photosensitization of sheep in New Zealand (*29,46*). Diosgenin is the only sapogenin that has been found in *P. dichotomiflorum* and the biliary crystals of sheep photosensitized after grazing the plant contain epismilagenin β-D-glucuronide (*46,50*).

Panicum schinzii has been implicated in photosensitization of sheep in Australia (*6,50,38*). Diosgenin is the only sapogenin reported to be found in *P. schinzii* and, as in *P. dichotomiflorum* intoxicated sheep, the biliary crystals of sheep affected by *P. schinzii* toxicosis contain epismilagenin β-D-glucuronide.

Panicum miliaceum contains the two sapogenins, diosgenin and yamogenin, and photosensitization is known to occur in sheep grazing the plant in New Zealand (*7,29,45,46*).

Panicum coloratum is known to cause photosensitization of sheep in the United States and in South Africa (*5,35*). Two steroidal sapogenins have been found in *P. coloratum*, diosgenin and yamogenin (*61*).

Panicum virgatum was reported to have caused photosensitization of sheep in the United States (*62*). Grass from this outbreak contained diosgenin (Barry Smith, personal communication).

Nolina texana and ***Nolina microcarpa***, of the lily family, are perennial desert shrubs reported to have caused photosensitization of sheep, cattle and goats in Texas, USA (*43,64*). No reports appear on the steroidal sapogenins *of N. texana* or *N microcarpa*, but *N. recurvata*, a Nolina species indigenous to Mexico, is known to contain 7 different steroidal sapogenins (*51*).

Plants Lacking Steroidal Saponins

Lantana camara of the verbena family is native to central America and Africa but has become distributed throughout many tropical and subtropical parts of the world. It grows as a bush up to 3 m high, and can become an impenetrable thicket (*60*). It proliferates readily in cleared land and has become a noxious weed in many places. The small, trumpet-shaped, yellow to orange, red and mauve to white flowers are borne in dense terminal clusters, usually with flowers of two different colours occurring in one cluster (*35*).

L. camara is potentially toxic to all ruminants grazing on it and poisoning has been reported from Australia, India, New Zealand, South Africa, and the Americas (*35,60*). Toxicity is not cumulative but occurs when enough of the plant is consumed at once. Poisonings normally occur when hungry stock are introduced to a pasture containing *L. camara* or at times when other food is scarce. Under natural conditions, *L.camara* poisoning occurs almost exclusively in cattle, although sheep and goats but not, apparently, horses are similarly susceptible to the toxins.

Pentacyclic triterpene acids, of which lantadene A and B are the most important, are the toxic components of *L. camara* (Figure 5) (*35,60*). The molecular structure of lantadene A is 22β-angeloyl-oxy-3-oxoolean-12-en-28-oic acid and lantadene B is 22β-dimethylacryloyloxy-3-oxoolean-12-en-28-oic acid (*28*). Both of these pentacyclic triterpene acids produce the same clinical disease.

Normally only a small proportion of the ingested lantadenes is absorbed, but ruminal stasis, resulting from inappetence combined with reflex inhibition due to activation of a hepatoruminal reflex, causes large amounts of toxins to be retained in the rumen which worsens the poisoning (*60*).

Fungal Pathogens

Pithomyces chartarum is a saprophytic fungus, worldwide in distribution, but more commonly found in tropical and warm temperate than in cooler zones. Under specific weather conditions, when hot dry periods are followed by warm rains and high humidity with grass minimum temperatures of 12°C or greater on two or more consecutive nights, the fungus grows rapidly and sporulates freely (*52*). The spores are quite characteristic; rough, dark, barrel-shaped, septate aleurospores, 8-20 X 10-30 μm borne singly on short conidiophores (*12*), and contain sporidesmin, the toxin causing liver damage resulting in phylloerythrin retention and photosensitization.

The disease produced, facial eczema, regularly causes severe problems in animals grazing improved pastures in New Zealand, especially in sheep and cattle, although cases in both goats and fallow deer are also often seen (*52*). Facial eczema has been also diagnosed in Australia, South Africa, the United States, Argentina, France, Uruguay and Paraguay (*14,41,52*). The incidence and severity of an outbreak of facial eczema is related to the number of spores ingested, their sporidesmin content, and the susceptibility of the animals exposed (*52*). Not all strains of *P. chartarum* produce spores containing sporidesmin, and the feasibility of using atoxigenic strains of *P. chartarum* for the biological control of toxigenic strains of *P. chartarum* has been studied. Results from a preliminary trial have shown that sporidesmin concentrations in pasture, can be reduced by the application of atoxogenic isolates of the fungus before rapid sporulation of resident *P. chartarum* occurs (*16*). However, further studies are needed to determine the effectiveness of this biocontrol method under field conditions.

P. chartarum produces several sporidesmins, sporidesmin A being the most important (*52*) and representing more than 90% of the total sporidesmins (Figure 6), the remainder being small amounts of sporidesmins B, C, D, E, F, G, H, and J. In the animal, sporidesmin undergoes a cyclic reduction/autoxidation reaction, involving the disulphide bridge, generating the toxic free-radical, superoxide which causes the intracellular lesions of the liver (*53, 54, 55, 56*). Zinc has been found to inhibit the generation of superoxide, thus protecting animals against sporidesmin toxicity. In New Zealand, zinc has been used a long time, either as a drench or in drinking water, to protect dairy cows during dangerous periods and, more recently, a bolus for slow release of zinc in the rumen has been developed to protect sheep (*57,67,72*).

Prediction of dangerous periods (From meterological observations) and counting of spores are important tools for controlling the disease. Spraying of pastures with fungicides to reduce the number of fungi and keeping the animals on pastures that are not overgrazed both

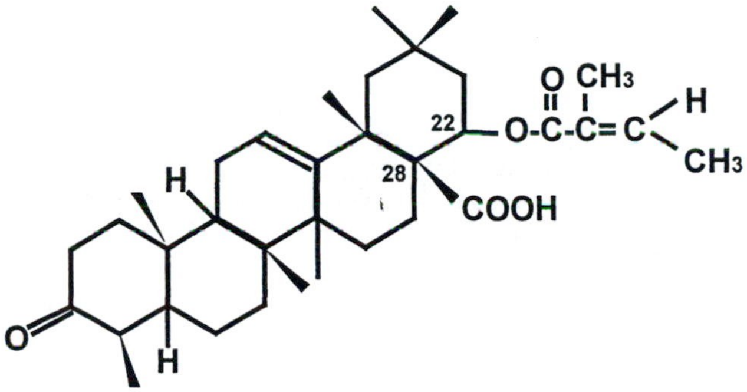

Figure 5. Chemical structure of lantadene A.

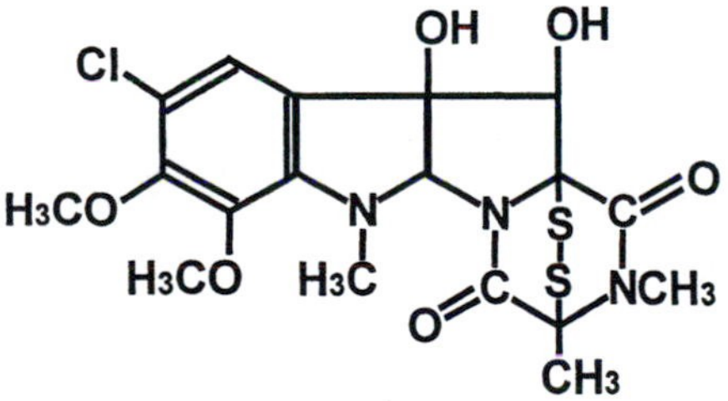

Figure 6. Chemical structure of sporidesmin A.

reduce the toxic hazard. In sheep, breeding for resistance to sporidesmin has been practiced for many years (*52*).

Phomopsis leptostromiformis is a pathogen of certain *Lupinus* spp. and can also grow saprophytically on its dead host. On the stems, pods, and seeds of infected lupin plants it can be distinguished by its stromatic pycnidia (*35,52*).

The fungus cause lupinosis, a disease most commonly affecting sheep, occasionally cattle, and rarely horses and pigs. It is also includes liver damage and it has got its name because it occurs only when animals graze on Lupins. The disease has been reported in several countries including Germany, Australia, New Zealand and South Africa (*35,52*).

A cyclic hexapeptide, Phomopsin A (Figure 7) has been identified to be the toxic principle of *P. leptostromiformis* (*9*).

Algae Pathogens

Microcystis aeruginosa is a cosmopolitan fresh water, blue-green algae known to have caused poisonings in cattle, sheep, horses, dogs, turkeys, ducks and fish (*11,35*), but ruminants are the species most commonly affected. Blue green algae are phototrophic cyanobacteria which differ from other groups of algae in that they possess no nuclear membrane. Factors promoting growth of the alga and increasing the risk of intoxications are high water temperatures, eutrophication with high water nutrient concentrations, high concentrations of electrolytes and winds which aggregate the resulting algal blooms.

The algal toxins causing the liver damage are cyclic heptapeptides (Figure 8) which are released into the water when the algae die or disintegrate (*11,35*). High doses of the toxin cause death within 24 hours, before signs of photosensitization occur.

Potential Medical Applications and Future Research Directions

Due to their photosensitizing properties, molecules of the porphyrine type are used extensively in human cancer therapy. A thorough study of the effects of phylloerythrin on skin cells under irradiation, may provide relevant results for those involved in cancer therapy, as well as further information on the pathogenesis of photosensitization diseases. A non-invasive fiberoptic method for measuring the concentration of phylloerythrin in the skin should be developed that diagnoses animals at risk of being photosensitized. These animals could be removed from the dangerous pastures and placed in the dark before clinical signs occur.

The uncertainty of the role of steroidal saponins in the pathogenesis of the liver lesions in photosentization diseases justifies a more thorough study of the hepatotoxicity of steroidal saponins. The importance of performing toxicity studies is further increased by the widespread use of steroidal saponins in the food and feed industry.

Variations, throughout and between seasons, in the concentration of saponins in grazed plants should be studied, because the plants appear at certain times to be non-toxic. A relevant study would show whether the incidence of photosentization rises in periods of high concentrations of saponins in the plants. Warning systems, based on results from chemical analyses of plants, could be develop

For those photosensitization diseases caused by known toxins, more thorough studies on the toxin's liver metabolism should be performed. The phase I- and II enzymes metabolizing the

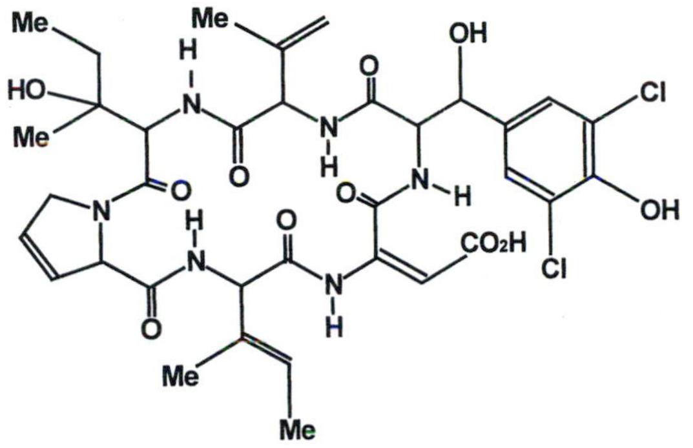

Figure 7. Chemical structure of phomopsin A.

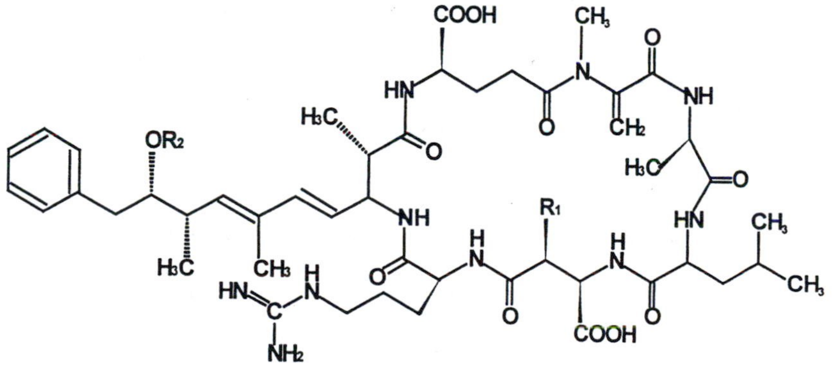

Figure 8. Chemical structure of 3 different mycrocystins.

	R_1	R_2
Structure 1:	CH_3	CH_3
Structure 2:	H	CH_3
Structure 3:	CH_3	H

Table I. Higher Plants causing Hepatogenous Photosensitization Diseases of Minor Importance

Plant	Toxic principle	Affected Species	Countries where disease has been reported
Tetradymia glabrata	Unknown	Sheep and goats	United States (*32,33*)
Tetradymia canescens	Unknown	Sheep	United States (32,33)
Myoporum aff. *insulare*	Unknown	Cattle	Australia (*31*)
Asaemia axillaris	Unknown	Sheep	South Africa (*35*)
Lasiospermum bipinnatum	Furanosesqui-terpenoids	Sheep and cattle	South Africa (*35*)
Athanasia trifurcata L.	Unknown	Sheep	South Africa (35)
Nidorella foetida	Unknown	Sheep	South Africa (66)
Kochia scoparia	Unknown	Cattle	United States (*13*)

toxins should be identified and their activities measured in animals of different age and breed. Such information could permit breeding for resistance, by selecting animals with the optimal activity of the phase I- and II enzymes involved in the metabolism and excretion of the toxins.

References

1. Abas Mazni, O.; Mohd Khusahry, Y.; Sheikh Omar, A.R. Jaundice and photosensitization in indigenous sheep of Malaysia grazing on *Brachiaria decumbens*. *Malaysian Vet.J.* **1983**, *7*, 254-263.
2. Abdelkader, S.V.; Ceh, L.; Dishington, I.W.; Hauge, J.G. Alveld-producing saponins. II. Toxicological studies. *Acta.Vet.Scand.* **1984**, *25*, 76-85.
3. Abdullah, A.S.; Noordin, M.M.; Rajion, M.A. Neurological disorders in sheep during signal grass (*Brachiaria decumbens*) toxicity. *Vet.Hum.Toxicol.* **1989**, *31*, 128-129.
4. Amjadi, A.R.; Ahourai, P.; Baharsefat, M. First report of geeldikkop in sheep in Iran. *Archives de l'Institut Razi*, **1977**, *29*, 71-78.
5. Bridges, C.H.; Camp, B.J.; Livingston, C.W.; Bailey, E.M. Kleingrass (*Panicum coloratum* L.) poisoning in sheep. *Vet.Pathol.* **1987**, *24*, 525-531.
6. Button, C.; Paynter, D.I.; Shiel, M.J.; Colson, A.R.; Paterson, P.J.; Lyford, R.L. Crystal-associated cholangiohepatopathy and photosensitisation in lambs. *Aust.Vet.J.* **1987**, *64*, 176-180.
7. Clare, N.T. *Photosensitization in diseases of domestic animals*; CAB: Farnham Royal, **1952**.
8. Coetzer, J.A.W.; Kellerman, T.S.; Sadler, W.; Bath, G.F. Photosensitivity in South Africa. V. A comparative study of the pathology of the ovine hepatogenous photosensitivity diseases, facial eczema and geeldikkop (*Tribulosis ovis*), with special reference to their pathogenesis. *Onderstepoort J.Vet.Res.* **1983**, *50*, 59-71.
9. Culvenor, C.C.J.; Cockrum, P.A.; Edgar, J.A.; Frahn, J.L.; Gorst-Allman, C.P.; Jones, A.J.; Marasas, W.F.O.; Murray, K.E.; Smith, L.W.; Steyn, P.S.; Vleggaar, R.; Wessels, P.L. Structure elucidation of phomopsin A, a novel cyclic hexapeptide mycotoxin produces by Phomopsis leptostromiformis. *J.Chem.Soc., Chem.Commun.* **1983**, 1259-1262.
10. Dahlgren, R.M.T.; Clifford, H.T.; Yeo, P.F. The families of the monocotyledons. *Structure, evolution, and taxonomy*; Springer-Verlag : Berlin, **1985**.
11. DeVries, S.E.; Namikoshi, M.; Galey, F.D.; Merritt, J.E.; Rinehart, K.L.; Beasley, V.R. Chemical study of the hepatotoxins from *Microcystis aeruginosa* collected in California. *J.Vet.Diagn.Invest.* **1993**, *5*; 409-412.
12. di Menna, M.E.; Mortimer, P.H.; White, E.P. The genus *Pithomyces*. In *Mycotoxic fungi, mycotoxins, mycotoxicoses;* Wyllie, T.D.; Morehouse L.G., Eds.; Marcel Dekker: New York, **1977**, 99-103.
13. Dickie, C.W.; James, L.F. *Kochia scoparia* poisoning in cattle. *J.Am.Vet.Med.Assoc.* **1983**, *183*, 765-768.
14. Edwards, J. Facing up to facial eczema. *J.Agric.W.Aust.* **1980**, *21*, 19-20.
15. Ender, F. Undersøkelser over alveldsykens etiologi. [Eng.: Etiological studies on "alveld"- a disease involving photosensitization and icterus in lambs]. *Nor.Vet-Med.* **1955**, *7*, 329-377.

16. Fitzgerald, J.M.; Collin, R.G.; Towers, N.R. Biological control of sporidesmin-producing strains of *Pithomyces chartarum* by biocompetitive exclusion. *Lett.Appl.Microbiol.* **1998**, *26*, 17-21.

17. Flåøyen, A. *Studies on the aetiology and pathology of alveld with some comparisons to sporidesmin intoxication.* Dr.med.vet.-thesis; Norges veterinærhøgskole: Oslo, **1993**.

18. Flåøyen, A. Do steroidal saponins have a role in hepatogenous photosensitization diseases of sheep? *Adv.Exp.Med.Biol.* **1996**, *405*, 395-403.

19. Flåøyen, A.; Frøslie, A. Photosensitization disorders. In *Handbook of plant and fungal toxicants*, D'Mello. J.P.F., Ed.; Boca Raton, Fl, **1997**, 191-204.

20. Flåøyen, A.; Hjorth Tønnesen, H.; Grønstøl, H.; Karlsen, J. Failure to induce toxicity in lambs by administering saponins from *Narthecium ossifragum*. *Vet.Res.Commun.* **1991**, *15*, 483-487.

21. Flåøyen, A.; Jóhansen, J.; Olsen, J. *Narthecium ossifragum*-associated photosensitization in sheep in the Faroe Islands. *Frodskaparrit*. **1994**, *41*, 103-106.

22. Flåøyen, A.; Smith, B.L.; Miles, C.O. An attempt to reproduce crystal-associated cholangitis in lambs by the experimental dosing of sarsasapogenin or diosgenin alone and in combination with sporidesmin. *N.Z.vet.J.* **1993** *41*, 171-174.

23. Flåøyen, A.; Wilkins, A.L. Metabolism of saponins from *Narthecium ossifragum* - a plant implicated in the aetiology of alveld, a hepatogenous photosensitization of sheep. *Vet.Res.Commun.* **1997**, *21*, 335-345.

24. Ford, E.J.H. A preliminary investigation of photosensitization in Scottish sheep. *J.Comp.Pathol.* **1964**, *74*: 37-44.

25. Glastonbury, J.R.W.; Boal, G.K. Geeldikkop in goats. *Aust.Vet.J.* **1985**, *62*, 62-63.

26. Glastonbury, J.R.W.; Doughty, F.R.; Whitaker, S.J.; Sergeant, E. A syndrome of hepatogenous photosensitisation, resembling geeldikkop, in sheep grazing *Tribulus terrestris*. *Aust.Vet.J.* **1984**, *61*, 314-316.

27. Graydon, R.J.; Hamid, H.; Zahari, P.; Gardiner, C. Photosensitisation and crystal-associated cholangiohepatopathy in sheep grazing *Brachiaria decumbens*. *Aust.vet.J.* **1991**, *68*, 234-236.

28. Hart, N.K.; Lamberton, J.A.; Sioumis, A.A.; Suares, H. New triterpenes of *Lantana camara*. A comparative study of the constituents of several taxa. *Aust.J.Chem.* **1976**, *29*, 655-671.

29. Holland, P.T.; Miles, C.O.; Mortimer, P.H.; Wilkins, A.L.; Hawkes, A.D.; Smith, B.L. Isolation of the steroidal sapogenin epismilagenin from the bile of sheep affected by *Panicum dichotomiflorum* toxicosis. *J.Agric.Food Chem.* **1991**, *39*, 1963-1965.

30. Hostettmann, K.; Marston, A. *Saponins*; Cambridge University Press: Cambridge, **1995**.

31. Jerrett, I.V.; Chinnock, R.J. Outbreaks of photosensitisation and deaths in cattle due to *Myoporum* aff. *insulare* R. Br. toxicity. *Aust.Vet.J.* **1983**, *60*, 183-186.

32. Johnson, A.E. Predisposing influence of range plants on Tetradymia-related photosensitization in sheep: Work of Drs. A. B. Clawson and W. T. Huffman. *Am.J.Vet.Res.* **1974**, *35*, 1583-1585.

33. Johnson, A.E. Tetradymia toxicity - a new look at an old problem. In *Effects of poisonous plants on livestock* Keeler, R.F.; Van Kampen, K.R.; James, L.F., Eds.; Academic Press: New York, **1978**, 209-216.

34. Johnson, A.E. Photosensitizing toxins from plants and their biological effects. In *Handbook of natural toxins. Plant and fungal toxins*; Keeler, R.F.; Tu, A.T., Eds.; Marcel Dekker : New York, 1983, 345-359.

35. Kellerman, T.S.; Coetzer, J.A.W.; Naudé, T.W. ; Kellerman TS. *Plant poisonings and mycotoxicoses of livestock in Southern Africa*; Oxford: Cape Town, **1988**.
36. Kellerman, T.S.; Erasmus, G.L.; Coetzer, J.A.W.; Brown, J.M.M.; Maartens, B.P. Photosensitivity in South Africa. VI. The experimental induction of geeldikkop in sheep with crude steroidal saponins from *Tribulus terrestris*. *Onderstepoort J.Vet.Res.* **1991**, *58*, 47-53.
37. Lajis, N.H.; Abdullah, S.H.; Salim, S.J.S.; Bremner, J.B.; Khan, M.N. Epi-Sarsasapogenin and epi-smilagenin: two sapogenins isolated from the rumen of sheep intoxicated by *Brachiaria decumbens*. *Steroids* **1993**, *58*, 387-389.
38. Lancaster, M.J.; Vit, I.; Lyford, R.L. Analysis of bile crystals from sheep grazing *Panicum schinzii* (sweet grass). *Aust.Vet.J.* **1991**, *68*, 281.
39. Lemos, R.A.; Salvador, S.C.; Nakazato, L. Photosensitization and crystal-associated cholangiohepatopathy in cattle grazing *Brachiaria decumbens* in Brazil.*Vet.Hum.Toxicol.***1997**, *39*, 376-377.
40. Low, S.G.; Bryden, W.L.; Jephcott, S.B.; Grant, I.M. Photosensitization of cattle grazing Signal grass (*Brachiaria decumbens*) in Papua New Guinea. *N.Z.Vet.J.* **1993**, *41*, 220-221.
41. Marasas, W.F.O.; Adelaar, T.F.; Kellerman, T.S.; Minné, J.A.; van Rensburg, I.B.J.; Borroughs, G.W. First report of facial eczema in sheep in South Africa. *Onderstepoort J.Vet.Res.* **1972**, *39*, 107-112.
42. Mathews, F.P. Lechuguilla (*Agave lecheguilla*) poisoning in sheep, goats, and laboratory animals. Texas Agricultural Experiment Station.1937. Bulletin 554.
43. Mathews, F.P. Poisoning in sheep and goats by sacahuiste (*Nolina texana*) buds and blooms. Texas Agricultural Experiment Station.1940. Bulletin 585.
44. McDonough, S.P.; Woodbury, A.H.; Galey, F.D.; Wilson, D.W.; East, N.; Bracken, E. Hepatogenous photosensitization of sheep in California associated with ingestion of *Tribulus terrestris* (puncture vine). *J.Vet.Diagn.Invest.* **1994**, *6*, 392-395.
45. Miles, C.O. A role for steroidal saponins in hepatogenous photosensitisation. *N.Z.Vet.J.* **1993**, *41*, 221.
46. Miles, C.O.; Munday, S.C.; Holland, P.T.; Smith, B.L.; Embling, P.P.; Wilkins, A.L. Identification of a sapogenin glucuronide in the bile of sheep affected by *Panicum dichotomiflorum* toxicosis. *N.Z.Vet.J.* **1991**, *39*, 150-152.
47. Miles, C.O.; Wilkins, A.L.; Erasmus, G.L.; Kellerman, T.S. Photosensitivity in South Africa. VII. Ovine metabolism of *Tribulus terrestris* saponins during experimentally induced geeldikkop. *Onderstepoort J. Vet.Res.* **1994**, *61*, 351-359.
48. Miles, C.O.; Wilkins, A.L.; Erasmus, G.L.; Kellerman, T.S.; Coetzer, J.A.W. Photosensitivity in South Africa. VII. Chemical composition of biliary crystals from a sheep with experimental induced geeldikkop. *Onderstepoort J. Vet.Res.* **1994**, *61*, 215-222.
49. Miles, C.O.; Wilkins, A.L.; Munday, S.C.; Flåøyen, A.; Holland, P.T.; Smith, B.L. Identification of insoluble salts of the β-D-glucuronides of episarsasapogenin and epismilagenin in the bile of lambs with alveld and examination of *Narthecium ossifragum*, *Tribulus terrestris*, and *Panicum miliaceum* for sapogenins. *J.Agric.Food Chem.* **1993**, *41*, 914-917.
50. Miles, C.O.; Wilkins, A.L.; Munday, S.C.; Holland, P.T.; Smith, B.L.; Lancaster, M.J.; Embling, P.P. Identification of the calcium salts of epismilagenin β-D- glucuronides in the

bile crystals of sheep affected by *Panicum dichotomiflorum* and *Panicum schinzii* toxicoses. *J.Agric.Food.Chem.* **1992**, *40*, 1606-1609.

51. Mimaki, Y.; Takaashi, Y.; Kuroda, M.; Sashida, Y.; Nikaido, T. Steroidal saponins from *Nolina recurvata* stems and their inhibitory activity on cyclic AMP phosphodiesterase. *Phytochemistry* **1996**, *42*, 1609-1615.

52. Mortimer, P.H.; Ronaldson, J.W. Fungal-toxin-induced photosensitization. In *Handbook of natural toxins. Plant and fungal toxins*; Keeler R.F.; Tu, A.T., Eds.; Marcel Dekker: New York, **1983**; 361-419.

53. Munday, R. Studies on the mechanism of toxicity of the mycotoxin, sporidesmin. I. Generation of superoxide radical by sporidesmin. *Chem.Biol.Interact.* **1982**, *41*, 361-374.

54. Munday, R. Studies on the mechanism of toxicity of the mycotoxin sporidesmin. 2-Evidence for intracellular generation of superoxide radical from sporidesmin. *J.Appl.Toxicol.* **1984**, *4*, 176-181.

55. Munday, R. Studies on the mechanism of toxicity of the mycotoxin sporidesmin. 3-Inhibition by metals of the generation of superoxide radical by sporidesmin. *J.Appl.Toxicol.* **1984**, *4*, 182-186.

56. Munday, R. Studies on the mechanism of toxicity of mycotoxin, sporidesmin. V. Generation of hydroxyl radical by sporidesmin. *J.Appl.Toxicol.* **1987**, *7*, 17-22.

57. Munday, R.; Thompson, A.M.; Fowke, E.A.; Wesselink, C.; McDonald, R.M.; Ford, A.J. A slow release device for facial eczema control in sheep. *N.Z.Vet.J.* **1993**, *41*, 220.

58. Opasina, B.A. Photosensitization jaundice syndrome in West African Dwarf Sheep and goats grazed on *Brachiaria decumbens*. *Tropical Grasslands* **1985**, *19*, 120-123.

59. Osbourn, A. Saponins and plant defence - a soap story. *Trends in Plant Science* **1996**, *1*, 4-9.

60. Pass, M.A. Poisoning of livestock by Lantana plants. In *Handbook of natural toxins. Toxicology of plant and fungal compounds*; Keeler, R.F.; Tu, A.T., Eds.; Marcel Dekker: New York, **1991**, 297-311.

61. Patamalai, B.; Hejtmancik, E.; Bridges, C.H.; Hill, D.W.; Camp, B.J. The isolation and identification of steroidal sapogenins in kleingrass. *Vet.Hum.Toxicol.* **1990**, *32*, 314-318.

62. Puoli, J.R.; Reid, R.L.; Belesky, D.P. Photosensitization in lambs grazing switchgrass. *Agronomy Journal* **1992**, *84*, 1077-1080.

63. Quin, J.I.; Rimington, C.; Roets, G.C.S. Studies on the photosensitisation of animals in South Africa. VIII. The biological formation of phylloerythrin in the digestive tracts of various domesticated animals. *Onderstepoort J.Vet.Sci.Anim.Ind.* **1935**, *4*, 463-471.

64. Rankins, D.L.; Smith, G.S.; Ross, T.T.; Caton, J.S.; Kloppenburg, P. Characterization of toxicosis in sheep dosed with blossoms of sacahuiste (*Nolina microcarpa*). *J.Anim.Sci.* **1993**, *71*, 2489-2498.

65. Rimington, C.; Quin, J.I. Photosensitising agent in 'geel-dikkop' phylloerythrin. *Nature.* **1933**, *132*, 178-179.

66. Schneider, D.J.; Green, J.R.; Collett, M.G. Ovine hepatogenous photosensitivity caused by the plant *Nidorella foetida* (Thunb.) dc. (asteraceae). *Onderstepoort J.Vet.Res.* **1987**, *54*, 53-57.

67. Smith, B.L.; Embling, P.P.; Towers, N.R.; Wright, D.E.; Payne, E. The protective effect of zinc sulphate in experimental sporidesmin poisoning of sheep. *N.Z.Vet.J.* **1977**, *25*, 124-127.

68. Smith, B.L.; Miles, C.O. A role for *Brachiaria decumbens* in hepatogenous photosensitization of ruminants? *Vet.Hum.Toxicol.* **1993,** *35,* 256-257.

69. Steyn, D.G. In *Vergiftiging van mens en dier*; Van Schaik: Pretoria, **1949,** 88.

70. Summerfield, R.J. Biological flora of British Isles. *J.Ecol.* **1974,** *162*, 325-339.

71. Tapia, M.O.; Giordano, M.A.; Gueper, H.G. An outbreak of hepatogenous photosensitization in sheep grazing *Tribulus terrestris* in Argentina. *Vet.Hum.Toxicol.* **1994,** *36*, 311-313.

72. Towers, N.R.; Smith, B.L. The protective effect of zinc sulphate in experimental sporidesmin intoxication of lactating dairy cows. *N.Z.Vet.J.* **1978,** *26*, 199-202.

73. Wall, M.E.; Warnock, B.H.; Willaman, J.J. Steroidal sapogenins. LXVIII. Their occurence in *Agave lecheguilla. Economic Botany* **1962,** 16, 266-269.

ANIMAL TOXINS

Chapter 16

Probing the Structure, Function, Dynamics, and Folding of Snake Venom Cardiotoxins

T. K. S. Kumar, T. Sivaraman, and C. Yu

Department of Chemistry, National Tsing Hua University,
Hsinchu 30043, Taiwan, Republic of China

Snake cardiotoxins are small molecular weight (6.5 - 7.0 kDa), highly basic (pI > 10), all β-sheet proteins. This class of toxins exhibit a wide array of interesting biological properties. The secondary structural elements in these proteins include antiparallel double and triple stranded β-sheets. Three-dimensional structures of cardiotoxins consistently reveal a cluster of cationic residues encircling, well organized hydrophobic patches. Such an asymmetric distribution of the positively charged and the non-polar residues is believed to important cytolytic activity exhibited by the class of toxins. The aim of this comprehensive review is to summarize and critically evaluate the progress made in research on the structure, function, dynamics and folding of snake venom cardiotoxins.

Venoms of snakes belonging to the elapidae family contain many toxic constituents (*1-3*). The lethality of the snake bites is primarily due to the presence of toxic principles such as the cardiotoxins and neurotoxins (*4*). Both cardiotoxins and neurotoxins are small molecular weight (6.5 - 7.0 kDa), homologous proteins (*5, 6*). Members belonging to these two classes of toxins (cardio - and neurotoxins) share more than 50% homology in their amino acid sequences (*7*). Interestingly, the three-dimensional structures of these toxins (*8*) resemble one another (Figure 1). Surprisingly, despite the structural similarities, cardiotoxins and neurotoxins exhibit drastically different biological properties (*9*). Neurotoxins block nerve transmission by selectively binding to the acetylcholine receptor at the post-synaptic junction (*4, 10*). Cardiotoxins, on the other hand, show a wide variety of biological properties like the contracture of the cardiac muscle, lysis of various types of cells such as the erythrocytes, epithelial cells, fetal lung cells and selective cytotoxicity to certain types of tumor cells, such as the Yoshida sarcoma cells (*11*). Cardiotoxins also are

Cardiotoxin III

Cobrotoxin

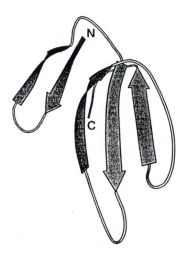

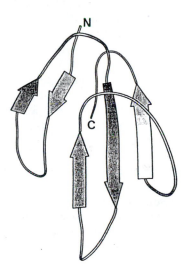

Figure 1: MOLSCRIPT representation of the backbone folding of cardiotoxin toxin analogue III and cobrotoxin (a neurotoxin analogue) isolated from the Taiwan cobra *(Naja naja atra)*. The ribbon arrows depict regions of the protein in the β-sheet conformation.

known to cause the depolarization and contracture of muscle cells and also prevent platelet aggregation (4). In addition, cardiotoxins from certain snake venom sources have been shown to inhibit the activity of key enzymes, such as the Na^+ - K^+ ATPase (12, 13) and Protein kinase C (14, 15). In the recent years, there has been intensive research to understand the structural basis for the diverse and interesting biological properties exhibited by snake venom cardiotoxins (16). In the present paper, we attempt to provide a comprehensive review of the progress made in the understanding of the structure-function relationship in cardiotoxins isolated from the Taiwan cobra (*Naja naja atra*).

Venom from the Taiwan cobra (*Naja naja atra*), is known to contain five different isoforms of cardiotoxins (17). They are highly conserved proteins and share more than 90% homology in their amino acid sequences (Figure 2). In addition to cardiotoxins, the Taiwan cobra possess a highly basic and virtually non-toxic, component with an amino acid sequence very similar to that of the cardiotoxins isoforms (18, 19). Owing to these reasons, this constituent of the venom has been labeled as the cardiotoxin-like basic protein (CLBP, 19). It is worth mentioning that CLBP lacks most of the biological properties shown by the cardiotoxin isoforms (20). Close scrutiny of the amino acid sequence of CLBP with the other cardiotoxin isoforms isolated from the same venom source (*Naja naja atra*) reveals subtle but important differences (17, 21). The cardiotoxin sequences including those isolated from the Taiwan cobra consistently reveal presence of a 'signature' tripeptide sequence, I-D-V, spanning residues 39-41 (Figure 2) in their amino acid sequences (7, 17). In contrast, CLBP lacks this signature tripeptide sequence. In addition, the amino acid sequence of CLBP shows the presence of glutamic acid at positions which is found to be conspicuously missing in the primary sequences of the cardiotoxin isoforms (Figure 2). The amino acid sequences of CLBP also shows the absence of methonine residues. In sharp contrast, all the cardiotoxin sequences reported to-date contain atleast two methionine residues and are strongly implicated in their biological activity (22, 23). Despite the significant differences in the amino acid sequences of CLBP and cardiotoxins, there are instances in the literature, wherein the CLBP's have been mistakenly classified as cardiotoxins (7, 17).

Chemistry and the Structure of Cardiotoxins

Cardiotoxins, in general, are single chain, sixty amino acid, highly basic (pI > 10.0), small molecular weight (6.5 kDa) proteins. There are four disulfide bonds located at identical positions in the sequences of the five cardiotoxins isoforms isolated from the Taiwan cobra (*Naja naja atra*). The disulfide bonds are located at positions 3-21, 14-38, 42-53 and 54-59 (Figure 2). The amino acid sequence of cardiotoxins are characterized by the absence of glutamic acid (Figure 2).

The structure of cardiotoxins have been probed by a variety of biological techniques (24-29). It should be of interest to note that three-dimensional structures of all the cardiotoxins from the Taiwan cobra (*Naja naja atra*) have been elucidated using multidimensional NMR techniques (22, 26, 30, 31).

10	20	30	40	50	60	
LKC-NKLIPI	ASKTCPAGKN	LCYKM-FMMS	DLTIPVKRGC	IDVCPKSNLL	VKYVCCNTDRCN	CTX I
LKC-NKLVPL	FYKTCPAGKN	LCYKM-FMVS	NLTVPVKRGC	IDVCPKNSAL	VKYVCCNTDRCN	CTX II
LKC-NKLVPL	FYKTCPAGKN	LCYKM-FMVA	TPKVPVKRGC	IDVCPKSSLL	VKYVCCNTDRCN	CTX III
RKC-NKLVPL	FYKTCPAGKN	LCYKM-FMVS	NLTVPVKRGC	IDVCPKNSAL	VKYVCCNTDRCN	CTX IV
LKC-NKLVPL	FYKTCPAGKN	LCYKM-FMVS	NKMVPVKRGC	IDVCPKSSLL	VKYVCCNTDRCN	CTX V
LKCHNTQLPF	IYKTCPEGKN	LCFKATLKKF	PLKFPVKRGC	ADNCPKNSAL	LKYVCCSTDKCN	CLBP

Figure 2: Representation of the amino acid sequences of the various cardiotoxin isoforms (CTX) and cardiotoxin-like basic protein (CLBP) isolated from the Taiwan cobra *(Naja naja atra)*.

Barring some subtle variations, in the orientation of a few side chain groups, the overall backbone folding of all the cardiotoxin isoforms appears to be similar. Hence, for the sake of convenience, we describe the general architecture of cardiotoxins, taking the solution structure of cardiotoxin analogue III (CTX III) isolated from the Taiwan cobra (*Naja naja atra*) as a model (*30*).

The gross three-dimensional, solution structure of CTX III depicts that the backbone of the toxin fold into three major loops projecting from a globular head (*30*). In analogy, the 3D structure presents a picture of three-fingers projecting from the palm of a hand and hence, they are aptly called the 'three finger' proteins (Figure 1). The secondary structure of the toxin (CTX III) is predominantly β-sheet, with no discernible helical segments. Secondary structure prediction analysis on the amino acid sequence reveals that CTX III has very little or no propensity to adopt helical conformation (*4*). There are five β-strands which make up the β-sheet conformation in the protein. These five β-strands are found to align themselves into double and triple stranded β-sheet segments. The double stranded β-sheet segment comprising of β-strand I (residues 1-5) and II (residues 10-14) is fairly flexible (*30*).

The triple-stranded β-sheet domain constitutes the central core of the molecule and it comprises of β-strand III (residues, 21-25), IV (residues, 35-39) and V (residues, 50-55). These three strands are placed antiparallel to each other. Equilibrium hydrogen-deuterium exchange experiments on CTX III unambiguously indicate that the triple stranded β-sheet segment is more stable than the double stranded β-sheet domain (Sivaraman, T. *et al.*, *Arch. Biochem. Biophys.*, in press). The head portions of the molecule is rigid and extensively crosslinked by four disulfide bridges. It is believed that the extraordinary structural stability of CTX III arises due to the high density of disulfide bonds in the protein (*32-34*). It should be mentioned that under neutral pH conditions, cardiotoxins in general, exhibit very high thermal stability. They are stable even upto a temperature of 70 ^{0}C (*35*). In contrast, neurotoxins despite possessing a similar three-dimensional structural fold, show much lower stability than the cardiotoxins (Sivaraman, T. *et al.*, *Arch. Biochem. Biophys.*, in press). It is believed that the disparity in the stability among the cardio- and neurotoxins stems from the differential packing of the hydrophobic residues in both the toxins (*36*). Comparison of the fractional solvent exposed surface accessible area in CTX III and cobrotoxin (a short neurotoin analogue from the Taiwan cobra (*Naja naja atra*) reveals that degree of exposure of the residues to the water molecules in CTX III is much lower when compared to that in cobrotoxin (Figure 3). This aspects implies that the three-dimensional structure of CTX III has higher spread of the non-polar residues which tend to get buried in the interior of the molecule. The net effect causes the promotion of stronger hydrophobic interactions among the buried, non-polar residues in the protein (*37*). It is well known that such interactions among the non-polar residues contributes significantly to the stability and dynamics of the protein (*37*).

The three-dimensional structure of CTX III as referred earlier, depicts the backbone of the protein to fold into three major loops, loops I, II and III. The loop I comprises of residues 2 to 14 and lodges the double stranded β-sheet segments. The studies based on cardiotoxins isolated from other snake venom sources reveal that the douple stranded, β-sheet domain is crucial for the erythrocyte lytic activity of the toxin (*4, 38-41*). The tip position of loop I is poorly defined in the three-dimensional solution structures for most of the cardiotoxins. Loop I shows an asymmetric distribution of hydrophobic and hydrophilic amino acids. It is interesting to note that most of the residues comprising the loop I are conserved and are believed to play an important role in the biological activity of cardiotoxins. The cationic residues in cardiotoxins are believed to have a major role in the cytolytic activity of the toxin (*42-44*). There are three lysine residues (Lys 2, Lys 5 and Lys 12) located in loop I. Chemical modification studies indicate that selective modification of the positive charges on Lys 12 by acetylation, knocks out most of the erythrocyte lytic activity of the cardiotoxin, implying that this residue (Lys 12) is functionally important (*39*). Since, no structural changes accompany the modification of Lys 12, it can be presumed that the positive charge on this residue (Lys 12) *per se* is a crucial factor in the lytic activity of the toxin. A similar selective modification of the positively charged side chains of Lys 2 and Lys 5 has no effect on the hemolytic activity of the cardiotoxins (*45*). The results of this study clearly indicates that Lys 12 is a functionally important residue. In addition, it is now more prevalently believed that the asymmetric distribution of the polar and non-polar residues located in loop I has a profound influence on the cytolytic activity exhibited by these toxins (*46*).

A major portions of the triple-stranded β-sheet domain resides in Loop II (Figure 1). Loop II lodges β-strands III (residues, 20-26) and IV (residues, 35-39). The solution structure of CTX III shows a typical G1 type β-bulge formed by the well conserved residues, Pro 33 and Val 34. The β-bulge seems to be a structural adjustment to compensate for the disruption of the regularity of the β-sheet due to the presence of proline at position 33. Occurrence of the β-bulge is believed to increase the twist of the triple-stranded β-sheet (*47*). It is opined that the bend by the presence of the β-sheet at the tip of loop II strongly influence the biological activity of cardiotoxins (*4*). In addition, the higher flexibility of the backbone at the distal end of the loop II is believed to produce a conducive effect for binding of these toxins to their cellular 'receptor(s)'. A lot of controversy exists on the existence of specific cardiotoxin 'receptors' (*4*). Dufton and Hider made an interesting observation that cardiotoxins binds specifically to 'band-3' protein located on the erythrocyte membrane (*7*). Similar observations were made by Condrea *et al.* (*48*) wherein the direct hemolytic factor (other name of cardiotoxins) from the Ringhal (the *Hemachetus hemachatus*) was shown to bind to limited number of sites on the RBC membrane. The results of these studies probably hint the existence of specific cellular receptors for cardiotoxins.

Loop II of the CTX III molecule consists of many residues which have been demonstrated by site-specific chemical modification to play a crucial role in biological activity of cardiotoxin (*4, 39*). Using toxin γ from *Naja melanoleuca,* Carlsson and Louw (*23*) had unambiguously demonstrated that the methionine residues in cardiotoxins are important for the lethal activity. It is reported that modification of the well conserved tyrosine at position 22, results in a drastic loss in both the structure and erythrocyte lytic activity of cardiotoxins. This possibly implies that Tyr 22 has a structure maintenance role in the toxin.

The backbone from loop II crosses over into the hind side in a right handed crossover to constitute the first half of the loop III. The loop III lodges the β-strand V (residues, 50-55) which antiparallel to the β-strand III (residues, 20-25) to complete the triple-stranded β-sheet segment (Figure 1). The disulfide link between Cys 54 and Cys 59 creates a compact auxiliary loop. This disulfide bond renders the otherwise unstructured, flexible C-terminal to be compact. Hydrogen-deuterium exchange studies in conjunction with multi-dimensional NMR techniques reveals that many of the amide protons are strongly protected from exchange (Sivaraman, T. *et al., Arch. Biochem. Biophys.,* in press). This is quite surprising bearing the fact that the residues in the C-terminal loop are involved in any secondary structural interactions. Closer examination of the three-dimensional structure revealed that the backbone of the protein spanning the C-terminal tail is involved in long-range interactions with the residues at the N-terminal end of the molecule. In addition, in the loop III, a type I-β turn is found to occur between residues 46 and 49. This β-turn is consistently found in the solution and crystal structures of all the cardiotoxin isoform(s). Recently, Sivaraman *et al.* (*49*) tracing the events in the folding pathway of CTX III found that this β-turn could be one of the earliest structure forming events during the refolding process of the protein.

Although at the gross level, the three-dimensional solution structure of CTX III appears to be flat, a clear curvature could be discerned in the molecule. Thus, the CTX III molecule has distinct concave and convex surfaces (Figure 4). The concave side is contributed by residues located at the junction of the N-and C-terminal end of the molecule. On the other hand, the convex side is located at a place wherein the backbone crosses over the β-sheet between loops II and III. The convex surface of the molecule is largely traversed by a large hydrophobic area distributed into two distinct patches (Figure 5). One of the hydrophobic patches is contributed by residues located in Loop I. This hydrophobic cluster is spatially connected through van der Waals interactions with the other hydrophobic patch located in the Loops II and III. The NOE's identified between the side chain of Leu 1 and the alpha protons of Lys 35 and Arg 36 characterize the van der Waal interactions contemplated between the hydrophobic patches. Thus, it appears that the hydrophobic patch is continuously spread on the convex surface. The hydrophobic patch located in the triple-stranded β-sheet segment basically occurs in two clusters. Each of the clusters encircle the most conserved tyrosines in the protein-namely, Tyr 22 and Tyr 51. Tyr 22 is encircled by Met 24, Gly 37, Ile 39, Pro 43 and Cys 53 . The second satellite hydrophobic cluster comprises of Met 24, Met 26, Val 32, Val 34, Gly 37 and Val 52 surrounding Tyr 51. In general, the solvent accessibility

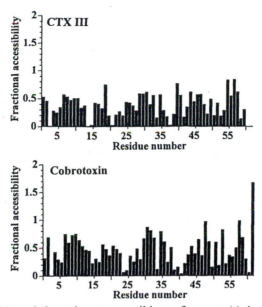

Figure 3: Plot of the solvent accessible surface area(s) in the solution structures of CTX III and Cobrotoxin.

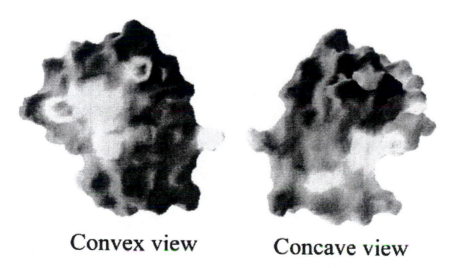

Convex view   Concave view

Figure 4: GRASP representation of the cationic residues on the convex and concave surfaces of the solution structure of CTX III. The darkened area in the figure represent the location of the cationic residues in the protein.

of the residues involved in the hydrophobic clusters is very poor. It is believed that the strong network of hydrophobic interactions together with the extensive disulfide crosslinks, bestows extraordinary structural stability to the cardiotoxins.

As mentioned earlier, the concave surface of the molecule is built-up from the residues located at the junction of the N- and C- terminal ends of the cardiotoxin molecules. The interactions among the residues in the concave surface are basically stabilized by the presence of three disulfide bonds. For example, the solution structure of CTX III indicates that the side chain of Leu 1 presents several NOE's with the residues located at the C-terminal such as Thr 56, Asp 57 and Arg 58. In addition, the side chains of the conserved Arg 36 is juxtaposed by Lys 23 and Asn 60. This ionic cluster is proximal to loop I and probably strengthens the tethering of the N- and C- terminal ends. One of the characteristic features of the three-dimensional structure of snake venom cardiotoxins has been the asymmetric spatial distribution of the polar and non-polar amino acids (44). This structural feature presents a sort of 'pseudo-amphipathic' character to the molecule. Such type of amphipathicity of the molecule is commonly seen in most of the cytolytic peptides such as mellitin, paradaxin and γ-toxin (4, 50-52). It could be discerned from the structures that the hydrophobic patches spread on the convex surface of the molecule are conspicuously surrounded by positively charged residues, for example, the hydrophobic patch in the loop I region is encircled by Lys 5, Lys 18 and Lys 35. Additionally, it is known that chemical modification of any one of these lysine residues results in substantial loss is the cytotoxic activity of the cardiotoxins (39).

Cardiotoxins, in general belong to the class A type of proteins and characteristically contain only tyrosine residues and lack in tryptophan and phenylalanine residues (7). The two conserved tyrosine residues located at positions 22 and 51 are positioned in the lower end of the molecule and lie close to each other in space (4). The phenolic rings of each of these tyrosine are tilted to about 80^0 with respect to each other in space. It is believed that tyrosine 22 is in a spatial restricted environment and thus opined to contribute to the near UV CD signal of cardiotoxins. In addition, chemical modification studies have indicated that substantial loss in activity occurs when the conserved tyrosine moieties are modified (7). However, the loss in the biological activity upon modification of the tyrosine residues is also accompanied by a substantial loss in the structure of the toxin, implying that the conserved tyrosine residues are probably involved in maintaining the structural integrity of the toxin molecule.

Venom of snakes including that of the Taiwan cobra contain multiple isoforms of cardiotoxins. However, the physiological relevance is not clearly understood. Jang et al. (31) recently addressed this question by comparing the structures and functional properties of two cardiotoxin analogues, namely, CTX II and CTX IV, isolated from Taiwan cobra (Naja naja atra). The amino acid sequences of CTX II and CTX IV are identical except for the N-terminal amino acid. The N-terminal amino acid, leucine in CTX II is replaced by a cationic residue, arginine. Comparison of the amino acid sequences of about 60 cardiotoxin sequences (known to date) shows that CTX IV is the only cardiotoxin analogue whose N-terminal amino acid is a positively charged residue (Arg). The amino acid sequences of all other cardiotoxin analogues have a hydrophobic amino acid as their N-terminal amino acid such as Leu/Ile. Jang *et al.* (*31*) compared the erythrocyte lytic activities of CTX II and CTX IV. Interestingly, it was found that erythrocyte lytic activity of CTX IV was atleast more than twice that of the lytic activity exhibited by CTX II, in the concentration range of the toxins used (Figure 6). These authors, also did a comparative study of the solution structures of these two cardiotoxin isoforms. Solution structures of CTX II and CTX IV, at a gross level, appeared to be the same (*31*). The secondary structural elements in both the toxin isoforms consisted of five β-strands arranged in the form of anti-parallel double- and triple-stranded β-sheets. The location and amino acid involved in the β-sheet region were found to be identical in both CTX II and CTX IV. The most prominent difference in the backbone folding of CTX IV and CTX II is found to occur at the N- and the C- terminal ends. The carboxyl end in CTX IV was shown too be spatially close to the amino-terminal end. This was substantiated by the present of several NOEs between the N- and C-terminal residues in CTX IV such as Asn 4NH - Cys 59NH, Cys 3Hα - Arg 58NH, and Asp 57NH - Arg 58Hα. In addition, the N- and C-terminal ends in CTX IV were found to connected *via* a salt bridge between the side chain. The guanido group of Arg 1 and the carboxylate group of Asp 57 (Figure 7). In CTX II, this crucial salt bridge was found missing (Figure 7). It was contemplated that this salt bridge aids in bringing together the N-and C-terminal ends in CTX IV. Hydrogen-deuterium amide proton exchange monitored by two-dimensional NMR techniques revealed that the amide protons of residues at the N-terminal ends of the CTX IV molecule were shown to be protected from hydrogen-deuterium exchange (*31*). A similar protection against exchange of the residues in the terminal region of the molecule were not observed in CTX II (Figure 7). The increased long range interactions observed in the solution structure of CTX IV is probably reflected in the higher thermal stability of CTX IV than that of CTX II (Figure 8). Critical evaluation of the degree of exposure of hydrophobic residues in CTX IV and CTX II owed that both the toxin isoforms showed almost similar disposition of the hydrophobic groups indicating that the observed difference in the erythrocyte lytic activity of CTX II and CTX IV could not be accounted by the disparity in the orientations of the non-polar groups in the three-dimensional structures of these toxins. Closer examination of the solution structure of CTX IV revealed that, the carbonyl and the amino groups of Lys 2 and Cys 59 (located at the C-terminal) are hydrogen bonded in the toxin molecule (*31*). As a consequence, of these long range structural interactions, a 'dense' cationic cluster comprising of Arg 1, Lys 2, Lys 5, Lys 23, Lys 50 and Arg 58 is formed in the N-terminal end of the molecule. The presence of cationic cluster is believed to be responsible for the enhanced erythrocyte lytic activity of CTX IV. Cardiotoxins are known to be membrane lytic peptides (*7*)

232

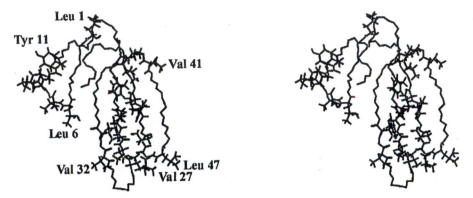

Figure 5: Depiction of the distribution of the hydrophobic 'patches' on the convex surface of the solution structure of CTX III.

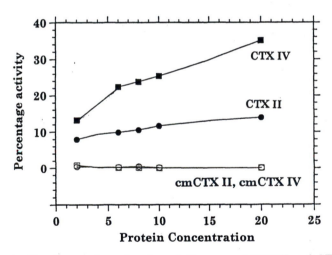

Figure 6: The dosage dependent hemolytic assay of CTX II and CTX IV. Cm CTX II and cm CTX IV represent the S-carboximethylated derivatives of CTX II and CTX IV, respectively (adapted from ref. 31).

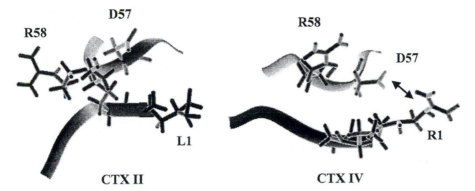

Figure 7: Representation of the salt-bridge formation between the side chain groups of Arg 1 and Asp 57 in CTX IV. The salt bridge is absent in CTX II (adapted from ref. 31).
Copyright 1997 American Chemical Society.

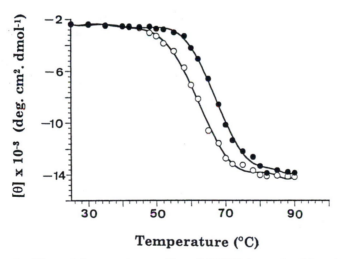

Figure 8: Thermal denaturation profiles of CTX II (open simple) and CTX IV (closed symbol) in 10 mM acetate buffer at pH 3.0 (adapted from ref. 31). Copyright 1997 American Chemical Society.

and hence they are contemplated to manifest their lytic activity by prior binding to the membrane surface. It is presumed that the 'dense' cationic cluster intensifies the binding of the toxin molecule onto the negatively charged erythrocyte membrane thereby resulting in an enhanced RBC lytic activity for CTX IV. In contrast, in favor the development of the cationic cluster (observed in CTX IV) and hence possibly accounting for the lower erythrocytes lytic activity of CTX II as compared to CTX IV.

Dynamics of Cardiotoxin Structures

Protein function is intricately linked to the molecular motion (*53-55*). For example, the recognition process involved in protein-receptor, antigen-antibody and enzyme-substrate interactions must involve protein flexibility in order to permit both fast recognition and filling to enable the appropriate conformational changes required for a function to take place (*56, 57*). Intramolecular motion can be understood using a variety of experimental and theoretical approaches. Examples of the former include techniques such as NMR spectroscopy (*58, 59*), fluorescence quenching and life time measurements (*60-63*), isotope exchange (*64, 65*), Mossbauer spectroscopy (*66*), Neutron scattering (*67, 68*) and X-ray crystallography (*69*). The theoretical methods include the use of molecular dynamics simulations and normal mode analysis (*70*). Among the various experimental methods, only the NMR and X-ray crystallography could provide insights into the molecular motions at the atomic level.

A lot of useful information exists on the internal dynamics of snake venom neurotoxins, but very little is known about the structural dynamics of snake venom cardiotoxins (*71, 72*). Recently, Lee *et al.* (*73*) using carbon-13 NMR spectroscopy at natural abundance made a through study of the backbone dynamics of a cardiotoxin analogue, CTX II, from the Taiwan cobra (*Naja naja atra*). These authors successfully assigned all the carbon-13 resonance in CTX II (*73*). This development paved way for investigating the internal dynamics of this cardiotoxin analogue (*73*).

The α-helical and β-sheet conformation in proteins could be effectively gauged from the conformation-dependent chemical shifts of the α-helical carbon atoms (*74, 75*). Using this strategy, Lee *et al.* (*73*) obtained some interesting results on the α-carbon chemical shifts of CTX II. They found that one of β-sheet segments in the protein spanning residues 35-39 (β-strand IV) exhibits a positive α-carbon secondary chemical shift instead of the anticipated negative value (Figure 9). It was convincingly argued that the anomalous secondary chemical shift values observed for the residues in the β-strand IV was due to the presence of a β-bulge just ahead of the β-strand IV (*73*). The β-bulge (as discussed earlier) is believed to distort the topology of the β-strand IV. Such structural distortions are expected to cause changes in the Ψ values to which the C^α and C^β secondary shifts are sensitive.

Investigations of the main-chain dynamics of CTX II using the T_1, T_2 and NOE values of 50 out of the 58 non-glycine residues, revealed that more than

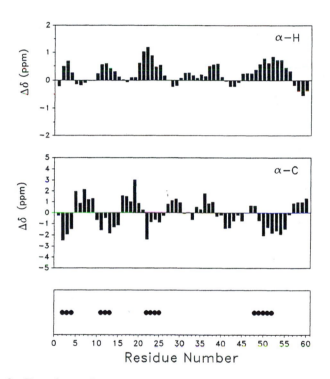

Figure 9: Experimental secondary chemical shift plots of H$^{\alpha}$ (top box) and C$^{\alpha}$ (bottom box) atoms of CTX II. Each bar indicates the secondary chemical shift value of each of these atoms in CTX II (adapted from ref. 73). Copyright 1997 American Chemical Society.

50% of the residues in CTX II exhibited very fast internal motions (<20 ps). Only Leu 30, Val 32 and Tyr 51 were found to perform slower motions in the nanosecond time scales (Figure 10). The generalized order parameters (S^2) which provides information on the relative flexibility of the residues, shows that the five β-strands, as expected, have high S^2 values (0.78 - 0.87) implying that the backbone of the protein in these regions is relatively rigid. Interestingly, the β-strands, located in the segment of residues, 20-25 (β-strand III) and 34-39 (β-strand IV) have higher S^2 values (Table I). Incidentally, these β-strands are constituents of the triple-stranded β-sheet segment. Based on the results of this study, it appears that the β-strand III and IV constituting the central are the most stable core of the cardiotoxins. The notion was confirmed by Sivaraman *et al.* (*Arch. Biochem. Biophys.*, in press) who using hydrogen-deuterium exchange technique conclusively demonstrated that the residues located in the β-strands III and IV are the most resistant to exchange and thus implying that the triple-stranded β-sheet domain is the most stable structural core of the protein. Interestingly, the partially structured states characterized along the acid and the alcohol induced unfolding pathways of CTX III (*76, 77*), showed the β-strands located between residues 20-25 and 35-39 were intact.

The backbone study on CTX II also provided useful clues on the flexibility of residues which are implicated in the biological activity of the toxin. There are many models proposed describing the interaction of cardiotoxin with the erythrocyte membrane (*4*). Interestingly, the internal dynamics study of Lee *et al.* (*73*) revealed that the average order parameter values of residues located at the tips of the three loops namely, 5-10 (Loop I), 26-33 (Loop II) and 40-49 (Loop III) are 0.78, 0.77 and 0.79, respectively (Table I). Thus, it appears that the residues located at the tips of the loops are relatively more flexible and hence could constitute the multipoint 'receptor' binding sites. It is possible that the higher flexibility of residues at the tips of the loops could have significant effect on the thermodynamics and kinetics of ligand -'receptor' binding of cardiotoxins. The higher flexibility of the residues at the tips of the three loops could facilitate an increase in the rate of association of the toxin to its 'receptor' and thus could favor complex formation by lowering the free energy barrier.

The backbone dynamics study of Lee *et al.*, also provided useful informations which helped settle long-standing debate on the functional role of a few residues (*7*). Carlson and Louw (*23*), examining the functional role of the well conserved methionine residues (Met 24 and Met 26) in snake venom cardiotoxins, found that non-selective chemical modifications of the conserved methionine resulted in the substantial loss in the lethal activity of the cardiotoxin analogue (*Naja melanoleuca*). Due to the non-selective nature of the modification reaction, it was unclear whether one or both the methionine residues are involved in the lethal activity of the toxins. Carlsson and Louw (*23*), had predicted that the more solvent accessible and flexible of the methionine residues could be responsible for the lethal action. Interestingly, the dynamic data of Lee *et al.* (*73*) showed that Met 26 (S^2 = 0.70) was more flexible than Met 24 (S^2 = 0.82). Thus, based on the prediction of Carlsson and

237

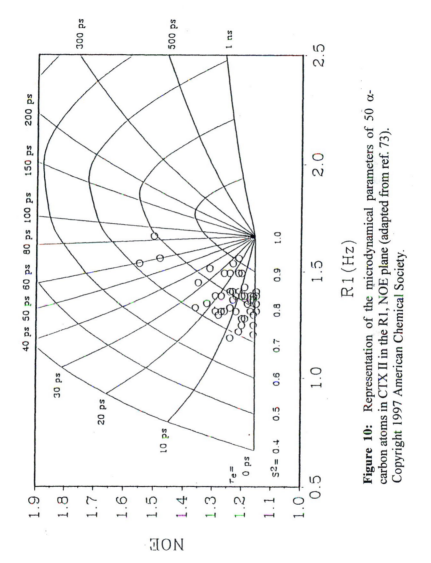

Figure 10: Representation of the microdynamical parameters of 50 α-carbon atoms in CTX II in the R1, NOE plane (adapted from ref. 73). Copyright 1997 American Chemical Society.

Table I: Average Order Parameter Values for Secondary Structure Elements

Structure elements	Residues considered	Components[*]	Average[*]
Strand I	2 – 4	[3] 0.85	
Strand II	11 – 13	[2] 0.75	
Strand III	20 – 25	[5] 0.85	[19] 0.84
Strand IV	34 – 39	[5] 0.89	
Strand V	50 – 54	[4] 0.80	
Loop I	5 – 10	[5] 0.78	
Loop II	26 – 33	[6] 0.77	[20] 0.78
Loop III	40 – 49	[9] 0.79	
N – terminal	1		[01] 0.72
C – terminal	55 - 60		[16] 0.76

[*]Number of residues for which data were available is noted in square bracket.

Louw (23), it appears that Met 26 could have a greater role in the lethal activity of the cardiotoxins than Met 24.

Mode of Action

Cardiotoxins, as stated earlier possesses a wide array of biological activities (4, 7). However, most of its actions are elicited upon binding surfaces. Although many models have been proposed to explain the cytolytic activity of cardiotoxins to-date there is no fool-proof model to explain the bio-activity of cardiotoxins. However the most popular and accepted of the proposed models is the binding and penetration model (78-80). The basic trend of the model is that the cytolytic activity of cardiotoxins is primarily promoted by the penetration of the toxin molecules into the membrane. The studies of Bougis and co-workers provide the strongest evidence for the binding and penetration model (78). Bougis et al. (78) suggest that following an electrostatic interaction step between the cationic centers in the protein molecule and the negatively charged phosphate groups on the surface of the membrane, the first hydrophobic loop of the CTX molecule penetrates the lipid phase of the membrane with the toxin molecule in an 'edgewise' orientation. The concomitant disorganization of the membrane is postulated to bring about a decrease in the surface pressure on the membrane (78). The decrease in the surface pressure is believed to promote the flip of the molecule from an 'edge wise' orientation to a 'flat' orientation. This molecular flipping is proposed to result in the amplification of the structural perturbation of the membrane resulting in the disorganization of the cell (78, 79). The weakening of the erythrocyte membrane upon cardiotoxin-erythrocyte interaction during the non-lytic period supports the flipping model proposed by Bougis et al. (78, 79). Interestingly there are various views on the extent to which the cardiotoxin penetrate into the membrane surface. It was proposed by Dufourcq et al. (81, 82) that only one of the three loops penetrate into the membrane. Loop I containing a high density of non-polar residues is proposed to be involved in membrane penetration (81). In contrast, Lauterwein et al. (83) proposed that all the three loops in the toxins are involved in the membrane insertion. They contemplate the middle loop traverses the bilayer thereby allowing the charged amino groups at the tip of the middle loop to be salt-bridged in the polar region of the lipid head groups. Thus, this model convincingly accounts for the high affinity, stoichiometry and specificity of binding. The serious demerit of the proposal of Lauterwein is the apparent disparity between the length of the hydrophobic region of the protein and the thickness of the hydrophobic sore of the lipid bilayer. Irrespective of the pros and cons of the models, it appears that cardiotoxins induce cell lysis by penetrating into the cell membrane upon initial binding. The cationic clusters on the CTX molecules appears to promote the conducive interactions for the initial biding of the toxin to the membrane.

Protein Folding Aspects

One of the challenging problems of the contemporary biochemistry and biophysics is the problem of protein folding. The astronomically large number of possible conformational states available for the protein's polypeptide chain leads us to believe that specific pathway of kinetic folding of a protein exists (84, 85). At the present juncture, it is believed that the folding of a protein is initiated by formation of isolated elements of the secondary structure which later diffuse together to adopt a kind of compact yet "molten globule" state (86, 87). Though many proteins in their "molten globule" and partially structured intermediate states have been characterized, it has not yet been possible to generalize the concept of the 'molten globule' state as an universal equilibrium protein folding intermediate(s) (88). This lacuna is primarily due to lack of evidence of the occurrence of 'molten globule' like intermediates in the folding pathway of proteins consisting of entirely of β-sheet structure. In this context, there is a spurt in the research activity focused on the investigation of protein folding of all β-sheet proteins, such as snake venom cardiotoxins (76, 77, 89). Cardiotoxins serve as ideal model proteins to understand the events in the folding/unfolding pathway(s) of all β-sheet protein. These toxins are small-sized proteins and the secondary structural elements in the protein are β-sheets, with no helical segments (4). In addition, the isolation procedures for these class of toxins are fairly simple and most importantly, the solution structures of cardiotoxins (from several venom sources) have been elucidated (4).

Recently, several partially structured intermediate(s) have been identified and characterized along unfolding pathway (s) of cardiotoxin analogue III (CTX III) from the Taiwan cobra (*Naja naja atra*) (76, 77). Jayaraman *et al.* (32) recently elucidated a partially structured intermediate in the thermal-induced. Unfolding pathway interestingly, these intermediate has been shown to be stable under high temperature conditions (90 ^{0}C). Yu and coworkers (76, 77), recently characterized the acid and alcohol-induced unfolding pathway(s) of the various cardiotoxin isoforms from the Taiwan cobra (*Naja naja atra*). These studies indicate the 2,2,2-trifluoroethanol (TFE) could induce helix-conformation in otherwise all β-sheet proteins. The results of these studies are significant because TFE per *se* is known to induce helical conformation in those protein which are either helical in the native state or have a strong sequence propensity to adopt helical conformation (90-93). In this context, it is rather surprising that the various cardiotoxin isoforms isolated from the Taiwan cobra have no helical segments in their native states and also posses no sequence propensity to adopt helical conformation (94). In this background the helical-induction by TFE appears to indicate the non-specific (helix - inducing) effect of TFE.

Sivaraman *et al.* (76) recently studied the acid-induced equilibrium unfolding pathway of CTX III from the Taiwan cobra. This study reported the identification of a stable 'molten globule' like state in the 2,2,2-Trichloroacetic acid (TCA) induced unfolding pathway of CTX III. The intermediate state was extensively characterized using two-dimensional NMR techniques (76). The protein in its 'molten globule' like state was shown to posses most of the native structural interactions. The structure of the intermediate state was found to be a stabilized by the native hydrophobic contacts. The aromatic side chain appeared to be mostly flexible as evidenced by the line broadening effects observed in the 1D ^{1}H NMR spectra of the protein (CTX III) in the 'molten globule'-like state. It was found that the triple-stranded β-sheet segment was intact in the intermediate state. However, the structural interactions stabilizing the double stranded β-sheet domain are either lost of considerable weakened in the TCA-induced 'molten globule'- like state in CTX III. It is believed that the identification and characterization of a 'molten globule' like state in the acid-induced unfolding pathway of an all β-sheet protein such as CTX III, could pave the way for the generalization of the 'molten globule' state as a universal protein-folding intermediate. Recently, partially structured intermediates with properties similar to that of a 'molten globule' state have been identified and characterized along the folding/unfolding pathways of several other β-sheet proteins (*95, 96*).

Folding of small, single domain protein is highly co-operative process (*97*). However, it is not clear at what stage the co-operativity in the folding reaction sets-in. To understand the co-operativity phenomenon observed in the folding of small proteins, characterization of early folding intermediates is a necessity (*98*). Characterization of transient intermediates become feasible with the advent of powerful techniques such as the quenched-flow deuterium-hydrogen exchange in conjunction with two-dimensional NMR experiments (*99-101*). The quenched-flow deuterium-hydrogen exchange is particularly powerful, because it permits many specific sites within a protein to be probed on the millisecond time scale (*102, 103*). Recently, this technique has been used successfully to investigate the folding pathway of several proteins (*104, 105*).

Sivaraman *et al.* (*49*), recently investigated the events in the kinetic folding pathway of cardiotoxin analogue III (CTX III) from the Taiwan cobra (*Naja naja atra*). The folding pathway of CTX III was probed using a variety of biophysical techniques including quenched-flow deuterium-hydrogen exchange. The results of this study indicated that the folding of the protein was very fast and complete within a time span of 200 ms. The folding pathway of CTX III was shown to proceed through the formation of a hydrophobic collapse. This clustering of hydrophobic amino acids was found to occur in the very early stages of folding within a time span of 20 ms (Table II). This is the first time the 'hydrophobic collapse' was observed in a very small protein such as CTX III. Sivaraman *et al.* (*49*), through the quenched-flow deuterium-hydrogen experiments unambiguously demonstrated that the segment spanning residues 51-55 (in the β-strand V), along with Lys 23, Ile 39, Val 49, Tyr 51 and Val 52 constituted the 'hydrophobic cluster' (Table II). The results of this study clearly revealed that the triple stranded β-sheet segment, which constitutes the 'central core' of the CTX III molecule folds faster than the double-stranded β-sheet segment (Figure 11).

Table II: Protection factors and Time constants of refolding residues in CTX III

Residue	Amplitude (%)	Time constant (ms)	Structural Context
K2	88.8±04.3	21.0±02.2	Strand I
K5	64.2±07.3	33.0±09.9	"
L6	96.3±02.6	120.5±08.7	Loop I
V7	76.7±05.4	52.4±10.5	"
F10	49.5±05.2	36.9±10.2	Strand II
Y11	61.9±08.3	35.2±11.7	"
K12	54.2±05.3	33.7±07.9	"
K18	91.8±06.5	80.8±12.4	Globular head
C21	79.7±03.3	27.6±02.8	Strand III
Y22	87.6±06.0	30.8±05.1	"
K23	42.4±03.2	13.5±02.0	"
A28	67.1±04.1	22.5±03.2	Loop II
T29	97.2±02.8	116.3±09.5	"
K31	71.8±07.4	55.9±13.1	"
V32	98.9±02.4	163.9±17.5	"
V34	59.6±21.4	50.5±41.3	"
K35	54.1±05.8	36.4±10.0	Strand IV
R36	64.0±04.7	32.1±06.2	"
C38	60.1±05.0	35.6±08.0	"
I39	58.2±02.1	12.4±00.9	"
V41	94.0±03.1	161.3±15.6	Loop III
K44	84.2±06.8	49.5±11.3	"
L48	94.2±24.9	129.9±69.1	"
V49	51.3±03.9	5.5±01.2	"
Y51	39.6±05.0	12.7±03.9	Strand V
V52	44.5±03.0	11.8±01.6	"
C53	37.5±03.4	13.6±02.7	"
N55	78.7±06.9	21.3±05.0	"
D57	56.7±01.3	9.8±00.5	C-Terminal
R58	38.7±04.6	12.9±03.0	"
C59	64.0±03.0	172.4±29.7	"
N60	63.9±03.8	51.8±09.4	"

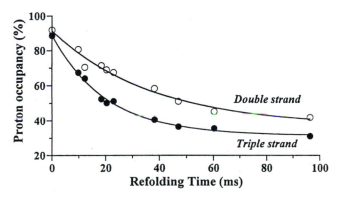

Figure 11: Time course for the protection of the amide protons from exchange of the residues involved in double- and triple stranded β-sheets during the refolding of CTX III (adapted from ref. 49).
Copyright 1998 American Society of Biological Chemistry.

Future Perspectives

Owing to the wide variety of interesting biological properties associated with snake venom cardiotoxin, they are and would remain as subject(s) of intensive research in the future. Recently, cardiotoxin analogue III (CTX III) from the Taiwan cobra has been cloned and expressed in large yields (*106*). This development is expected to pave way for the generation of site-specific mutants. The creation of site-specific mutants of cardiotoxins is expected to give strong clues on the structure-function relationship of cardiotoxins. Recently, cardiotoxin have been reported to block the activity of key enzymes such as Na^+-K^+ ATPase and Protein kinase C (*4*). Hectic research activity is expected to be triggered to understand the molecular basis for the cardiotoxin(s)-mediated enzyme(s) inhibition. Snake venom cardiotoxins have been recently demonstrated to act as potent inhibitors of platelet aggregation and co-aggregation (*107*). Owing to the immense biochemical applications, research on this aspect is expected to intensify in the coming years. Despite the small size and its all β-sheet nature, cardiotoxins, have not been popular models to understand protein folding. However, the recent report on the identification of the formation of a 'hydrophobic cluster' in the early stages of refolding of CTX III (*49*) is expected to draw the attention of the 'Protein folders' the world over. A detailed study on the understanding of the thermodynamic stability of the various kinetic and equilibrium intermediates that occur unfolding/folding pathway(s) of snake venom cardiotoxins using a appropriate site-specific mutants is expected to be carried out in the coming years. On the whole, a great deal of exciting research on snake venom cardiotoxins covering many aspects is expected in the near future.

Acknowledgments: We would like to thank all our coworkers and collaborators who have significantly contributed to research on snake venom cardiotoxins. We also express our thanks to National Science Council and Medical Research Advancement Foundations, Taiwan for funding various projects (NSC 88-2311-B007-021, NSC 88-2113-M007-028 & VGHTH-87-09-02) in our group pertaining to snake venom cardiotoxins. We also express our appreciation to the Regional Instrumentation Center at Hsinchu (Supported by the National Science Council, Taiwan) for the 600 MHz NMR facility.

Literature Cited

1. Dufton, M. J and Hider, R. C. in *Snake toxins*; Harvey, A. L., Ed.; Pergamon press, New York, **1991**; pp 259-302.
2. Harvey, A. L in *Handbook of Natural Toxins*; Tu, A. T., Ed.; Marcel Dekker, New York, **1991**; pp 85-106.
3. Harvey, A. L *Toxicol. Toxin. Rev.* **1983**, *4*, 41-69.
4. Kumar, T. K. S., Jayaraman, G., Lee, C. S., Arunkumar, A. I., Sivaraman, T., Samuel, D and C. Yu. *J. Biomolec. Stru. Dyn.* **1997**, *15*, 431-363.

5. Dufton, M. J. and Fider, R. C. J. *Toxicol. Toxin. Res.* **1988**, *4*, 41-69.
6. Dufton, M. J and Hider, R. C. *Trends in Biochem. Sci.* **1980**, *5*, 53-66.
7. Dufton, M. J and Hider, R. C. CRC *Crit. Rev. Biochem.* **1983**, *114*, 113-171.
8. Kumar, T. K. S., Lee, C. S. and Yu, C. in *Natural toxins II* (B. R. Singh and Tu, A. T. Eds), Plenum Press, New York, **1995**, 114-129.
9. Kumar, T. K. S., Pandian, S. K., Sailam, S. and Yu, C. J. *Toxicol. Toxin. Rev.,* **1998**, *17*, 183-212.
10. Betzel, C., Lange, G., Pal, G., P., Wilson, K., s., Maelicke, A. and Seagner, W. *J. Biol. Chem.* **1991**, *269*, 21530-21536.
11. Kumar, T. K. S., Pandian, S. K., Jayaraman, G. and Yu, C. *Proc. Natl. Acad. Sci (ROC),* **1999**, *23*, 1-19.
12. Bougis, P. E., Marcheot, P. and Rochat, H. *Biochemistry,* **1989**, *28*, 3637-3642.
13. Harvey, A. L., Marshall, R. J., and Karlsson, E. *Toxicon,* **1982**, *20*, 379-396.
14. Raynor, R. L., Rin, Z., Kuo, J. F. *J. Biol. Chem.,* **1986**, *266*, 2753-2758.1.
15. Chiou, S. H., Raynor, R. L., Zhang, B., Chamber, T. C and Kuo, J. F. *Biochemistry,* **1993**, *32*, 2062-2067.
16. Yu, C., Bhaskaran, R and Yang, C. C. *J. Toxicol. Toxin Rev.* **1994**, *13*, 291-315.
17. Sivaraman, T., Kumar, T. K. S., Yang, P. W. and Yu, C. *Toxicon,* **1997**, *35*, 1367-1371.
18. Takechi, M., Tanaka, Y. and Hayashi, K. *Biochem. Int.,* **1985**, *11*, 795-800.
19. Takechi, M., Tanaka, Y. and Hayashi, K. 1986, *FEBS Lett.,* **1986**, *205*, 143-146.
20. Chien, K. Y., Chiang, W. N., Jean, J. H. and Wu, W. G. *J. Biol. Chem.,* **1991**, *266*, 3232-3239.
21. Dufton, M. J. *J. Mol. Evol.,* **1984**, *20*, 128-134.
22. Sivaraman. T., Kumar, T. K. S., Huang, C. C. and Yu, C. *Biochem. Mol. Biol. Int.,* **1998**, *44*, 29-39.
23. Carlsson, F. H. and Louw, A. I, *Biochem. Biophys. Acta,* **1978**, *534*, 325-329.
24. Bilwes, A., Rees, D., Moras, D, Menez, R. and Menez, A. *J. Mol. Biol.,* **1991**, *239*, 122-136.
25. Hung, M. C. and Chen, Y. H. Intl. *J. Pept. Prot. Res.,* **1977**, *10*, 277-285.
26. Bhaskaran, R., Huang, C. C., Tsai, Y. C., Jayaraman, G., Chang, D. K. and Yu, C. *J. Biol. Chem.,* **1994**, *269*, 23500-23508.
27. Hider, R. C., Drake, A. F. and Tamiya, N. *Biopolymers,* **1988**, *27*, 113-122.
28. Bhaskaran, R., Yu, C. and Yang, C. C. *J. Prot. Chem.,* 1994, *13*, 503-504.
29. Williams, R. J. P. *Biol. Rev.,* **1979**, *54*, 389-390.
30. Bhaskaran, R., Huang, C. C., Chang, D. K. and Yu, C. *J. Mol. Biol.,* **1994**, *235*, 1291-1301.
31. Jang, J. Y., Kumar, T. K. S. Jayaraman, G. Yang, P. W and Yu, C. *Biochemistry,* **1997**, *36*, 14635-14641.
32. Jayaraman, G., Kumar, T. K. S., Sivaraman, T., Lin, W. Y., Chang, D. K. and Yu, C. *Int. J. Biol. Macromol.* **1995**, *18*, 303-306.

33. Sivaraman, T., Kumar, T. K. S. and Yu, C. *Int. J. Biol. Macromol.* **1996**, *19*, 235-239.
34. Arunkumar, A. I., Kumar, T. K. S., Jayaraman, G., Samuel, D and Yu, C. *J. Biomol. Strut. Dyn.* **1996**, *14*, 381-385.
35. Grognet, J. M., Menez, A., Drake, K., Hayashi, K., Morrisson, E. G. and Hider, R. C. *Eur. J. Biochemistry.* **1988**, *172*, 383-388.
36. Rees, B. and Bilwer, A. *Chem. Res. Toxicol.* **1993**, *6*, 385-406.
37. Behe, M. J., Lattman, E. E. and Rose, G. D. Proc. Natl. Acad. Sci. (USA). **1991**, *88*, 4195-4199.
38. Bilwes, A., Rees, B., Moras, D., Menez, R. and Menez, A. *J. Mol. Biol.* **1994**, *239*, 122-136.
39. Menez, A., Gatineau, E., Roumestand, C., Harvey, A. T., Mauwad, L., Gilquin, B. and Toma, F. *Biochemie.* **1990**, *72*, 575-588.
40. Roumestand, C., Gilquin, B., Tremeau, O., Gatineau, E., Mouwad, L., Menez, A. and Toma, F. *J. Mol. Biol.* **1994**, *243*, 719-735.
41. Marchot, P., Bougis, P. E., Ceard, B., Riefchoten, J. and Rochat, H. *Biochem. Biophys. Res. Commun.* **1988**, *153*, 642-647.
42. Shashidharan, P. and Ramachandran, L. K. *Ind. J. Biochem. Biophys.* **1986**, *29*, 232-247.
43. Kumar, T. K. S., Rao, C. N. and Reddy, R. in *Recent advances in Toxinology*; Gopalakrishnakone. P., Ed.; Venom and toxin research group, Singapore, **1991**, pp 514 -520.
44. Kini, R. M. and Evans, H. J. *Int. J. Pept. Prot. Res.* **1989**, *4*, 277-286.
45. Gilquin, B., Roumestand, C., Zinn-Justin, S., Menez, A. and Toma, F. *Biopolymers,* **1993**, *33*, 1659-1675.
46. Kini, R. M. and Evans, H. J. *Biochemistry,* **1989**, *28*, 9209-9215.
47. Richardson, J. S. *Adv. Protein. Chem.* **1981**, *34*, 167-339.
48. Condrea, E., Barzilay, M. and De Vries, A. *Naunyn Schmeidebergs. Arch. Pharmacol.* **1971**, *265*, 442-449.
49. Sivaraman, T., Kumar, T. K. S., Chang, D. K., Lin, W. Y. and Yu, C. *J. Biol. Chem.* **1998**, *273*, 10181-10189.
50. Teeter, M. M., Ma, S., Rao, U and Whitlow, M. *Proteins,* **1990**, *8*, 118-132.
51. Saberwal, G. and Nagaraj, R. *Biochim. Biophys. Acta.* **1994**, *1197*, 109-137.
52. Ho, C. L. and Hwang, L. L. *Biochem. J.* **1991**, *274*, 453-456.
53. Debrucner, P. G. and Frauenfeider, H. *Ann. Rev. Phy. Chem.* **1982**, *33*, 283-299.
54. Wodak, S. J., de Crombrugghe, M. and Janin, J. *Prog. Biophys. Mol. Biol.* **1987**, *49*, 29-63.
55. Williams, R. J. P. *Eur. J. Biochem.* **1989**, *183*, 479-497.
56. Ringe, D. and Pefsko, G. *Prog. Biophys. Mol. Biol.* **1985**, *45*, 197-235.
57. Brooks, B. R. and Karplus, M. *Proc. Natl Acad. Sci. (USA).* **1983**, *80*, 6571-6575.
58. Wagner, G. *Q. Rev. Biophys.* **1983**, *16*, 1-57.
59. Karplus, M. and Dobson, C. M. *Methods in Enzymol.* **1986**, *131*, 362-389.

60. Grallon, E. and Lakowicz, J. R. in *Structure and Motion: Nucleic acids and Proteins;* Clementi, G., Sarma, M. H. and Sarma, R. H., Ed.; Adenine press, New York, **1985**, 155-168.

61. Bucci, E. and Steiner, R. F. *Biophys. Chem.* **1988**, *30*, 199-224.

62. Merola, F., Rigler, R., Holmgren, A. and Brochon, J. C. *Biochemistry,* **1989**, *28*, 3383-3398.

63. Bismoto, E., Irace, G. and Grathan, 6. *Biochemistry,* **1989**, *28*, 1508-1512.

64. Woodward, C. K. and Hilton, B. D. *Annu. Rev. Biophys. Bioeng.* **1979**, *8*, 99-127.

65. Englander, S. W. and Kallenbalch, N. R. *Q. Rev. Biophys.* **1984**, *16*, 521-655.

66. Nadler, W. and Schulten, K. *Proc. Natl. Acad. Sci. (USA).* **1984**, *81*, 5719-5723.

67. Cussack, S., Smith, J., Finney, J., Tider, B. and Karplus, M. *J. Mol. Biol.* **1988**, *202*, 903-908.

68. Dorster, W., Cursack, S. and Petzy, W. *Nature,* **1989**, *337*, 754-756.

69. Hartmann, H., Parak, F., Steigemann, W., Petsko, G. A. and Ringe, D. *Proc. Natl. Acad. Sci. (USA).* **1982,** *79*, 4967-4971.

70. Brooks, C. L., Karplus, M. and Peltitt, B. *Adv. Chem. Phys.* **1988**, *71*, 1-259.

71. Inagaki, F., Boyd, J., Campbell, I. D., Clayden, W. J., Miyazawa, T., Hull, W. E., Tamiya, N. and Williams, R. J. P. *Eur. J. Biochem.* **1982**, *121*, 609-622.

72. Inagaki, F., Miyazawa, T. and Williams, R. J. P. *Bioscience Reports*, **1981**, *1*, 743-756.

73. Lee, C. S., Kumar, T. K. S., Cheng, J, W., Lian, W. J. and Yu, C. *Biochemistry,* **1998**, *37*, 155-164.

74. Wishart, D. S., Sykes, B. D. and Ricchards, F. M. *J. Mol. Biol.* **1991**, *222*, 311-333

75. Wishart, D. S., Sykes, B. D. *J. Biomolec. NMR.* **1994**, *4*, 171-180.

76. Sivaraman, T., Kumar, T. K. S., Jayaraman, G., Han, C. C. and Yu, C. *Biochem. J.* **1997**, *321*, 457- 464.

77. Kumar, T. K. S., Jayaraman, G., Lee, C. S., Sivaraman, T., Lin, W. Y. and Yu, C. *Biochem. Biophy. Res. Commun.* **1995**, *207*, 536-543.

78. Bougis, P. E., Tiessie, J., Rochat, H., Pieroni, G. and Verger, R. **1987,** *Biochem. Biophys. Res. Commun.* 1987, *143*, 506-511.

79. Bougis, P. E., Rochat, H., Pieroni, G. and Verger, R. *Biochemistry,* **1981**, *20*, 4915-49920.

80. Bougis, P. E., Marchot, P. and Rochat, H., *Biochemistry,* **1986**, *25*, 7235-7243

81. Dufourcq, J. and Faucan, J. *Biochemistry,* **1978**, *17*, 1170-1176.

82. Dufourcq, J., Faucan, J. F., Bernard, E., Pezolet, M., Tessier, M., Bougis, P. E., Van Rietschotenm, Delori, P. and Rochat, H. *Toxicon,* **1982**, *29*, 165-174.

83. Lauterwein, J., Lazdweski, M. and Wuthrich, K. *Eur. J. Biochem.* **1978**, *90*, 361-371.

84. Schimid, F. X. In *Protein folding*; Creighton, T. E., Ed.; Freeman Press, New York, **1992**, 197-238.

85. Christensen, H., Pain, R. H. in *Mechanisms of Protein folding;* Pain. R. H., Ed.; Oxford University press, New York, **1994**, 26-54.

86. Semisotnov, G. V., Rodionova, N. A., Razgalyaev., Uvenskey, V. N., Gripas, A. F., Gilmanshin, R. I. *Biopolymers,* **1991**, *31*, 119-128.

87. Ptitsyn, O. B., Pain, R. H., Semisotnov, G. V., Zeronik, E., Razgulyaev, O. I. *FEBS Lett.* **1990**, *262*, 20-24.

88. Barrick, D., Hughson, F. M. and Baldwin, R. L. *J. Mol. Biol.* **1994**, *237*, 588-601.

89. Sivaraman, T., Kumar, T. K. S., Jayaraman, G. and Yu, C. *J. Prot. Chem.* **1997**, *16*, 291-297.

90. Nelson, J. W. and Kallenbalch, N. R. *Biochemistry,* **1989**, *28*, 5256-5261.

91. Shiraki, K., Nishikawa, K. and Goto, Y. *J. Mol. Biol.* **1995**, *245*, 180-184.

92. Dyson, H. J., Merutka, G., Waltho, J. P., Lerner, R. A. and Wright, P. E. *J. Mol. Biol.* **1992**, *226*, 795-817.

93. Yang, J.J., Buck, M., Pitkeathly, M., Kotik, M., Haynie, D. T., dobson, C. M. and Radford, S. E. *J. Mol. Biol.* **1995**, *252*, 483-491.

94. Jayaraman, G., Kumar, T. K. S., Arunkumar, A. I. and Yu, C. *Biochem. Biophys. Res. Commun.* **1996**, *222*, 33-37.

95. Fan, P., Bracken, C. and Baum, J. *Biochemistry,* **1993**, *32*, 1573-1582.

96. Ptitsyn, O. B. in *Protein Folding*; Creighton, T. E., Ed.; W. H. Freeman and Co., New York, **1992**, *358*, pp 302-307.

97. Radford, S. E., Dobson, C. M. and Evans, P. A. *Nature,* **1992**, *358*, 302-307.

98. Kuwajima, K., Yamaya, h., Miwa, S., Sugai, S., Naggyamura, T. *FEBS Lett.,* **1987**, *221*, 115-118.

99. Kiefhaber, T. and Baldwin, R. L. *Proc. Natl. Acad. Sci., (USA).* **1995**, *92*, 2657-2661

100. Kiefhaber, T., Labhardt, A. M., Baldwin, R. L. *Nature,* **1995**, *375*, 513-515.

101. Bai, Y., Sosnick, T. R., Mayne, L. and Englander, S. W. *Science,* **1995**, *269*, 192-196.

102. Loftus, D., Gbenle, G. O., Kim, P. S. and Baldwin, R. L. *Biochemistry,* **1986**, *25*, 1428-1436.

103. Roder, H., Elove, G. A. and Englander, S. W. *Nature,* **1988**, *335*, 700-704.

104. Loh, S. N., Rohl, C. A., Kiefhaber, T. and Baldwin, R. L. *Biochemistry,* **1993**, *32*, 11022-11028.

105. Clarke, J. and Fersht, A. R. *Folding and Design,* **1998**, *1*, 243-254.

106. Kumar, T. K. S., Yang, P. W., Lin, S. H., Wu, C. Y., Lei, B., Lo, S. J., Tu, S. C. and Yu, C. *Biochem. Biophys. Res. Commun.* **1996**, *219*, 450-456.

107. Kini, R. M. and Evanns, H. J. *Toxicon,* **1990**, *28*, 1387-1422.

Chapter 17

Structures and Pharmacological Activities of Phospholipase A_2s from *Agkistrodon halys* Pallas

Z.-J. Lin[1], L. Tang[1], K.-H. Zhao[1], H.-Y. Zhao[1], X.-Q. Wang[1], W.-Y. Meng[1],
L.-L. Gui[1], S.-Y. Song[1], Y.-C. Chen[2], and Y.-C. Zhou[2]

[1]National Laboratory of Biomacromolecules, Institute of Biophysics,
Academia Sinica, Beijing 100101, China
[2]Shanghai Institute of Biochemistry, Academia Sinica, Shanghai 200031, China

In addition to esterase activity, phospholipase A_2 (PLA_2) from snake venom possesses a wide variety of pharmacological activities. The venom of *Agkistrodon halys* Pallas (i.e. *Agkistrodon blomhoffii brevicaudus*) contains three PLA_2 species, i.e. acidic, neutral and basic PLA_2s based on their isoelectric points, and they differ broadly in pharmacological effects. The neutral PLA_2 is a potent presynaptic neurotoxin, designated as agkistrodotoxin, the acidic PLA_2 displays the ability to inhibit platelet aggregation and the basic PLA_2 possesses the ability to cause hemolysis. The structures of all three PLA_2 isoforms have been determined in six crystal forms by molecular replacement to intermediate or high resolution. As expected, the overall foldings of these PLA_2 isoforms are similar to each other and to those from other species. However, the finer details of the structures and the molecular association patterns are different. The putative neurotoxic, hemolytic and inhibiting platelet aggregation sites of these PLA_2s are proposed by comparing the structures, amino-acid sequences, biochemical and pharmacological data.

Phospholipase A_2 (PLA_2) catalyzes specifically the hydrolysis of the C2 ester bond of 3-*sn*-phosphoglycerides. They are traditionally classified as intracellular or extracellular. Extracellular PLA_2s, abundant in mammalian pancreas and snake venom, have been shown to be mostly a small single-chain protein with seven disulfide bonds (MW range of 13-15 KD). These can be further classified into two groups. The venom enzymes from *Elapidae* and *Hydrophidae* belong to group I, while those from *Crotalidae* and *Viperidae* belong to group II (*1*). More than 150 amino acid sequences and about 20 structures of group I and II PLA_2s have been determined. The PLA_2 structure (Figure 1) is dominated by three long helices (1-15,

39-55, 89-110), a β-wing (74-85), the Ca^{2+}-binding loop (25-35), and a C-terminal ridge, connected by several short helices, loops or turns (2,3). A catalytic triad consisting of Asp99, His48 and a water molecule hydrogen bonding to Nδ atom of His48, were found in almost every known PLA_2 structures. The catalytic mechanism of PLA_2 has been proposed based on the structural information (4,5). The enzyme displays much higher hydrolytic activity towards aggregated phospholipids at lipid/water interface (6). It was also demonstrated that in addition to catalytic site, the enzyme possesses a topographically distinct region for the interaction with lipid/water interfaces, the so-called interface recognition site, IRS (5).

In addition to esterase activity, snake venom phospholipases A_2 have been shown to possess a wide variety of pharmacological activities, such as neurotoxic, cardiotoxic, hemolytic, anticoagulant, and myotoxic actions (7). It was found that each PLA_2 can exhibit more than one pharmacological activity, and PLA_2s from different venom or even same venom can display pharmacological activities of different types or potencies. A small protein with a high degree of homology in sequence and structure can display diverse pharmacological activities, indicating that the relationship between structure and pharmacological activity of this enzyme is complicated.

Recently, several structures of snake venom phospholipase A_2 with specific pharmacological activities were determined by X-ray crystallography in other laboratories. This includes myotoxin from *Bothrops asper* (3), presynaptic neurotoxins from the Australian tiger snake (notexin, 8), *Bungarus multicinctus* (β-bungarotoxin, 9) and *Vipera ammodytes ammodytes* (vipoxin, 10,11) and anticoagulant PLA_2 from *Vipera russelli russelli* (12). A series of structures of PLA_2 isoforms from *Agikistrodon halys* Pallas have been determined in our laboratory for correlating the structural basis for neurotoxic, platelet and hemolytic activities.

PLA$_2$s from *Agkistrodon halys* Pallas

The snake, *Agkistrodon halys* Pallas or *Agkistrodon blomhoffii* brevicaudus lives in Jiangsu and Zhejiang province of China. Phospholipase A_2 from this venom has been chosen for the study because it has been well isolated, purified and characterized by biochemists and physiologists in Shanghai of China as well as in foreign countries (13-17).

The venom contains three homologous PLA_2s (13). They were designated as acidic, neutral and basic PLA_2s according to their isoelectric points 4.5, 6.9, and 9.3, respectively. The sequence homology between these isoenzymes ranges from 51.6 % to 62.3 % (Table 1). They show different enzymatic properties and toxicities. The lethal activities of these isoenzymes do not correlate with their enzymatic activities. Among them, the neutral enzyme has the highest toxicity (LD_{50} =0.055mg/kg mice), thus it is the major toxic component of the venom and designated as agkistrodotoxin (ATX) by Prof. Ke Xu in Shanghai Institute of Physiology (18). The three isoenzymes also show quite different pharmacological activities: the acidic PLA_2 inhibits platelet aggregation, the basic PLA_2 causes hemolysis, and the neutral PLA_2 possesses presynaptic neurotoxicity. Agkistrodotoxin has some distinctive features among the presynaptic neurotoxins (15). It is the first reported instance in which a single chain, non-basic PLA_2 exhibits potent presynaptic neurotoxicity.

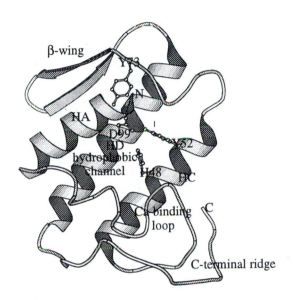

Figure 1. Overall structure of a typical group II PLA$_2$

Table I. Sequence comparison of the PLA$_2$s from *Agkistrodon halys* Pallas

	1	10	20	30	40	50	60	70
BPLA$_2$[a]	HLLQFRKMIK-KMTGKEPVVSYAFYGCYCGSGGRGKPKDATDRCCFVHDCCYEKVT--G-C-----DPKW
APLA$_2$	S.I...TL.M-.VAK.SGMFW.SN......W.'...R.Q.............G...--.-.-----...M
NPLA$_2$	S.L..N....-EE...NAIPF.................G...........GRLV--N-.-----NT.S

	71	80	90	100	110	120	130
BPLA$_2$	DDYTFSWEN.TIVCEG-DDPCKKEVCECDKAAAICFRDNLKTYKKRKYAAPG-IL.S SKSEKC
APLA$_2$	.V.SF.E.N.D...E.-.............R..........TL.NDK..W.F.AKN.PQEE..P.
NPLA$_2$	.I.S..LKE.Y.T.GK-GTN.EEQI....RV..E...R..D...NGYMFYRD-SK.T .T..E.

[a] BPLA$_2$, APLA$_2$ and NPLA$_2$ indicate the basic, acidic and neutral PLA$_2$ enzymes, respectively. Data taken from references (*13*), (*15*), and (*16*).

Crotoxin, a well-known presynaptic neurotoxin from *Crotalus durissus terrificus,* is a heterodimer of a basic PLA_2 type subunit with weak neurotoxicity and a chaperonic acidic subunit with no toxicity and no enzymatic activity (*19*). It is of interest that neutral PLA_2 is highly homologous with the basic subunit of crotoxin (sequence identity of 80%), while acidic PLA_2 is homologous with the precursor of the acidic subunit of crotoxin (*15,19*).

Structure determination by X-ray crystallography

The PLA_2 isoenzymes from *Agkistrodon halys* Pallas have been crystallized in several crystal forms. The photographs of these crystals are shown in Figure 2. The crystallization conditions and crystal parameters for six crystal forms (*20-24*) are listed in Table II. As can be seen from the Table, these crystallization conditions significantly differ from each other. Although the sequences and structures of these PLA_2 isoforms are highly homologous with each other, the structural details are different, especially on the molecular surface, thus resulting in different crystal forms. In the cases of neutral and basic enzymes, two or three crystal forms were obtained in different crystallization conditions, and these crystal forms contain multiple copies of the molecule in the asymmetric unit of the unit cell.

The crystal structures of all six crystal forms have been determined by the molecular replacement technique to high or medium resolutions. The atomic coordinates of four structures have been deposited with the Brookhaven Protein Data Bank (entry codes: 1psj for the acidic PLA_2, 1jia for $P2_12_12_1$ of the basic PLA_2, 1a2a for $P2_1$ of ATX, and 1bjj for $P2_12_12_1$ of ATX, respectively). Some of them have been published (*25-27*). The first structure determination of agkistrodotoxin ($P2_1$ crystal form) was delayed for several years due to the complexity of the structure (presence of eight molecules in the asymmetric unit). This difficulty was solved by using the pseudo-symmetry of the crystal and painstakingly designing the search model with the help of program *AMoRe* (*28*). The final atomic models of all six crystal forms have acceptable crystallographic *R*-factors, reasonable stereochemistry, and good electron density. Table III shows that the root-mean-square deviations of Cα atom positions and the correlation of the structural similarity with the sequence homology between any two isoforms. The low Cα root-mean-square deviations (range from 0.8Å to 1.5Å) and good correlation indicate the quality of the models of these three PLA_2 isoforms. The structural analyses and comparisons of these PLA_2s from *Agkistrodon halys* Pallas in different crystal forms may improve our understanding of the structural features and structure-pharmacological activity relationships of the enzyme.

As expected, the overall folding of the three isoforms is similar to each other and to other group II PLA_2s such as PLA_2 from *Crotalus atrox* (*29*). The three long α-helices, catalytic site, Ca^{2+}-binding site, etc. show greatest structural conservation. Atomic root-mean-square diviations of four essential residues His48, Asp99, Tyr73 and Tyr52 in the catalytic site are within 0.90Å. In basic PLA_2, in spite of the enhancement of positive charges at N-terminal (His1, Arg6, Lys7, Lys10, Lys11), the N-terminal hydrogen bonding network and α-helix remains perfect (*26*). However the finer details of the structure are different. As shown in Figure 3, two significant

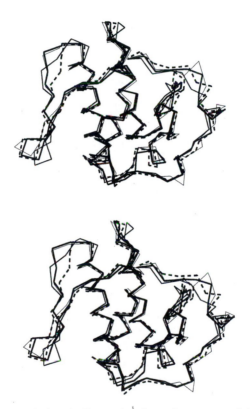

Figure 2. Crystals of phospholipase A_2s from the venom of *Agkistrodon halys* Pallas. A) acidic PLA_2; B) and C) monoclinic and orthorhombic crystal forms of neutral PLA_2; D) and E) monoclinic and orthorhombic crystal forms of basic PLA_2, respectively.

Table II. The crystallization conditions and crystal parameters of the three PLA₂s in six crystal forms.

Species	Acidic	Basic		Neutral		
Space group	$P6_1$	$P2_12_12_1$	$C2$	$P2_1$	$P2_12_12_1$	$R3$
Crystallizing	Na(CH₃)₂AsO₂	CHES	Tris-HCl	NaAc-HAc	Tris-HCl	Tris-HCl
Buffer	pH6.0	pH9.5	pH8.5	pH4.5	pH8.5	pH8.5
Precipitant	2,5-hexanediol	PEG4K	PEG4K	MPD	MPD	MPD
a/Å	83.57	97.13	100.45	108.46	87.75	125.20
b/Å	83.57	103.69	54.29	84.86	105.80	125.20
c/Å	32.72	23.27	108.39	70.82	110.30	49.69
β			111.76°	109.87°		
No. of mol per a.u.	1	2	4	8	6	2
Resolution/Å	2.0	2.0	2.6	2.6	2.8	2.8
Ref.	22	23	24	26	20	-

Table III. The correlation between Cα rms deviations and sequence homology of three PLA₂s

	APLA₂	NPLA₂	BPLA₂
APLA₂	--	51.6%[a]	62.3%
NPLA₂	1.365Å	--	59.0%
BPLA₂	0.795Å	1.053Å	--

[a] Structure deviations were calculated by least square fit of Cα atoms from residue 1 to residue 114.

conformational differences occur in β-wing (74-85) and in C-terminal ridge. In β-wing, the difference may be a reflection of the flexibility of this region, where no covalent link between the wing and the main-body of the molecule in group II PLA$_2$s was observed; and in C-terminal ridge, the difference is apparently due to two extra insertions in acidic PLA$_2$. The differences in other sites may have relevance to the pharmacological activities, which will be discussed below.

Putative pharmacological sites

As a small protein with similar structural scaffold, the presence of diverse pharmacological activities in PLA$_2$ should be related to the finer differences of the molecular structure. It was suggested by Kini & Evans (*30*) that in addition to the catalytic site, there are pharmacological sites on the surface of PLA$_2$ molecule. These pharmacological sites are usually separated from the catalytic site; however, in some cases, they overlap with each other. With these pharmacological sites the enzymes recognize and bind to the target sites of target molecules (membrane proteins or glycoproteins) in a particular organ or tissue, then induce diverse pharmacological effects. In some cases, PLA$_2$ activity is required for expressing the pharmacological activity, while it is not required in other cases. The model can explain the diversity of pharmacological activity of PLA$_2$.

Based on the data of structures, sequences, chemical modifications and mutations of PLA$_2$ isoforms from *Agkistrodon halys* Pallas, we proposed some putative pharmacological sites (*25-27*) as indicated in Figure 4.

Site for inhibiting platelet aggregation. Acidic PLA$_2$ has inhibitory effect on blood platelet aggregation caused by several inducers. It was suggested that the acidic PLA$_2$ has interaction with platelet membrane, leading to the increase of cyclic AMP level, then affecting the normal function of cytoskeleton and an inhibition on platelet aggregation (*31*).

It was found, by structure analysis of acidic PLA$_2$, that an aromatic cluster occurs on a face of the molecule. The cluster comprises Phe20 ,Trp21, Tyr113 and Trp119 and two acidic residues (Glu6 and Asp115) surrounding the cluster. This unique structure does not overlap with the catalytic site, but does slightly with interface recognition site. We propose that this unusual structure may be responsible for inhibiting platelet aggregation for the following reasons (*25*): (1) The specific sequence forming such unique structure is absent in neutral and basic PLA$_2$s without platelet activity. However it is in common in two acidic PLA$_2$s from other species possessing platelet activity with the exception of the acidic residue Asp115, which is displayed from Asp115 to Asp114; (2) The site contains two acidic residues. This is consistent with the observation that most PLA$_2$s with activity of inhibiting platelet aggregation are acidic proteins and thus negatively charged groups may be important for the platelet function; Recently, it has been found that the mutation of the two acidic residues completely inhibited the platelet activity of the acidic PLA$_2$ (Zhou, Y.-C., et al., unpublished result).

Hemolytic site.   Phospholipase A$_2$s show a great variety of behavior against

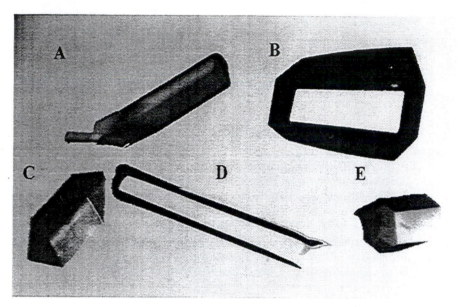

Figure 3. Cα backbone superposition of three PLA₂ isoforms from *Agkistrodon halys* Pallas and PLA₂ from *Crotalus. atrox*

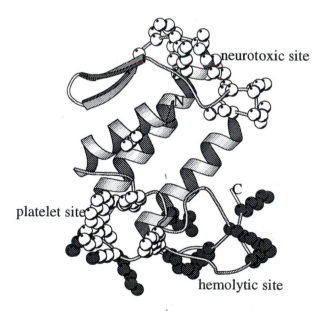

Figure 4. Summary of three pharmacological sites

biological membranes and model systems. Some PLA$_2$s can hydrolyze the phospholipids of the intact erythrocyte membrane which can lead to hemolysis of the cell (*32*). Some others cannot hydrolyze this kind of phospholipids. Among venom PLA$_2$s with hemolytic activity, the potency of hemolytic activity differs significantly from one enzyme to the other. The difference in the potency may result from the ability of the binding of the enzyme to the receptor on the membrane followed by enzymatic hydrolysis of the membrane phospholipids. The former may be related to the structural features of the hemolytic site in PLA$_2$ molecule. The hemolytic property is unique for the basic PLA$_2$ from *Agkistrodon halys* Pallas, whereas the acidic, neutral and basic-acidic PLA$_2$s (a new protein cloned and expressed by a natural mutant, Pan, H. et al., unpublished result) from the same venom show no such activity. Recently it was shown that when lysine and arginine were modified by cyclohexanedione and maleic anhydride respectively, the hemolytic activity of the basic PLA$_2$ would only retain 40%, indicating further that the positively charged residues are required for the pharmacological activity (Zhou, Y. -C., et al., unpublished data).

The structure analysis of basic PLA$_2$ reveals two specific basic-residue-rich surfaces: one is at the N-terminal of the enzyme, and the other at the C-terminal (*26*). Only basic residues Lys111, Lys114, Lys115, Lys129 and Lys132 as well as Arg34 are unique for basic PLA$_2$ from sequence comparison of the isoenzymes of this venom (Pan, H., et al., unpublished data). They are distributed over the two ends of the C-terminal basic-residue-rich specific surface. Three other basic residues (Lys36, Lys38 and Arg43) join the two parts, forming a continually positively charged surface which is remote from IRS. It was proposed that this specific face may be responsible for the hemolytic activity (Zhao, K. -H., et al., unpublished results).

It was demonstrated by Forst, et al., (*33*) that, in contrast to most of other PLA$_2$s, a few basic PLA$_2$s such as those from *Agkistrodon halys* blomhoffii and *Agkistrodon halys* Pallas have a special hydrolytic property: the ability to degrade the phospholipids of *E. coli* in the presence of a bactericidal/permeability-increasing (BPI) protein and that the unusual concentration of basic residues in functionally important N-terminal -helix may be responsible for this specific property (*34*). The crystallographic analysis of basic PLA$_2$ from *Agkistrodon halys* Pallas reveals the location of these basic residues (Lys7,10,11,16) to be close to the IRS. They are exposed to the solvent environment and do not interact with the other structural elements of the molecule, thus permitting interactions with *E. coli* membrane as well as the activator agent (e.g. BPI) that facilitates enzyme action.

Neurotoxic site. Agkistrodotoxin causes a typical triphasic modulation of neurotransmitter (acetylcholine) release on isolated nerve-muscle preparations, i.e. initial depression, subsequent facilitation and final blockade of neurotransmission. It was found (*27*) that two spatially adjacent regions of agkistrodontoxin molecules, turn 55-61 and stretch 85-91, are remarkably different from those of the two non-neurotoxic isoforms (3.61 and 2.96 for C backbone comparison in these two regions). In agkistrodotoxin, stretch 85-91 forms a compactly organized local structure, which is sharply different from the loosely extended conformations in non-neurotoxic PLA$_2$s, and segment 55-61 forms a unique turn, where a random conformation exists in non-neurotoxic PLA$_2$s.

This remarkable conformational changes correspond with the notable sequence changes. For examples, when group II neurotoxic PLA$_2$s are compared with non-neurotoxic PLA$_2$s, we can see the changes from lysine--glycine to glycine--aspardic acid in two consecutive positions 86 and 88, and the changes from acidic residues to basic residues at positions immediately following the stretch, i.e. 92 and 93. Accordingly, the electrostatic distribution in the regions is quite different from agkistrodontoxin to non-neurotoxic isoforms. In conclusion, these sequence, conformational and charge differences may be contributing to the neurotoxicity. We propose that these two spatially adjacent regions of agkistrodontoxin molecules, turn 55-61 and stretch 85-91, form the neurotoxic site, functioning as recognizing and binding site for the receptor on the presynaptic membrane.

Lys86 has been predicted as one of important residues involved in presynaptic neurotoxicity of group II PLA$_2$ (15). A natural isotoxin of agkistrodotoxin isolated from the same source mutates only at the two positions: the substitution of proline for valine 56 and the substitution of tryptophan for aspartigine 90, but its neurotoxicity remains one seventh of the original (35). It supports the proposal of the neurotoxic site of agkistrodotoxin.

It was reported (8) that the structure of notexin, a presynaptic neurotoxin belonging to single chain group I PLA$_2$s, deviates from non-neurotoxic PLA$_2$s at residues 86-89, which is located within the putative neurotoxic site of agkistrodotoxin. This suggests a possibility for a common location of neurotoxic site on the molecular surface of single chain presynaptic neurotoxins. However, the structural features observed in agkistrodotoxin differ from the others, which may cause different neurotoxins to recognize different receptors on the presynaptic membrane.

It was reported that the sequence comparison between ammodytoxin isoforms and antipeptide antibody experiments indicate that C-terminal part of ammodytoxin A and subunit B of crotoxin is likely to be involved in the neurotoxic action (36). In agkistrodotoxin, the C-terminal ridge does not show significant conformational differences from those of non-toxic basic PLA$_2$ or C. atrox PLA$_2$ (Figure 3, the unique conformation in acidic PLA$_2$ results from two insertions in this region).

Further structural, biochemical and pharmacological studies on venom PLA$_2$s and their receptors are required to confirm the proposed pharmacological sites.

The intermolecular interactions in the crystals

Neutral PLA$_2$ and basic PLA$_2$ have been shown to have a tendency for aggregation. The measurements of the molecular weights in the crystallization conditions support the presence of dimers in the two enzymes (20, Zhao, K. -H., et al., unpublished result). The presence of multicopies of molecule in the asymmetric unit of the two enzymes is consistent with this tendency. The structure analyses reported here reveal the molecular packing patterns in the asymmetric units of the crystals.

In neutral PLA$_2$, eight molecules in $P2_1$ form two tetramers or two "dimer of dimer" structure (27). The intermolecular interactions within the dimer are extensive and asymmetric, including as many as 16 hydrogen bonds, involving the two long - helices and -wing. A unique four-stranded antiparrallel -sheet was formed by two juxtaposed β-wings from two dimers with two pairs of mainchain-mainchain interactions. The C-terminal carboxylate group makes a salt bridge with basic residue Arg54. The six molecules in $P2_12_12_1$ form three Ca^{2+} linked dimers (Tang, L., et al., unpublished result). The two molecules in $R3$ form a similar Ca^{2+} linked dimer (Tang, L., et al., Photon Factory Activity Report, 1997, 106). In spite of different association modes observed in three crystal forms of agkistrodotoxin, a common packing mode remains unchanged, which involves the hydrophobic interactions between the hydrophobic patches at the interface recognition site (IRS), consisting of Ile19, Pro20, Phe21, Ala23, Phe24, Met118, and Phe119. In basic PLA_2, the two molecules in $P2_12_12_1$ form a dimer, and the four molecules in $C2$ form "dimer of dimers" structure with a dimer of different type (Zhao, K. -H., et al., unpublished result). The dimer in $C2$ seems to be more stable, and may exist in the solution.

The presence of invariant hydrophobic patch interactions in agkistrodotoxin and the formation of dimer in basic PLA_2 may have some biological implications. The phospholipid is almost ubiquitous in the cell, thus some mechanisms may exist to avoid the interaction of non-specific substrates with the enzyme (9). The substrate binding sites that are occluded by oligomeric interfaces have been observed in the crystal structures of several venom phospholipase A_2 (27-29). This occlusion may provide a way to avoid the binding of PLA_2 to non-target membrane before the enzyme diffuses from the injection place to the target sites. The structure analyses reported here provide new examples for self-protection mechanism of this kind. The contact between the hydrophobic patches in agkistrodotoxin and the dimer interface in basic PLA_2 may partially occlude the IRS, the site for interaction with aggregated substrates.

pBPB modified PLA₂ structure and its implication

Venom PLA_2s display both enzymatic and pharmacological activities. In some cases, the enzymatic activity is required for expressing pharmacological activity. For assessing the role of enzymatic activity in pharmacological effect, the inhibition of the enzyme active center (His48) by pBPB (p-bromo-phenacyl-bromide) is frequently used (37). If the inhibition leads to the depression of pharmacological activity, it means that the catalytic activity plays some roles in pharmacological effect. However, the decrease in pharmacological activity may also be due to the conformational changes caused by the pBPB modification. In order to resolve this issue, we determined acidic PLA_2 structure complexed with pBPB and compared with native one (38). The results reveal that the inhibitor molecule fits well with the active site and hydrophobic channel in conformation and hydrophobicity, and the pBPB modification does not lead to observable conformational changes on the surface of the molecule, where the pharmacological sites are located (least square superposition of all C atoms of native and modified acidic PLA_2s results in root-mean-square deviations of only 0.243). In other words, His48 pBPB modification of PLA_2 does not lead to significant conformational changes at the pharmacological site of the enzyme. Hence, the depression of a pharmacological activity of PLA_2 by pBPB inhibition may indicate the dependence of a pharmacological activity on the enzymatic activity, rather than the conformational change at the pharmacological site.

This conclusion may be applicable to the isozymes from same source and some other snake venom PLA_2s which have structure features common to acidic PLA_2 .

Acknowledgments

The research grants were supported by the Chinese Academy of Sciences and China National Natural Science Foundation. The travel grant for attending Third International Symposium on Natural Toxins at the 216th ACS meeting was provided by China National Natural Science Foundation.

Literature Cited

1. Dennis, E.A. *TIBS* . 1997, *22*, 1-2.
2. Renetseder, R., Brunie, S., Dijkstra, B.W., Drenth, J. & Sigler, P.B. *J. Biol. Chem.* 1985, *260*, 11627-11634.
3. Arni, R.K. & Ward, R.J. *Toxicon* 1996, *34*, 827-841.
4. Scott, D. L., White, S. P., Otwinowski, Z., Yuan, W., Gelb, M.H. & Sigler, P.B. *Science* 1990, *250*, 1541-1546.
5. Dijkstra, B. W., Drenth, J. & Kalk, K. H. *Nature* 1981, 289, 604-606.
6. Heinrikson, R.L. & Kezdy, F.J. *Adv. Exp. Med. Biol.* 1990, *279*, 37-47.
7. Yang, C.-C. *J. Toxicol.-Toxin Reviews* 1994, *13*, 125-177.
8. Westerlund, B., Nordlund, P., Uhlin, U., Eaker, D. & Eklund, H. *FEBS Letters* 1992, *301*, 159-164.
9. Kwong, P. D., McDonald, N. Q., Sigler, P.B. & Hendrickson, W. A. *Structure* 1995, *3*, 1109-1119.
10. Devedjiev, Y., Popov, A., Atanasov, B. & Bartunik, H. D. *J. Mol. Biol.* 1997, *266*, 160-172.
11. Perbandt, M., Wilson, J.C., Mancheva, I., Aleksiev, B., Genov, N., Eschenburg, S., Willingmann, P., Weber, W., Singh, T.P. & Betzel, C.H. *FEBS Letters* 1997, *412*, 573-577.
12. Carredando, E., Westerlund, B., Persson, B., Saarinen, M., Ramaswamy, S., Eaker, D. & Eklund, H. *Toxin* 1998, *36*, 75-92.
13. Chen, Y.-C., Maraganore, J.M., Reardon, I., & Heinrikson, R.L. *Toxicon* 1987, *25*, 401-409.
14. Chen, Y.-C., Wu, X.-F., Zhang, J.-K., Jiang, M.-S. & Hsu, K. *Acta. Biochim. Biophys. Sinica* 1981,*13*, 205-212.
15. Kondo, K., Zhang, J.-K., Xu, K. & Kagamiyama, H. *J. Biochem.* 1989, *105*,196-203.
16. Pan, H., Ouyang, L., Yang, G., Zhou, Y.-C & Wu, X.-F *Acta. Biochim. Biophys. Sinica* 1996, *28*, 579-582.
17. Pan, H., Liu, X.-L., Ou-Yang, L.-L., Yang, G.-Z., Zhou, Y.-C., Li, Z.-P. & Wu, X.-F. *Toxicon* 1998, *36*,1155-1163.
18. Xu, K. *Chinese Sci. Bull.* 1990, *35*,1-7
19. Chang, C.-C. *Proc. Natl. Sci. Counc. ROC.* 1985, *B9*, 126-142.
20. Jin, L., Gui, L.-L., Bi, R.-C., Lin, Z.-J. & Chen, Y.-C. *Chinese Journal of Biochemistry and Biophysics.*1990, *23*, 270-276.

21. Li, D.-N., Gui, L.-L., Song, S.-Y., Lin, Z.-J., Qian, R. & Zhou, Y.-C. *Chinese Sci. Bull.*1995, *40*, 664-667.
22. Gui, L.-L., Niu, X.-T., Bi, R.-C., Lin Z.-J., & Chen, Y.-C. *Chinese Sci. Bull.* 1992, *37*, 1394-1396
23. Meng, W.-Y., Niu, X.-T., Gui, L.-L., Lin, Z.-J., Zhu H., Zhou Y.-C. & Li G.-P. *Acta Physico-Chimica Sinica* 1996, *12*, 946-949.
24. Niu X.-T., Meng, W.-Y., Gui L.-L., Wang X.-Q., Lin Z.-J., Gu, P.-G. & Zhou, Y.-C. *Acta Biochem. Biophys. Sinica* 1996, *28*, 206-209.
25. Wang, X.Q., Yang, J., Gui, L.-L., Zhou, Y.-C. & Lin, Z.-J. *J. Mol. Biol.* 1996, *255*, 669-676.
26. Zhao, K.-H., Song, S.-Y. , Lin Z.-J. & Zhou Y.C. *Acta Cryst.,* 1998, D54, 510-521.
27. Tang, L., Zhou, Y.-C. & Lin Z.-J. *J.Mol.Biol.,* 1998, 282,1-11.
28. Tang, L., Lin Z.-J. & Zhou Y.-C. *Science in China (Series C)* 1997, *40*, 481-487.
29. Brunie, S., Bolin, J., Gewirth, D. & Sigler, P.B. *J. Biol. Chem.* 1985, *260*, 9742-9749.
30. Kini, R.M. & Evans, H.J. *Toxicon,* 1989, *27*, 613-635.
31. Zhang, X., Zhou, Y.-C. & Zhuang, Q.-Q. *Acta Biochem. Biophys. Sinica* 1995, *27*, 361-366
32. Shukla, S. & hanahan, D. J. *Arch. Biochem. Biophys.* 1981, *209*, 668-676
33. Forst, S., Weiss, J., Maraganore, J. M., Heinrikson, R. L. & Elsbach, P. *Biochim. Biophy. Acta* 1987, *920*, 221-225.
34. Forst, S., Weiss, J. Blackburn, P., Frangione, B., Goni, F. & Elsbach, P. *Biochemistry* 1986. *25*, 4309-4314.
35. Samejima, Y. Yanagisawa, M., Tomomatsu, Y., Ji, Y.-H., Lu, L.-F. & Xu, K. *Third congress of federation of Asian and Oceanian physiological societies,* Nov.7-10, 1994, Shanghai, 139.
36. Curin-Serbec, V. & Choumet, V. *Toxicon,* 1994, *32*, 1337-1348.
37. Renetseder, R., Dijkstra, B.W., Huizinga, K., Kalk, K.H. & Drenth, J. *J. Mol. Biol.* 1988, *200*, 181-188.
38. Zhao, H.-Y., Tang, L., Wang, X.Q., Zhou, Y.-C. & Lin, Z.-J. *Toxicon* 1998, *36*, 875-886.

Chapter 18

Snake Venom Disintegrin: An Effective Inhibitor of Breast Cancer Growth and Dissemination

Francis S. Markland and Qing Zhou

Department of Biochemistry and Molecular Biology and USC/Norris
Comprehensive Cancer Center, School of Medicine,
University of Southern California, Los Angeles, CA 90033

Contortrostatin, a 13.5 kDa disulfide-linked, homodimeric polypeptide possessing an Arg-Gly-Asp sequence, was isolated from venom of the southern copperhead snake. Contortrostatin binds to integrins and blocks the adhesion of human breast cancer cells (MDA-MB-435) to extracellular matrix proteins including fibronectin and vitronectin. Contortrostatin also inhibits invasion of MDA-MB-435 cells. Daily intratumor injection of contortrostatin (5 µg/mouse/day) into MDA-MB-435 tumor, established in an orthotopic xenograft nude mouse model, significantly inhibits growth of the tumor and reduces the occurrence of pulmonary metastasis. Contortrostatin is not cytotoxic to MDA-MB-435 cells *in vitro*. Integrin αvβ3, which mediates cell motility and tumor invasion, is one of the binding sites for contortrostatin on MDA-MB-435 cells. On the chick embryo chorioallantoic membrane, contortrostatin inhibited angiogenesis induced by growth factors and by MDA-MB-435 cells. Contortrostatin effectively blocked adhesion of human umbilical vein endothelial cells (HUVEC) to immobilized vitronectin and significantly inhibited invasion of HUVEC. Integrin αvβ3 is one of the binding sites for contortrostatin on HUVEC. Detachment of HUVEC from vitronectin by contortrostatin induced apoptosis. It is concluded that contortrostatin blocks αvβ3, and perhaps other integrins on breast cancer and vascular endothelial cells, and thus inhibits *in vivo* progression by a combination of anti-tumor and anti-angiogenic activities.

Integrins are important cell surface receptors that are involved in both cell-cell and cell-extracellular matrix (ECM) protein interactions (1) All integrins are α/β heterodimers that require divalent cations for the proper noncovalent association of their subunits. Specificity toward distinct ECM proteins is accomplished by the various pairings of α and β subunits. Different α subunits may combine with the same β subunit, conversely different β chains are capable of pairing with a particular α subunit (1, 2). The interaction of ECM with integrins leads to integrin association with cytoskeleton elements, resulting in the formation of specialized adhesive junctions, such as focal adhesions and hemidesmosomes (3). Various cytoskeletal elements assemble at the cytoplasmic face of focal adhesions mediating stable cell attachment (3). Formation of focal adhesions triggers a series of intercellular signaling events. Experimental evidence suggests that through these intracellular signals, integrins regulate gene transcription, cell proliferation and cell differentiation (3). Several observations indicate that the expression and distribution of integrins may be affected by neoplastic transformation in human tumors (4). For example, the expression level of $\alpha v\beta 3$ in melanoma and several other cancers is proportional to its invasiveness (5-7). While changes in integrin expression promotes the acquisition of an invasive phenotype by tumor cells, there is strong evidence that tumor invasion is facilitated by proteolytic degradation of ECM components which is also controlled by integrin mediated signaling (8-11). Expression of $\alpha v\beta 3$ is required for vasculogenesis, wound healing, and angiogenesis. Tumor-induced angiogenesis is a critical step for cancer growth beyond the size of 1 to 2 mm^3 (12-16).

Several classes of integrins recognize the RGD sequence present in ECM proteins (1, 17). Previous studies by our group and others showed that peptides containing the RGD sequence bind competitively to integrins on the surface of tumor cells and inhibit binding of the cancer cells to the ECM (18, 19). Agents that disrupt interactions of these integrins should have significant anti-tumor activity. Disintegrins are a family of polypeptides found in the venom of viper and pit viper snakes. They were originally characterized by their ability to inhibit platelet aggregation by blockage of the fibrinogen receptor $\alpha IIb\beta 3$ (also known as GP IIb/IIIa) (20). Investigations have shown that disintegrins are potent functional blockers of multiple integrins including $\alpha v\beta 3$ and $\alpha 5\beta 1$ (18, 20). All disintegrins contain an R/KGD (Arg/Lys-Gly-Asp) sequence in the carboxyl-terminal half of the molecule that is essential to their ability to obstruct integrin interaction with the ECM (21-23). We have isolated a 13.5 kDa disintegrin, contortrostatin, from *Agkistrodon contortrix contortrix* (southern copperhead) venom by means of multiple step high performance liquid chromatography (HPLC) (24). Contortrostatin is distinct from other disintegrins by its homodimeric structure (19). Previous studies in our laboratory demonstrated that this disintegrin binds to $\beta 1$ integrin (most likely $\alpha 5\beta 1$) in M24met, human metastatic melanoma cells (19). Pretreatment of M24met cells with this disintegrin prevented the cells from forming pulmonary metastatic nodules after tail vein injection in a mouse model (19).

It is well established that the growth of a tumor depends on persistent neovascularization (25, 26). Since the initial report of this hypothesis by Folkman, many angiogenic inducers and inhibitors have been identified (reviewed in (27). It is believed that the balance of inhibitors and inducers governs the angiogenic switch in the cancer cell (28). Recently, the role of integrins in angiogenesis has been investigated (reviewed in (29). As already noted, $\alpha v\beta 3$ undergoes upregulation in endothelial cells during angiogenesis (12, 13). A monoclonal antibody (mAb) to integrin $\alpha v\beta 3$, as well as a cyclic RGD-containing peptide, perturbed angiogenesis and produced regression of human breast cancer growing on the chick embryo chorioallantoic membrane (CAM) (12, 13). Antagonists of $\alpha v\beta 3$ apparently cause apoptosis of vascular endothelial cells which results from the activation of p53 and the increase of *bcl*-2/bax ratio (30-32). In addition to $\alpha v\beta 3$, there is evidence that $\alpha v\beta 5$ is also involved in angiogenesis (33).

It is hypothesized that contortrostatin, by blocking integrins on breast cancer cells, interferes with adhesion of the cells to the ECM, and additionally inhibits cell motility and invasion. Disintegrins also inhibit angiogenesis by disrupting endothelial cell integrin function (34, 35). We present *in vitro* and *in vivo* evidence that contortrostatin has multiple functions in prevention of breast cancer progression.

Materials and Methods.

Materials. Venom of *Agkistrodon contortrix contortrix* was purchased from Biotoxins, Inc. (St. Cloud, FL). Contortrostatin was purified by multistep high performance liquid chromatography (HPLC) according to an established protocol (24). Human vitronectin, fibronectin, human and rat type I collagen, basic fibroblast growth factor (bFGF), vascular endothelial growth factor (VEGF) and Matrigel were purchased from Becton Dickinson (Bedford, MA). Peptide GRGDSP was purchased from Gibco Life Technologies (Gaithersburg, MD). Monoclonal antibody (mAb) LM609 (anti-αvβ3) was a gift from Dr. David A. Cheresh (Scripps Research Institute, La Jolla, CA). mAb 7E3 (anti- αIIbβ3, which also cross reacts with αvβ3), ^{125}I-c7E3 Fab and 10E5 (anti-αIIbβ3), as well as purified integrins αIIbβ3 and αvβ3 were kindly provided by Dr. Marian T. Nakada (Centocor, Malvern, PA). Goat anti-mouse IgG conjugated with fluorescein isothiocyanate (FITC) was purchased from Jackson ImmunoResearch (West Grove, PA). MDA-MB-435, an estrogen-receptor negative cell line established from isolated cancer cells from the pleural effusion of a woman with metastatic, ductal adenocarcinoma of the breast (36), was obtained from Dr. J. Price (MD Anderson Cancer Center, University of Texas, Houston, TX). Frozen HUVEC were obtained from Dr. Florence Hofman (Dept. of Pathology, University of Southern California). Other reagents used in these studies were from Fisher Scientific (Pittsburgh, PA).

Cell Adhesion Assay. MDA-MB-435 human breast cancer cells were cultured in Minimum Essential Medium containing 10% FBS, 2 mM L-glutamine, 1 mM sodium pyruvate, and penicillin/streptomycin. HUVEC were seeded in 1% gelatin coated tissue culture flask in RPMI-1640 medium containing 2mM L-glutamine, 20% fetal bovine serum, 0.1 mg/ml EndoGro (VEC Technology, Inc., NY, NY), 1% Nutridoma HU (Boehringer Mannheim, Indianapolis, IN), 20 units/ml heparin, 100 units/ml penicillin, 100 μg/ml streptomycin, and 2.5 μg/ml Fungizone. HUVEC were used between passage 4 and 7. All cells were incubated in a 37°C incubator with 5% CO_2. Contortrostatin or extracellular matrix proteins (100 μl) were immobilized on Immulon-II 96-well microtiter plate (Dynex Technologies, Inc., Chantilly, VA) by incubating the protein, dissolved in phosphate-buffered saline (PBS), in the well overnight at 4°C. The amount of contortrostatin, human fibronectin, vitronectin or Matrigel immobilized were 0.1 μg, 0.5 μg, 1μg or 1/100 dilution per well, respectively. Excess proteins were washed away with PBS. Unbound sites were blocked with 1% bovine serum albumin (BSA) in PBS. One hundred μl of MBA-MD-435 cells (7.5 x 10⁵ cells/ml) or HUVEC (5 x 10⁵ cells/ml) were incubated at 25°C for 20 minutes with various reagents before being applied to the coated wells. Seeded cells were allowed to adhere for one hour at 37°C. After unbound cells were washed away, the extent of cells adhesion was determined by CellTiter 96™ AQ$_{ueous}$ Non-Radioactive Cell Proliferation Assay kit (Promega, Madison, WI). The tests were performed in triplicate, and the assays were repeated at least three times to confirm results.

Cell Invasion Assay. Modified Boyden chambers were employed for this assay (37). Twelve-well Transwell chambers (Corning Costar, Cambridge, MA) with 12 μm pores were coated with 150 μl of 1:50 diluted Matrigel in serum free medium. Coated wells were allowed to air dry in a sterile hood overnight and rehydrated with serum free medium for 2 hours at room temperature prior to use. For MDA-MB-435 cells, the cells were mixed with various concentrations of contortrostatin or other inhibitors in serum free medium and 2.5 x 10⁵ cells were applied to the chambers and cells were allowed to invade across the Matrigel-coated membrane for 12 hours. Conditioned medium collected from fibrosarcoma cell (HT1080) culture was added to the lower chambers as chemoattractant. For HUVEC, 2.5 x 10⁵ cells in 200 μl of medium were pre-treated with contortrostatin (1 μM), 7E3 (200 μg/ml), or vehicle (control), and then applied to the upper wells. Medium containing bFGF (20 ng/ml) as chemoattractant was added in the bottom well. HUVEC were incubated at 37°C for 6 hr. For both cell types, after non-invaded cells were removed with wet cotton swabs, invaded cells attached to the bottom of the membrane were fixed and stained with Diff Quick™ staining kit (Dade Diagnostics of P. R. Inc., Aguada, Puerto Rico). The number of invaded cells were quantitated microscopically by finding the mean cell number of three randomly selected high power vision fields. Treatments were tested in duplicate, and the assays were repeated to verify results.

Competitive Binding Assay. MDA-MB-435 cells or HUVEC were resuspended in 1% BSA/PBS at a density of 1×10^7/ml. Aliquots of 100 µl were incubated with different concentrations of contortrostatin at room temperature for 30 minutes, followed by addition of 7E3, 10E5, or LM609 (final concentration 5 µg/ml). Incubations were continued for another 30 minutes. The cells were washed twice and resuspended in 1% BSA/PBS. Goat anti-mouse IgG conjugated with FITC was added to the suspension at a final titer of 1:200. After 30 minutes incubation at room temperature in the dark, unbound FITC-conjugated IgG was washed off, and the fluorescent intensity of the cells was analyzed using flow cytometry (FACScan, Becton Dickinson, Bedford, MA). Tests were performed in duplicate and the experiments were repeated three times.

Effect of Contortrostatin on the Proliferation of MDA-MB-435 Cells *In Vitro*. Six-well cell culture plates (Corning Costar, Cambridge, MA) were coated with 0.5 ml of 1% (v/v) Matrigel in serum free medium. MDA-MB-435 cell suspension at a density of 0.3×10^6/ml in 3 ml of complete DMEM medium containing contortrostatin at various concentrations were added to the wells. Cell densities were determined every 24 hours with a Coulter counter. Assays at each contortrostatin concentration were performed in duplicate, and each assay was repeated twice.

Effect of Contortrostatin on Metastatic Breast Cancer in Nude Mouse Model. For the humane treatment of experimental animals, a protocol approved by the Institutional Animal Care and Use Committee, University of Southern California was strictly followed. Female nude mice (BALB/c/nu/nu) at 4-weeks of age were purchased from Simonsen Lab (Gilroy, CA). Animals were kept in a pathogen-free environment, and fed sterilized food and water. Orthotopic xenograft nude mouse models were established according to Price *et al* (36). All procedures were performed with aseptic techniques. Nude mice were anesthetized by inhalation of Metofane (Pitman-Moore, Mundelein, IL) in a closed chamber. Mammary fat pad under the second nipple from the rostral side was selected for implantation of cancer cells. A longitudinal incision of 1 cm was made on the lateral side of the nipple where the mammary fat tissue beneath the skin was carefully exposed. MDA-MB-435 cells (5×10^5) resuspended in 0.1 ml PBS were injected into the mammary fat pad. The wound was then sutured. Palpable tumor masses appeared 14 days post-implantation. At that time the mice were randomized into control and treatment groups. Diameters of the tumor masses (subcutaneous) were measured weekly with a caliper. Volumes of the masses were calculated as tumor volume = (width)^2 x (length) x 0.5 (mm^3) (38). Daily infiltrative injections into the established tumor masses were carried out with 5 µg contortrostatin in 0.1 ml of 0.9% NaCl over a period of 7 weeks. An equal volume of 0.9% NaCl was injected into control animals. Upon termination of the daily administration, all tumor masses were removed surgically. Each tumor mass was weighed and sections were immediately fixed for pathological examination. No effort was made to eliminate the residual tumor or regional lymph nodes. Animals were allowed to survive for another 2 weeks without intervention. The animals were then euthanized and autopsied. The lungs were removed and perfused with buffered formaldehyde. The number of metastatic nodules on the surface of lungs were counted under stereo microscope. Lung metastases were also quantitated by microscopic examination of the hematoxylin-eosin-stained tissue sections.

Angiogenesis Assay on Chick Embryo Chorioallantoic Membrane (CAM). Ten-day old chick embryos were purchased from a local poultry farm (AA Lab, Westminster, CA). The embryos were incubated at 37°C with 60% humidity. CAM preparation was described in detail elsewhere (13). To grow tumor masses, 1 x 10^7 MDA-MB-435 cells in 30 μl were applied on the CAM. After 7 days, the tumor masses and surrounding CAM were cut out. For the angiogenesis assay, growth factor (200 ng) impregnated filter discs or carefully trimmed tumor masses (about 10 mg each) were transplanted on freshly prepared 10-day-old CAMs. Contortrostatin treatment, either i.v. injection or topical administration, was performed 24 hours after transplantation. CAM around the filter discs or tumors were cut off 48 hours after treatment, and angiogenesis on the inner side of the CAM was examined under stereomicroscope and documented photographically.

Apoptosis Assay. HUVEC (6 x10^5) were incubated in the presence or absence of contortrostatin for 20 minutes at room temperature prior to applying to 6-well plates coated with either vitronectin (5 μg/ml), fibronectin (2.5 μg/ml), or contortrostatin (2.5 μg/ml). After 18 to 20 hours incubation at 37°C, cells were collected. DNA fragmentation was analyzed by electrophoresis on 1.5% agarose gel using DNA extracted from HUVEC samples. Fragmented DNA extraction method is described elsewhere (39).

Results.

Contortrostatin Binds to MDA-MB-435 Cells Exclusively via Integrin(s). To confirm that contortrostatin binds to MDA-MB-435 cells via their integrins, contortrostatin (0.5 μg/well) was immobilized in the wells of 96-well microtiter plates. Contortrostatin can support binding of MDA-MB-435 cells in a dose-dependent manner (data not shown). Figure 1 shows that binding of MDA-MB-435 cells to immobilized contortrostatin was blocked by the soluble GRGDSP peptide (IC$_{50}$=0.4 mM) (Fig. 1A), and by EDTA (IC$_{50}$=0.8 mM) (Fig. 1B), suggesting that contortrostatin interacts with MDA-MB-435 cells solely through a specific RGD-dependent mechanism (ie., integrin-mediated).

Contortrostatin Inhibits Adhesion of MDA-MB-435 Cells to Fibronectin and Vitronectin. ECM proteins fibronectin, vitronectin and collagen support adhesion of MDA-MB-435 cells. By pretreating MDA-MB-435 cells with various concentrations of contortrostatin, we tested contortrostatin's ability to interfere with adhesion to the immobilized ECM proteins. Our results indicated that contortrostatin inhibits adhesion of MDA-MB-435 cells to fibronectin and vitronectin in a dose-dependent manner (Fig. 2). IC$_{50}$ for contortrostatin inhibition of adhesion to fibronectin is 18 nM (Fig. 2A), and for vitronectin it is 1.5 nM (Fig. 2B). Contortrostatin had minimal effect on the adhesion of MDA-MB-435 cells to human or rat type I collagen (data not shown). Contortrostatin failed to inhibit the adhesion of MDA-MB-435 cells to Matrigel (data not shown).

Contortrostatin inhibits MDA-MB-435 cell invasion through Matrigel. Because contortrostatin did not inhibit adhesion and spreading of the MDA-MB-435 cells to a Matrigel surface, we determined whether contortrostatin would inhibit the invasion of the cells through a Matrigel barrier. A layer of three-dimensional Matrigel was applied on the porous membrane of a Transwell insert. Conditioned medium collected from fibrosarcoma HT-1080 when placed below the Transwell, strongly induced the invasion of MDA-MB-435 cells within 8 hours. However, as seen in Figure 3A, invasion of the breast cancer cells was inhibited by contortrostatin in a dose-dependent manner. This inhibitory effect is most likely due to the functional interruption of αvβ3, since pretreatment of the cells with 7E3 was observed to have a similar inhibitory effect on invasion (Fig. 3B).

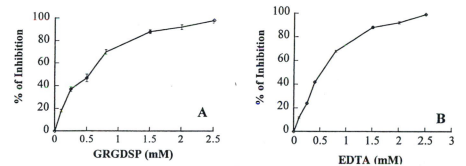

Figure 1. RGD-containing peptide and EDTA inhibit adhesion of MDA-MB-435 cells to immobilized contortrostatin. One hundred µl of MDA-MB-435 cells (7.5 x 10^5/ml) pretreated with various concentrations of the synthetic hexapeptide, GRGDSP (A), or EDTA (B), were allowed to adhere to microtiter plates coated with contortrostatin (0.1 µg/well). Extent of adhesion of MDA-MB-435 cells was determined after unbound cells were washed away. Adhesion of the cells to contortrostatin was blocked by the soluble GRGDSP peptide (IC$_{50}$=0.4 mM), and by EDTA (IC$_{50}$=0.8 mM). Each point is expressed as mean percentage of inhibition from triplicate analyses (error bars show SD). The assay was repeated three times.

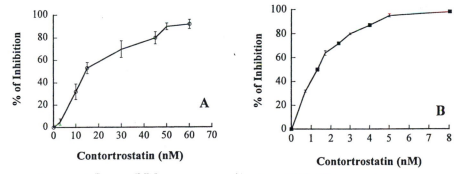

Figure 2. Contortrostatin inhibits MDA-MB-435 cell adhesion to fibronectin and vitronectin. One hundred µl of MDA-MB-435 cells (7.5 x 10^5/ml) were seeded in microtiter plates coated with human fibronectin (0.5 µg /well) or vitronectin (1 µg /well). Cells were treated with various concentration of contortrostatin prior to seeding and were allowed to adhere for one hour at 37°C. Extent of adhesion of MDA-MB-435 cells was determined after unbound cells were washed away. The results show that contortrostatin inhibited adhesion of MDA-MD-435 to both ECM proteins in a dose dependent manner. (A) IC$_{50}$ for adhesion to fibronectin is 18 nM. (B) IC$_{50}$ for adhesion to vitronectin is 1.5 nM. Each point is expressed as mean percentage of inhibition from triplicate analyses (error bars show SD). The experiment was repeated three times with identical results.

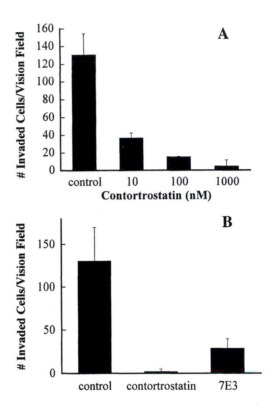

Figure 3. Antagonists of αvβ3 inhibit MDA-MB-435 cell invasion through Matrigel. (A) MDA-MB-435 cells were treated with various concentrations of contortrostatin. Cells (2.5×10^5) were applied to each Transwell chamber and allowed to migrate across the Matrigel layer towards HT1080 conditioned medium for 8 hours. After non-invaded cells were removed, invaded cells attached on the bottom of the membrane were fixed and stained. The number of cells were determined microscopically. (B) MDA-MB-435 cells treated with contortrostatin (0.5 μM) or 7E3 (200 μg/ml) were subjected to an identical assay. Both antagonists inhibit invasion of the cells through Matrigel basement membrane. Assays were performed in triplicate, and repeated three times. Error bars show SD.

αvβ3 is Identified as One of the Binding Sites for Contortrostatin on MDA-MB-435 Cells. Contortrostatin was previously shown to bind to α IIbβ3 on human platelets (24). Since contortrostatin inhibits the binding of MDA-MB-435 cells to vitronectin, we determined whether contortrostatin binds to the vitronectin receptor αvβ3, like other disintegrins (18). Using flow cytometry, we were able to detect 7E3 (mAb against αvβ3 and αIIbβ3) (40, 41) binding to MDA-MB-435 cells (Fig. 4A). αIIbβ3 is mainly expressed in platelets and megakaryocytes (42), but is also found in some highly metastatic melanoma cells (43, 44). Binding of 10E5, a highly specific mAb to αIIbβ3 (45), to MDA-MB-435 cells was not detected (Fig. 4C), suggesting that 7E3 binds solely to αvβ3 on these cells. When pretreated with contortrostatin, binding of 7E3 to MDA-MB-435 cells was significantly inhibited (Fig. 4B), suggesting it binds to αvβ3 on MDA-MB-435 cells.

Contortrostatin is not cytotoxic to MDA-MB-435 cells *in vitro*. We tested whether contortrostatin is cytotoxic to the breast cancer cells. To avoid detachment by contortrostatin, MDA-MB-435 cells were cultured in dishes coated with Matrigel. The growth curve (Fig. 5) indicates that cells in the presence of contortrostatin at concentrations of 100 nM or 500 nM, proliferate equally as well as those without contortrostatin pretreatment. The results suggest that contortrostatin is not cytotoxic to MDA-MB-435 cells *in vitro* at the concentrations tested.

Contortrostatin inhibits growth and spontaneous metastasis of human breast cancer (MDA-MB-435) in a nude mouse model. An orthotopic xenograft metastatic model of human breast cancer in nude mice was established by implantation of MDA-MB-435 cells in the mammary fat pads of 4-week-old female nude mice as previously described (36). Palpable tumors appeared 2 weeks after implantation. Without therapeutic intervention, the implanted tumor grew to 1 cm^3, and metastasized to lungs and other organs in 8 weeks. Local injection of contortrostatin (5 μg/mouse/day) was started at the 14th day post-implantation when the tumor take was confirmed by a subcutaneous palpable mass (designated as "week 0" of injection, Fig. 6). Weekly measurements of tumor volume during treatment showed that local injection of contortrostatin substantially inhibited the growth rate of the tumor (Fig. 6). Typical appearance of tumors in the control group showed high nodularity in addition to their large size. In comparison, tumors in the treated group had a smoother surface and much smaller size. The weight of the removed tumor masses from treated animals were also significantly lower than those from untreated mice (Table 1). Despite platelet aggregation inhibitory activity of contortrostatin (24), no spontaneous hemorrhage was observed during the experiment. Slightly prolonged bleeding times at the injection sites in the treated animals, however, were noticed. No other side-effects were observed, suggesting that contortrostatin was well-tolerated. A two-week treatment-free period after the removal of primary tumors allowed possible micrometastases to develop into marcometastatic nodules on the lung surface. Upon termination of the experiment, autopsies were performed to determine the extent of metastases. It was found that the number of surface nodules in the contortrostatin treated group was reduced by at least 65%. The proportion of micrometastases was well correlated to the number of surface nodules. Table 1 summarizes the analysis of *in vivo* experiments.

Contortrostatin inhibits the adhesion of HUVEC to ECM proteins. We chose HUVEC to investigate the effect of contortrostatin on the interaction of vascular endothelium with the ECM. Different ECM proteins were immobilized on the bottom of 96-well plates. Contortrostatin was examined for its ability to inhibit the adhesion of HUVEC to the immobilized proteins. Fig. 7 demonstrates that contortrostatin inhibits adhesion of HUVEC to immobilized vitronectin (IC$_{50}$ approximately 3 nM). However, at the concentration examined, it is less effective on the adhesion of HUVEC to fibronectin and Matrigel.

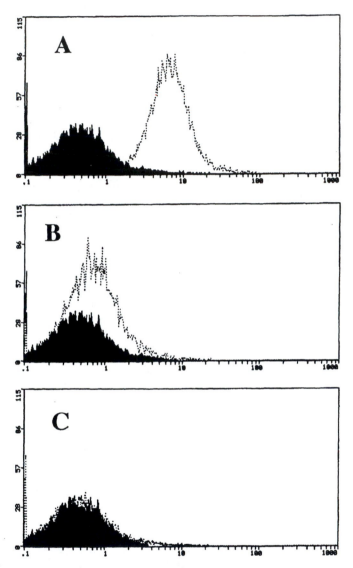

Figure 4. Contortrostatin specifically inhibits binding of 7E3 to αvβ3 in MDA-MB-435 cells. This figure shows the binding of antibodies to MDA-MB-435 integrins as detected by flow cytometry. (A) 7E3 (5 μg/ml), a mAb to αIIbβ3 which also cross reacts with αvβ3, binds to MDA-MB-435; (B) Contortrostatin (1 μM), added 20 minutes prior to addition of 7E3, effectively inhibits binding of 7E3; (C) Lack of binding of 10E5 (10 μg/ml), a specific mAb to αIIbβ3 to MDA-MB-435 cells. Solid area, background; dotted line, experimental. Data shown here is representative of three identical experiments.

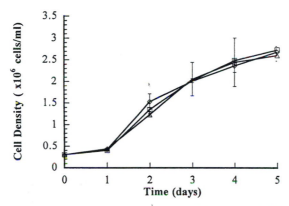

Figure 5. Effect of contortrostatin on proliferation of MDA-MB-435 cells *in vitro*. Six-well cell culture plates coated with Matrigel (1/100 dilution) were seeded with 3 ml of MDA-MB-435 cells (0.3 x 10^6 /ml) in complete DMEM medium. Cell density was determined every 24 hours. Growth curves of MDA-MB-435 cells *in vitro* without contortrostatin (diamonds), and with contortrostatin at 100 nM (squares) and 500 nM (triangles) are illustrated. Each point is expressed as mean (cells/ml) from duplicates (error bars show SD).

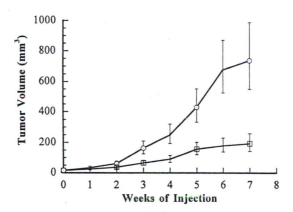

Figure 6. Effect of contortrostatin on the growth of MDA-MB-435 tumor in experimental nude mice. MDA-MB-435 (5×10^5) cells were implanted into mammary fat pads of 4-week old female nude mice. Palpable tumor masses appeared around the 14th day post-implantation. Daily local injection of contortrostatin was initiated on the 14th day post-implantation. The volumes of tumor masses (geometric mean ± 95% confidence interval) of control (circles) and contortrostatin-treated group (5 μg/mouse/day) (squares) are shown.

Contortrostatin inhibits invasion of HUVEC. The effect of contortrostatin on invasion of HUVEC was investigated using Matrigel-coated Boyden chambers. Basic FGF (20 ng/ml) in the lower chamber strongly induced the migration of HUVEC, whereas, pretreatment of HUVEC with contortrostatin (0.5 μM) effectively prevented invasion (Fig. 8). It is worthwhile to note that while the invasion of HUVEC was inhibited, adhesion of these cells to the Matrigel-coated filter was not affected. Therefore, the observed inhibitory effect of contortrostatin on HUVEC is presumably due to the suppression of cellular motility instead of prevention of adhesion. Pre-treatment of HUVEC with 7E3 (200 μg/ml) does not prevent the cells from adhering to the Matrigel-coated filter. However, this treatment does inhibit invasion of HUVEC (Fig. 8), suggesting that the functional blockade of integrin αvβ3 inhibits HUVEC motility.

Table 1: Daily Injection of Contortrostatin Inhibits Breast Cancer Growth and Metastasis in Nude Mice

	Tumor Volume (mm^3)	Tumor Weight (g)	Number of Metastatic Nodules In Lungs
Experiment 1			
Control (n=5)	1012.6	1.48	47.5
CN-Treated (n=6)	339.7	0.87	4.5
% Change	-66%	-41%	-90.5%
Experiment 2			
Control (n=14)	954.3	0.95	6.5
CN-Treated (n=14)	249.8	0.54	2.3
% Change	-74%	-44%	-65%

n= number of animals in experiment

Identification of the binding site for contortrostatin on HUVEC. Since contortrostatin blocks the adhesion of HUVEC to immobilized vitronectin, we determined whether contortrostatin binds to the vitronectin receptor, αvβ3, on HUVEC (46). Using flow cytometry, we were able to detect LM609 (mAb against αvβ3) (47) binding to HUVEC (data not shown), suggesting that integrin αvβ3 is expressed on HUVEC. FACS analysis indicates that 7E3 binds to HUVEC (Fig. 9A), and contortrostatin competitively inhibits 7E3 binding to αvβ3 (Fig. 9B). This competitive binding was not observed with LM609 (data not shown), because LM609 apparently binds to a different epitope than contortrostatin. We did not detect binding of 10E5 to HUVEC (data not shown). This finding excluded the possibility that binding of contortrostatin to HUVEC was due to αIIbβ3.

Detachment of HUVEC from immobilized vitronectin induces apoptosis. Endothelial cells are anchorage-dependent, detachment from their anchorage site results in apoptosis (39, 48). We determined whether detachment of HUVEC from a vitronectin-coated surface by contortrostatin causes apoptosis. DNA fragmentation, a marker of apoptosis, has been observed by electrophoretic analysis of DNA isolated from HUVEC. As shown in Fig. 10, DNA fragmentation was not detected in HUVEC 18-20 hours after plating on a vitronectin-coated surface. Contortrostatin-treated HUVEC did not adhere to vitronectin-coated plates and DNA fragmentation was obvious in these cells after 18-20 hours. The fragmentation pattern is similar to that observed by Meredith *et al.* in a similar experiment (39), and was observed only in cells detached from immobilized vitronectin by contortrostatin. Importantly, however, DNA fragmentation was not detected in HUVEC following 18-20 hours of

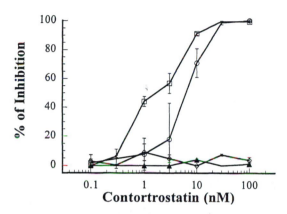

Figure 7.   Contortrostatin inhibits adhesion of HUVEC to immobilized vitronectin. Contortrostatin inhibits adhesion of HUVEC to immobilized vitronectin (1 μg/well) (squares) and contortrostatin (0.1 μg/well) (circles), but not to fibronectin (0.5 μg/well) (diamonds) or Matrigel (1:100 dilution) (solid triangles). Each point is the mean absorbance of triplicate wells ± SD. The experiment was repeated 3 times.

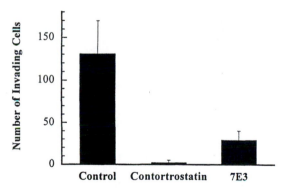

Figure 8.   Contortrostatin and 7E3 inhibit invasion of HUVEC through Matrigel. HUVEC pretreated with contortrostatin (0.5 μM) or 7E3 (200 μg/ml) were allowed to invade through the Matrigel-coated Boyden chamber. Basic FGF (20 ng/ml) served as chemoattractant in the lower chamber. 7E3 inhibits invasion by approximately 75%, while contortrostatin completely blocks invasion. The means of duplicates are presented, error bars indicate SD.

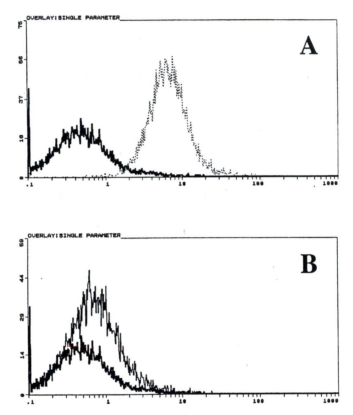

Figure 9. Contortrostatin specifically inhibits binding of 7E3 to αvβ3 in HUVEC. This figure shows the binding of antibodies to HUVEC integrins as detected by flow cytometry. (A) 7E3 (5 μg/ml), a mAb to αIIbβ3 which also cross reacts with αvβ3, binds to HUVEC; (B) Contortrostatin (1 μM), added 20 minutes prior to addition of 7E3, effectively inhibits binding of 7E3. 10E5 (10 μg/ml), a specific mAb to αIIbβ3, did not bind to HUVEC (not shown). Solid line, background; dotted line, experimental. Data shown is representative of three identical experiments.

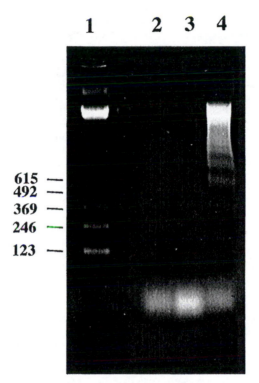

Figure 10. Contortrostatin induces HUVEC apoptosis. HUVEC were seeded on contortrostatin-coated (5 µg) (lane 2) 30 mm-plate, as well as vitronectin-coated (2.5 µg) plate in the absence (lane 3) or presence (lane 4) of contortrostatin (500 nM). Nuclear extracts from these cells were separated on 1.5% agarose gel. DNA fragmentation was only found in HUVEC treated with contortrostatin which were detached from immobilized vitronectin, as shown in lane 4. DNA ladder (123 bp) was loaded in lane 1.

adhesion to immobilized contortrostatin. Thus, it would appear that CN induces apoptosis by an indirect effect involving detachment of HUVEC from ECM proteins.

Contortrostatin inhibits angiogenesis induced by bFGF and VEGF in CAM.
VEGF and bFGF, expressed by multiple malignant cell lines, have been well documented as specific and potent endothelial mitogens and chemoattractants (49, 50). We tested whether contortrostatin blocks angiogenesis specifically induced by VEGF and bFGF using the CAM assay. Contortrostatin was applied topically on the filter discs 24 hours after application of the growth factors and angiogenesis was determined 48 hours later. Contortrostatin (5 µg/ embryo) effectively inhibited angiogenesis induced by bFGF and VEGF. Vessels that developed prior to contortrostatin treatment were not altered, indicating that contortrostatin only blocks angiogenic vessels.

Contortrostatin inhibits angiogenesis induced by MDA-MB-435 tumor in CAM.
We employed the CAM to test if the inhibitory effect of contortrostatin on breast cancer progression is partially due to its antiangiogenic effect. Angiogenesis was induced on the CAM by transplantation of MDA-MB-435 tumor mass (~50 mg) on an area devoid of blood vessels. These CAM models were treated on day 2 post-transplantation with contortrostatin by either i.v. injection (125 ng/embryo) or topical administration (1 µg/embryo). In the control CAM (saline treated), numerous blood vessels growing around and into the tumors were visible. The results clearly showed that both systemic and topical administration of contortrostatin significantly inhibited angiogenesis induced by the transplanted tumors. Vessels already existing before contortrostatin treatment were not affected.

Discussion.

Breast Cancer Adhesion, Invasion and Metastasis. At diagnosis, over 60% of breast cancer patients have metastatic disease. Dissemination of the cancer cells from the primary tumor to remote tissues or organs, and progressive growth of the metastatic cells at the distant site, is the most common cause of death in breast cancer patients. Therefore, the control of metastasis offers an important avenue for breast cancer treatment. Metastasis is a complex process which requires the participation of many tumor gene products and environmental factors (51). Local invasion and penetration of the cancer cells into the lumen of a blood vessel (intravasation) will allow tumor cells to disseminate. Other important events are the formation of tumor cell microthrombi, adherence, penetration through the endothelium and the adjacent basement membrane (extravasation), and ability to grow in a foreign environment and to stimulate angiogenesis. These processes must be successfully accomplished by the cancer cells to develop into metastatic tumors. In view of the complexity of this process, a combination of factors may be required to effectively arrest metastasis (52).

Integrins and Tumor Metastasis. The importance of integrins in tumor metastasis is now becoming apparent (3). Integrins mediate cell adhesion, migration, matrix degradation, proliferation, gene regulation, and angiogenesis. Contortrostatin, a unique dimeric disintegrin, is a potent antagonist of several integrins in cancer and vascular endothelial cells, and therefore has an effect on multiple steps crucial to tumor metastasis. Invading cancer cells express various adhesion molecules to facilitate their interaction with the matrix. These enable the tumor cells to migrate and escape from the primary site and ultimately attach to the matrix of the vessel wall leading to extravasation (3). We have shown that contortrostatin binds to MDA-MB-435 cells solely via integrins, since a soluble RGD-containing peptide or a cation chelator (EDTA) can abrogate adhesion of the cells to contortrostatin. Conversely, by binding to specific integrins, contortrostatin potently inhibits human breast cancer cell adhesion to immobilized ECM proteins.

Cellular adhesion to fibronectin is mediated primarily by integrin α5β1 through an RGD-mediated interaction (17). Using a mAb against α5β1 (P1D6), we detected the expression of α5β1 on MDA-MB-435 cells (data not shown). Anti-α5β1 antibody inhibited the adhesion of M24 met human melanoma cells to contortrostatin-coated plates, suggesting that contortrostatin binds to integrin α5β1 (19). The inhibitory effect of contortrostatin on MDA-MB-435 cell adhesion to fibronectin is most likely due to its specificity for α5β1, however, the role of other RGD-dependent integrins, such as αvβ3, can not be excluded (46). Cellular adhesion to vitronectin is mediated by αvβ3 (47) and other vitronectin receptors, such as αvβ1 and αvβ5 (46). Our data indicate that contortrostatin competes with 7E3 binding to MDA-MB-435 cells. 7E3 is a mAb to αIIbβ3, which was recently found in melanoma cells (43, 44). However, 7E3 also cross-reacts with αvβ3. Using a highly specific αIIbβ3 mAb, 10E5 (17), expression of αIIbβ3 was not detected in MDA-MB-435 cells, supporting the conclusion that contortrostatin binds to αvβ3, not αIIbβ3, on these cells. The specificity of contortrostatin to this vitronectin receptor probably accounts for its inhibitory effect on adhesion of MDA-MB-435 cells to vitronectin.

Contortrostatin Effects on Breast Cancer. Contortrostatin has also been shown to affect MDA-MB-435 cell invasive capacity, as evidenced by its ability to reduce invasion through a Matrigel membrane. Matrigel is a solubilized basement membrane preparation extracted from the Engelbreth-Holm-Swarm mouse sarcoma, a tumor rich in extracellular matrix proteins. Its major component is laminin, followed by collagen IV, heparin sulfate proteoglycans, entactin and nidogen. Since contortrostatin does not disrupt the adhesion of MDA-MB-435 breast cancer cells to Matrigel, the anti-invasive activity is not likely due to the action of contortrostatin on laminin receptor(s). In addition, 7E3 had no inhibitory effect on adhesion of MDA-MB-435 cells to immobilized Matrigel, however, it significantly inhibited the invasion of MDA-MB-435 cells through a Matrigel barrier. Therefore, we conclude that the anti-invasive activity observed here is due to the blockage of αvβ3 by contortrostatin and is independent of its anti-adhesive activity. Pathological studies strongly suggest that the invasiveness of cancer cells is proportional to the expression level of αvβ3 (5-7). Apart from mediating adhesion to vitronectin, αvβ3 transmits signals from the matrix to regulate cellular functions. A recent cell biology study demonstrated that ligation of αvβ3 activates myosin light chain kinase via the ras/mitogen activated protein (MAP) kinase signal transduction pathway, resulting in increased phosphorylation of myosin light chain, with subsequent enhancement of the myosin/actin interaction which provides locomotion of the invading cells (53). It is reasonable to postulate that obstruction of αvβ3 by contortrostatin stops the signal transduction pathway and inhibits the mobility of MDA-MB-435 cells.

The orthotopic metastatic model of human breast cancer, employing MDA-MB-435 cells in the nude mouse, has been extensively utilized in breast cancer research (36). Because of the rapid induction of metastasis and the relatively short latency period between implantation and formation of a visible tumor, this model is particularly valid for testing the inhibitory effect of contortrostatin on tumor growth, and for investigating the effect of contortrostatin on tumor dissemination. Using this model, we demonstrated that parenteral administration of contortrostatin effectively inhibits tumor growth and lung metastasis of the breast cancer cells. The anti-metastatic effect of contortrostatin can be explained by a combination of anti-adhesive and anti-invasive activities.

Angiogenesis and Breast Cancer. The role of angiogenesis in cancer growth and dissemination has long been known (26) and the hypothesis that tumor growth is angiogenesis dependent (25) has now became dogma. Folkman (54) proposed a two-compartment tumor model composed of vascular endothelial cell and tumor cell compartments. Tumor cells can stimulate the proliferation and migration of

278

endothelial cells by producing angiogenic factors such as bFGF and VEGF. On the other hand, endothelial cells can promote tumor growth by secretion of platelet-derived growth factor, insulin-like growth factor 1, bFGF, heparin-binding-epithelial growth factor, granulocyte colony-stimulating factor and interleukin-6 (54). The paracrine effect, together with blood perfusion supplied by angiogenesis, promotes growth and metastasis of the tumor. It has been reported recently that anti-angiogenic therapy does not induce drug resistance and is able to attain prolonged tumor dormancy (55). In the context of the two-compartment model, it is obvious that anti-angiogenic therapy could be a potentially effective mechanism of cancer therapy.

Anti-angiogenic therapies in a mouse model not only inhibited growth and metastasis of the xenograft tumors, but also sustained the tumors in a long-term dormant state (55-57). Importantly, due to the genetic stability of endothelial cells, drug resistance is an unlikely event for anti-angiogenic therapy (55). Therefore, anti-angiogenic therapy has developed into an attractive approach for cancer treatment. The important role of integrin $\alpha v\beta 3$ in angiogenesis has recently been revealed (12), and evidence indicates that blocking $\alpha v\beta 3$ effectively suppresses angiogenesis (13, 14). In coordination with the rapidly invading carcinoma cells, the expression of $\alpha v\beta 3$ in vascular endothelium is upregulated during vasculogenesis and angiogenesis (14, 16). Adhesion and spreading of the vascular endothelial cells on vitronectin, an important component of the basement membrane, is mediated by $\alpha v\beta 3$ (58). Adhesion mediated by $\alpha v\beta 3$ suppresses apoptosis caused by detachment (known as anoikis) (30, 58). Moreover, ligation of $\alpha v\beta 3$ activates cytoskeleton reorganization, and consequently confers endothelial cell motility (53, 59). We demonstrated that contortrostatin inhibits angiogenesis in CAM by blocking $\alpha v\beta 3$ on endothelial cells. Other disintegrins (35) and RGD-containing peptides (13), have also been shown to exert an anti-angiogenic effect in the CAM assay. It is reasonable to predict that, by blocking $\alpha v\beta 3$, contortrostatin disrupts the adhesion and spreading on vitronectin of newly generated vascular endothelial cells, which is essential in the microenvironment of angiogenesis, and induces apoptosis of the endothelial cells. In addition, contortrostatin interrupts $\alpha v\beta 3$-dependent signal transduction and consequent cell motility, which is critical for formation of new blood vessels. Anti-angiogenic agents induce a high rate of apoptosis of the cancer cells which counteracts the high level of cell proliferation (57). The net result of these effects is a tumor mass with a slower growth rate. Meanwhile, inhibition of angiogenesis impedes the escape of primary cancer cells into the circulation, therefore directly reducing metastasis.

Conclusion. Contortrostatin, by blocking integrins on breast cancer cells and HUVEC, not only interferes with adhesion of these cells to the extracellular matrix, but also inhibits cellular mobility which is essential for invasion. Contortrostatin is also a potent inhibitor of angiogenesis; it appears to block several critical steps in neovascularization by a direct effect on endothelial cell integrins. First, blockade of $\alpha v\beta 3$ prevents the adhesion of vascular endothelial cells to vitronectin; this in turn elicits a survival signal conflict and induces apoptosis of the vascular endothelial cells. Second, contortrostatin inhibits HUVEC mobility by interfering with the normal function of $\alpha v\beta 3$ (and possibly $\alpha v\beta 5$). These combined effects of contortrostatin dramatically inhibit the metastatic capability of MDA-MB-435 breast cancer cells. *In vivo* intratumor injection of contortrostatin, by repressing angiogenesis induced by the breast cancer, appears to shut down the blood supply and inhibit the growth rate of the breast cancer. Additionally, contortrostatin treatment significantly reduces pulmonary metastasis in the nude mouse model. Further studies on contortrostatin may lead to the development of a potent anti-angiogenic agent with clinical potential for cancer therapy.

Acknowledgements.

This work was supported by funds from the California Breast Cancer Research Program of the University of California, Grant Number 1RB-0052 (to F.S.M.). Q.Z. is a recipient of a Postdoctoral Fellowship from the California Breast Cancer Research Program of University of California, Grant Number 3FB-0125, and a Postdoctoral Fellowship Supplement Support from the University of Southern California/Norris Comprehensive Cancer Center. The authors would like to acknowledge the expert technical assistance of Ms. Catherine Arnold.

Literature Cited.

1.   Hynes, R. O. Integrins: Versatility Modulation, and Signaling in Cell Adhesion, Cell. *69:* 11-25, 1992.
2.   Cheresh, D. A. Structural and Biologic Properties of Integrin-Mediated Cell Adhesion, Clin. Lab. Med. *12:* 217-236, 1992.
3.   Giancotti, F. G. and Mainiero, F. Integrin-Mediated Adhesion and Signaling in Tumorgenesis, Biochim. Biophys. Acta. *1198:* 47-64, 1994.
4.   Juliano, R. L. and Varner, J. A. Adhesion Molecules in Cancer: the Role of Integrins, Current Opinion Cell Biol. *5:* 812-818, 1993.
5.   Jacob, K., Bosserhoff, A. K., Wach, F., Knuchel, R., Klein, E. C., Hein, R., and Buettner, R. Characterization of Selected Strongly and Weakly Invasive Sublines of a Primary Human Melanoma Cell Line and Isolation of Subtractive cDNA Clones, Int. J. Cancer. *60:* 668-675, 1995.
6.   Kawahara, E., Imai, K., Kumagai, S., Yamamoto, E., and Nakanishi, I. Inhibitory Effects of Adhesion Oligopeptides on the Invasion of Squamous Carcinoma Cells with Special Reference to Implication of av Integrins, J. Cancer Res. Clin. Oncol. *121:* 133-140, 1995.
7.   Liapis, H., Adler, L. M., Wick, M. R., and Rader, J. S. Expression of avb3 Integrin is Less Frequent in Ovarian Epithelial Tumors of Low Malignant Potential in Contrast to Ovarian Carcinomas, Human Pathology. *28:* 443-449, 1997.
8.   Huhtala, P., Humphries, M. J., McCarthy, J. B., Tremble, P. M., Werb, Z., and Damsky, C. Cooperative Signaling by a5b1 and a4b1 Integrins Regulates Metalloproteinase Gene Expression in Fibroblasts Adhering to Fibronectin, J. Cell Biol. *129:* 867-879, 1995.
9.   Seftor, R. E. B., Seftor, E. A., Gehlsen, K. R., Stetler-Stevenson, W. G., Brown, P. D., Ruoslahti, E., and Hendrix, M. J. C. Role of the avb3 Integrin in Human Melanoma Cell Invasion, Proc. Natl. Acad. Sci. *89:* 1557-1561, 1992.
10.   Seftor, R. E. B., Seftor, E. A., Stetler-Stevenson, W. G., and Hendrix, M. J. C. The 72kDa Type IV Collagenase Is Modulated via Differential Expression of avb3 and a5b1 Integrins during Human Melanoma Cell Invasion, Cancer Res. *53:* 3411-3415, 1993.
11.   Brooks, P. C., Stromblad, S., Sanders, L. C., von Schalscha, T. L., Quigley, J. P., and Cheresh, D. A. Localization of Matrix Metalloproteinase MMP-2 to the Surface of Invasive Cells by Interaction with Integrin avb3, Cell. *85:* 683-693, 1996.
12.   Brooks, P. C., Clark, R. A., and Cheresh, D. A. Requirement of Vascular Integrin avb3 for Angiogenesis, Science. *264:* 569-571, 1994.
13.   Brooks, P. C., Montgomery, A. M. P., Rosenfeld, M., Reisfeld, R. A., Hu, T., Klier, G., and Cheresh, D. A. Integrin avb3 Antagonists Promote Tumor Regression by Inducing Apoptosis of Angiogenic Blood Vessels, Cell. *79:* 1157-1164, 1994.
14.   Brooks, P. C., Stromblad, S., Klemke, R., Visscher, D., Sarkar, F. H., and Cheresh, D. A. Antiintegrin avb3 Blocks Human Breast Cancer Growth and Angiogenesis in Human Skin, J. Clin. Invest. *96:* 1815-1822, 1995.
15.   Clark, R. A. F., Tonnesen, M. G., Gailit, J., and Cheresh, D. A. Transient Functional Expression of avb3 on Vascular Cells during Wound Repair, Am. J. of Pathol. *148:* 1407-1421, 1996.

16.   Drake, C. J., Cheresh, D. A., and Little, C. D. An Antagonist of Integrin avb3 Prevents Maturation of Blood Vessels During Embryonic Neovascularization, J. Cell Sci. *108:* 2655-2661, 1995.
17.   Pierschbacher, M. D. and Rouslahti, E. The Cell Attachment Activity of Fibronectin Can be Duplicated by Small Fragments of the Molecule, Nature. *309:* 30-35, 1984.
18.   Knudsen, K. A., Tuszynski, G. P., Huang, T. F., and Niewiarowski, S. Trigramin, an RGD-Containing Peptide from Snake Venom, Inhibits Cell-Substratum Adhesion of Human Melanoma Cells, Exp. Cell Res. *179:* 42-49, 1988.
19.   Trikha, M., De Clerk, Y. A., and Markland, F. S. Contortrostatin, a Snake Venom Disintegrin, Inhibits β1 Integrin-Mediated Human Metastatic Melanoma Cell Adhesion, and Blocks Experimental Metastasis, Cancer Res. *54:* 4993-4998, 1994.
20.   Savage, B., Marzec, U. M., Chao, B. H., Harker, L. A., Maraganore, J. M., and Ruggeri, A. M. Binding of the Snake Venom-Derived Proteins Applaggin and Echistatin to the Arginine-Glycine-Aspartic Acid Recognition Site(s) on Platelet Glycoprotein IIb/IIIa Complex Inhibits Receptor Function, J. Biol. Chem. *265:* 11766-11772, 1990.
21.   Dennis, M. S., Henzel, W. J., Pitti, R. M., Lipari, M. T., Napier, M. A., Deisher, T. A., Bunting, S., and Lazarus, R. A. Platelet Glycoprotein IIb/IIIa Protein Antagonists from Snake Venoms: Evidence for a Family of Platelet-aggregation Inhibitors, Proc. Natl. Acad. Sci. *87:* 2471-2475, 1990.
22.   Gould, R. J., Polokoff, M. A., Friedman, P. A., Huang, T.-F., Holt, J. C., Cook, J. J., and Niewiarowski, S. Disintegrins: A Family of Integrin Inhibitory Proteins from Viper Venoms, Proc. Soc. Exper. Biol. Med. *195:* 168-171, 1990.
23.   Niewiarowski, S., McLane, M. A., Kloczewiak, M., and Stewart, G. J. Disintegrins and Other Naturally Ocurring Antagonists of Platelet Fibrinogen Receptors, Seminars in Hematology. *31:* 289-300, 1994.
24.   Trikha, M., Rote, W. E., Manley, P. J., Lucchesi, B. R., and Markland, F. S. Purification and Characterization of Platelet Aggregation Inhibitors From Snake Venoms, Thromb. Res. *73:* 39-52, 1994.
25.   Folkman, J. Tumor Angiogenesis: Therapeutic Implications, N. Engl. J. Med. *285:* 1182-1186, 1971.
26.   Folkman, J. Editorial: Angiogenesis and Breast Cancer, J. Clin. Oncol. *12:* 441-443, 1994.
27.   Folkman, J. Tumor Angiogenesis. *In:* J. F. Holland, R. C. Bast, D. L. Morton, E. Frei III, D. W. Kufe, and R. R. Weichselbaum (eds.), Cancer Medicine, Vol. 1, pp. 181-204. Baltimor: Williams &Wilkins, 1997.
28.   Hanahan, D. and Folkman, J. Patterns and Emerging Mechanisms of the Angiogenic Switch during Tumorigenesis, Cell. *86:* 353-364, 1996.
29.   Stromblad, S. and Cheresh, D. A. Cell Adhesion and Angiogenesis, Trends in Cell Biology. *6:* 462-468, 1996.
30.   Stromblad, S., Becker, J. C., Yebra, B., C., B. P., and Cheresh, D. A. Suppression of p53 Activity and p21WAF1/CIP1 Expression by Vascular Cell Integrin avb3 during Angiogenesis, J. Clin. Invest. *98:* 426-433, 1996.
31.   Miyashita, T., Harigai, M., Hanada, M., and Reed, J. C. Identification of a p53-dependent Negative Response Element in the *bcl*-2 Gene, Cancer Res. *54:* 3131-3135, 1994.
32.   El-Deiry, W. S., Harper, J. W., O'Connor, P. M., Velculescu, V. E., Canman, C. E., Jackman, J., Pietenpol, J. A., Burrell, M., Hill, D. E., Wang, Y., Wiman, K. G., Mercer, W. E., Kastan, M. B., Kohn, K. W., Elledge, S. J., Kinzler, K. W., and Vogelstein, B. WAF1/CIP1 is Induced in p53-mediated G1 Arrest and Apoptosis, Cancer Res. *54:* 1169-1174, 1994.
33.   Friedlander, M., Brooks, P. C., Shaffer, R. W., Kincaid, C. M., Varner, J. A., and Cheresh, D. A. Definition of Two Angiogenic Pathways by Distinct av Integrins, Science. *270:* 15001502, 1995.
34.   Juliano, D., Wang, Y., Marcinkiewcz, C., Rosenthal, L. A., Stewart, G. J., and Niewiarowski, S. Disintegrin Interaction with avb3 Integrin on Human Umbilical

Vein Endothelial Cells: Expression of Ligand-Induced Binding Site on b3 Subunit, Exp. Cell Res. *225:* 132-142, 1996.
35.   Sheu, J. R., Yen, M. H., Kan, Y. C., Hung, W. C., Chang, P. T., and Luk, H. N. Inhibition of Angiogenesis *in vitro* and *in vivo*: Comparison of the Relative Activities of Triflavin, an Arg-Gly-Asp-containing Peptide and Anti-avb3 Integrin Monoclonal Antibody, Biochimica et Biophysica Acta. *1336:* 445-454, 1997.
36.   Price, J. E., Polyzos, A., Zhang, R. D., and Daniels, L. M. Tumorigenicity and Metastasis of Human Breast Carcinoma Cell Lines in Nude Mice, Cancer Res. *50:* 717-721, 1990.
37.   Repesh, L. A. A New *in vitro* Assay for Quantitating Tumor Cell Invasion, Invasion and Metastasis. *9:* 192-208, 1989.
38.   Osborne, C. K., Hobbs, K., and Clark, G. M. Effect of Estrogens and Antiestrogens on Growth of Human Breast Cancer Cells in Athymic Nude Mice, Cancer Res. *45:* 584-590, 1985.
39.   Meredith, J. E., Fazeli, B., and Schwartz, M. A. The Extracellular Matrix as a Cell Survival Factor, Mol. Biol. Cell. *4:* 953-961, 1993.
40.   Coller, B. S., Peerschke, E. I., Scudder, L. E., and Sullivan, C. A. A Murine Monoclonal Antibody That Completely Blocks the Binding of Fibrinogen to Platelets Produces a Thrombasthenic-like State in Normal Platelets and Binds to Glycoproteins IIb and/or IIIa, J. Clin. Invest. *72:* 325-338, 1983.
41.   Charo, I. F., Bekeart, L. S., and Phillips, D. R. Platelet Glycoprotein IIb-IIIa Like Proteins Mediate Endothelial Cell Attachment to Adhesive Proteins and the Extracellular Matrix, J. Biol. Chem. *262:* 9935-9938, 1987.
42.   Phillips, D. R., Charo, I. F., and Scarborough, R. M. GP IIb-IIIa: The Responsive Integrin, Cell. *65:* 359-362, 1991.
43.   Chang, Y. S., Chen, Y. Q., Timar, J., Nelson, K. K., Grossi, I. M., Fitzgerald, L. A., Diglio, C. A., and Honn, K. V. Increased Expression of aIIbb3 Integrin in Subpopulations of Murine Melanoma Cells with High Lung-Colonizing Ability, Int. J. Cancer. *51:* 445-451, 1992.
44.   Chen, Y. Q., Gao, X., Timar, J., Tang, D., Grossi, I. M., Chelladurai, M., Kunicki, T. J., Fligiel, S. E. G., Taylor, J. D., and Honn, K. V. Identification of the aIIbb3 Integrin in Murine Tumor Cells, J. Biol. Chem. *267:* 17314-17320, 1992.
45.   Peerschke, E. I. and Coller, B. S. A Murine Monoclonal Antibody That Blocks Fibrinogen Binding to Normal Platelets Also Inhibits Fibrinogen Interactions with Chymotrypsin-treated Platelets, Blood. *64:* 59-63, 1984.
46.   Felding-Habermann, B. and Cheresh, D. A. Vitronectin and Its Receptors, Current Opinion in Cell Biology. *5:* 864-868, 1993.
47.   Wayner, E. A., Orlando, R. A., and Cheresh, D. A. Integrins avb3 and avb5 Contribute to Cell Attachment to Vitronectin but Differentially Distribute on the Cell Surface, J. Cell Biol. *113:* 919-929, 1991.
48.   Frisch, S. M. and Francis, H. Disruption of Epithelial Cell-Matrix Interactions Induces Apoptosis, J. Cell Biol. *124:* 619-629, 1994.
49.   Connolly, D. T., Heuvelman, D. M., Nelson, R., Olander, J. V., Eppley, B. L., Delfino, J. J., Siegel, N. R., Leimgruber, R. M., and Feder, J. Tumor Vascular Permeability Factor Stimulates Endothelial Cell Growth and Angiogenesis, J. Clin. Invest. *84:* 1470-1478, 1989.
50.   Ferrara, N., Houck, K., Jakeman, L., and Leung, D. W. Molecular and Biological Properties of the Vascular Endothelial Growth Factor Family of Proteins, Endocr. Rev. *13:* 18-32, 1992.
51.   Nicolson, G. L. Tumor Cell Instability, Diversification and progression to Metastatic Phenotype from Oncogene to Oncofetal Expression, Mol. Biol. Cell. *47:* 1473-1487, 1987.
52.   Liotta, L. A. Tumor Invasion and Metastasis - Role of the Extracellular Matrix :Rhoads Memorial Award Lecture, Cancer Res. *46:* 1-7, 1986.
53.   Klemke, R. L., Cai, S., Giannini, A. L., Gallagher, P. J., de Lanerolle, P., and Cheresh, D. A. Regulation of cell motility by mitogen-activated protein kinase, Journal of Cell Biology. *137:* 481-92, 1997.

54.   Folkman, J. Fighting Cancer by Attacking Its Blood Supply,   Scientific American. *Sept.:* 150-154, 1996.
55.   Boehm, T., Folkman, J., Browder, T., and O'Reilly, M. S. Antiangiogenic Therapy of Experimental Cancer Does Not Induce Acquired Drug Resistance, Nature. *390:* 404-407, 1997.
56.   O'Reilly, M. S., Holmgren, L., Chen, C., and Folkman, J. Angiostatin Induces and Sustains Dormancy of Human Primary Tumors in Mice, Nature Medicine. *2:* 689-692, 1996.
57.   O'Reilly, M. S., Boehm, T., Shing, Y., Fukai, N., Vasios, G., Lane, W. S., Flynn, E., R., B. J., Olsen, B. R., and Folkman, J. Endostatin: An Endogenous Inhibitor of Angiogenesis and Tumor Growth, Cell. *88:* 277-285, 1997.
58.   Chen, C. S., Mrksich, M., Huang, S., Whitesides, G. M., and Ingber, D. E. Geometric Control of Cell Life and Death., Science. *276:* 1425-1428, 1997.
59.   Eliceiri, B. P., Klemke, R., Stromblad, S., and Cheresh, D. A. Integrin avb3 Requirement for Sustained Mitogen-Activated Protein Kinase Activity During Angiogenesis, J. Cell Biol. *140:* 1255-63, 1998.

Chapter 19

Lethal Toxins of Lizard Venoms
That Possess Kallikrein-Like Activity

Anthony T. Tu

Department of Biochemistry and Molecular Biology, Colorado State University,
Fort Collins, CO 80523–1870

The venoms of the lizard genus *Heloderma* contain many different types of proteins. Two types of *Heloderma* toxins, horridum toxin and gila toxin, were isolated in our laboratory. There are some similarities and some differences between these two toxins. For example, both toxins have kallikrein enzymatic activity, and the primary structure is also similar; moreover, they both have hypotensive effects when injected into rats. Gila toxin has a higher molecular weight and consists of 245 amino acid residues, while horricum toxin contains 210 residues. Both horridum and gila toxins are hemorrhagic, yet only horridum toxin causes exophthalmia. In this review article the chemical and several biological activities of these two toxins are compared extensively.

Among lizards, only two species, *Heloderma horridum* and *Heloderma suspectum*, have been found to be venomous (Figure 1A and B). *Heloderma* venom, like most venoms, is a complex mixture of proteins with diverse biological activity. Several enzymes and bioactive compounds have been isolated. These properties include phospholipase A$_2$, kallikrein-like enzymes such as gila toxin and helodermatine, arginine ester hydrolases and hyaluronidase (*1*). Lizard venom proteins such as extendin-3 and extendin-4 interact with vasoactive intestinal peptide receptors (*2,3*). Other physiologically active proteins include helodermins and helospectins (*4, 5, 6*). Lethal activities have been reported for gila toxin, horridum toxin, and another toxin called "lethal toxin" (*7, 8, 9, 10, 11*). However, aside from being lethal, gila toxin possesses kallikrein-like activity, and horridum toxin has ester hydrolase activity, both of which can potentially be valuable pharmacological agents.

Isolation of Gila Toxin and Horridum Toxin

Analysis of venom fractions of *Heloderma suspectum suspectum* and *H. horridum horridum* venoms revealed that both gila toxin and horridum toxins were found in the second fraction of Sephadex G-75 fractionation (Figure 2A). Both toxins were eventually purified to a homogeneous state by subsequent three-step purifications. Molecular weight determination by SDS-PAGE indicates that homogeneous gila toxin has a higher molecular weight than that of horridum toxin (Figure 3).

Primary Structure

Both gila toxin and horridum toxins are glycoproteins (*7, 12*). The carbohydrate portion of horridum toxin has been better characterized than that of gila toxin. For horridum toxin, mannose, glucose, galactose, and N-acetylglucosamine are the monosaccharides present, with a total of 4 mol of carbohydrates/mol of protein. Horridum toxin is quite different from gila toxins isolated from *Heloderma* venoms in primary structure and exophthalmic effect. The primary structure of apoprotein portions of two toxins was identified and is shown in Figure 4. Horridum toxin has 210 residues, whereas gila toxin has 245, with the major differences being in the C-terminus. Within the first 210 residues, only seven sites of residue substitution exist, at positions 25, 55, 107, 111, 112, 144, and 209 (Figure 4).

CD spectroscopy demonstrated similar secondary structure between the two proteins, as follows: horridum toxin: 11% α-helix, 41% β-structure, 26% β-turns, and 25% unordered structure; gila toxin: 6% α-helix, 42% β-structure, 27% β-turns, and 25% unordered structure (Figure 5).

Proteolytic Activity

Both toxins possess proteolytic activity, and both substrate specificity and sequence homology studies indicate that they are both kallikrein-like proteases. Both proteins hydrolyze kininogen and produce bradykinin, a hypotensive peptide (Figures 6 and 7). Degradation products of high molecular weight kininogen following incubation with horridum toxin are shown in Figure 8.

Both gila toxin and horridum toxin hydrolyze arginine esters (BAEE and TAME) and do not hydrolyze tyrosine ester such as ATEE (Table 1). They do not hydrolyze chromophogenic substrates for thrombin, factor X activation enzyme, or plasmin. The fact that kallikrein-like activity is inhibited by DFP but not by EDTA suggests that both toxins are serine type protease rather than metalloenzyme.

Since bradykinin is a well known hypotensive agent, hypotensive activity should be observed following animal injection with gila toxin or horridum toxin; indeed, injection of gila toxin into a rat does show pronounced effect of hypotensive action (Figure 9). Bradykinin is also known to stimulate uterus smooth muscle and to induce contractions. Testing fraction B_{40} (isolated after incubation of kinogen with gila toxin) using rat uterus revealed that the production of rat uterus contractions is

Figure 1. A. *Heloderma horridum horridum* from Mexico.
B. *Heloderma suspectum* from southern Utah.

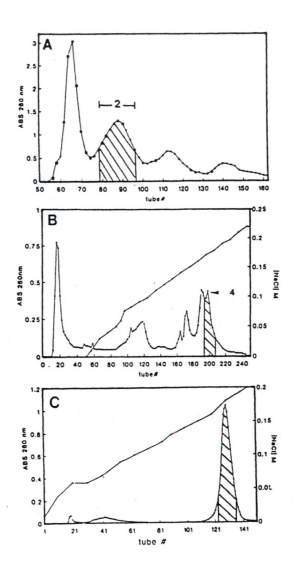

Gilatoxin

Figure 2. Comparison of fractionation patterns for gila toxin and horridum toxin.

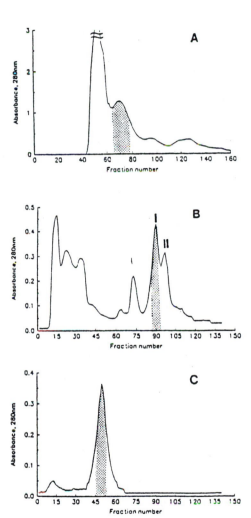

Horridum Toxin

Figure 2. *Continued.*

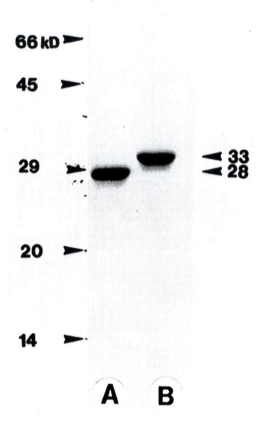

Figure 3. Homogeneity of (A) horridum toxin and (B) gila toxin on SDS-PAGE (10% gels).

```
                                10                    20                    30
Horridum toxin   I  I  G  G  Q  E  C  D  E  T  G  H  P  W  L  A  L  L  H  R  S  E  G  S  D  W  S  G  V  L
Gilatoxin        I  I  G  G  Q  E  C  D  E  T  G  H  P  W  L  A  L  L  H  R  S  E  G  S  T  W  S  G  V  L

                                40                    50                    60
Horridum toxin   L  N  R  D  W  I  L  T  A  A  H  C  E  E  L  G  P  M  K  I  C  F  G  M  H  N  R  N  V  L
Gilatoxin        L  N  R  D  W  I  L  T  A  A  H  C  E  E  L  G  P  M  K  I  C  F  G  M  K  N  R  N  V  L

                                70                    80                    90
Horridum toxin   R  G  D  E  Q  V  K  V  A  A  V  K  K  C  Y  P  A  T  A  G  T  I  Y  N  C  N  Y  V  N  T
Gilatoxin        R  G  D  E  Q  V  K  V  A  A  V  K  K  C  Y  P  A  T  A  G  T  I  Y  N  C  N  Y  V  N  T

                                100                   110                   120
Horridum Toxin   V  L  M  N  N  D  L  L  K  R  E  L  F  P  M  L  F  K  L  D  E  Y  V  D  Y  N  E  R  V  A
Gilatoxin        V  L  M  N  N  D  L  L  K  R  E  L  F  P  M  L  I  K  L  D  S  S  V  D  Y  N  E  R  V  A

                                130                   140                   150
Horridum Toxin   P  L  S  L  P  T  S  P  A  S  L  G  A  E  C  S  V  L  G  W  G  T  T  S  P  D  D  V  T  L
Gilatoxin        P  L  S  L  P  T  S  P  A  S  L  G  A  E  C  S  V  L  G  W  G  T  T  T  P  D  D  V  T  L

                                160                   170                   180
Horridum Toxin   P  D  V  P  V  C  V  N  I  E  I  F  N  N  A  V  C  Q  V  A  R  D  L  W  K  F  T  N  K  L
Gilatoxin        P  D  V  P  V  C  V  N  I  E  I  F  N  N  A  V  C  Q  V  A  R  D  L  W  K  F  T  N  K  L

                                190                   200                   210
Horridum Toxin   C  A  G  V  D  F  G  G  K  D  S  C  K  G  D  S  G  G  P  L  V  C  D  N  Q  L  T  G  I  V
Gilatoxin        C  A  G  V  D  F  G  G  K  D  S  C  K  G  D  S  G  G  P  L  V  C  D  N  Q  L  T  G  N  V

                                220                   230                   240
Gilatoxin        S  W  G  F  N  C  E  Q  G  E  K  Y  G  Y  I  K  L  I  K  F  N  F  W  I  Q  N  I  I  Q  G

                                245
Gilatoxin        G  T  T  C  P
```

Figure 4A. Alignment of amino acid residues of horridum toxin and gila toxin. Residues that differ are shaded.

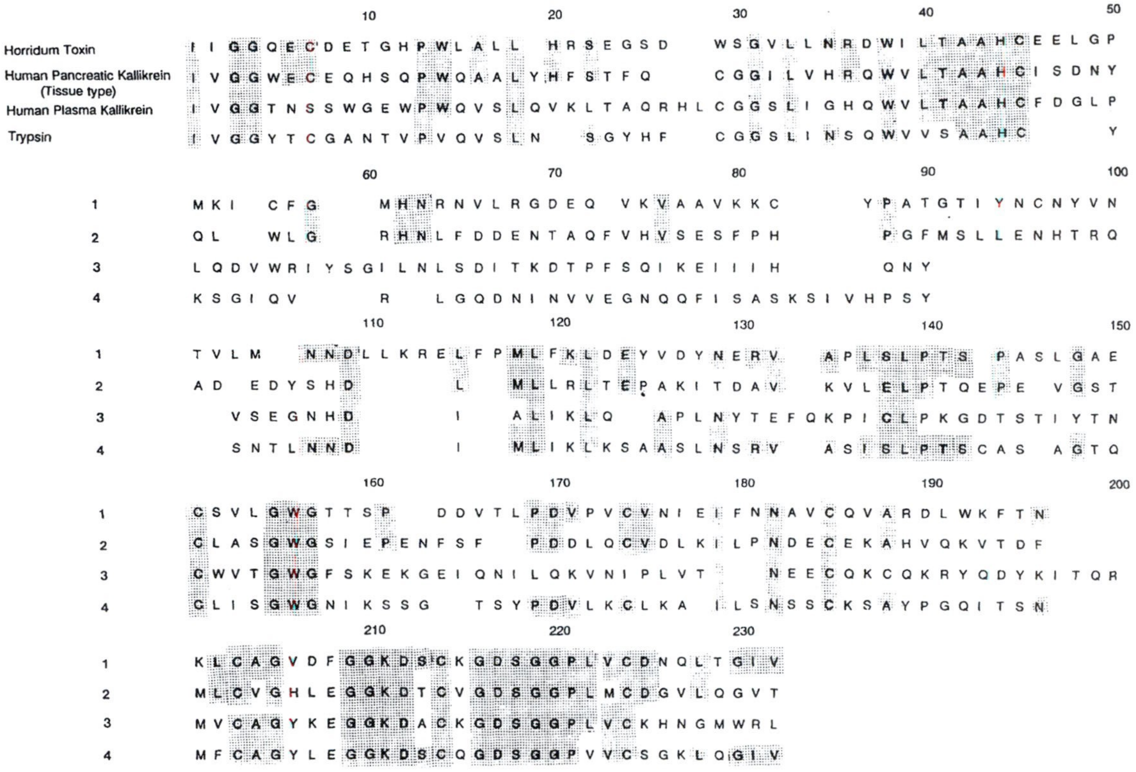

Horridum Toxin
Human Pancreatic Kallikrein (Tissue type)
Human Plasma Kallikrein
Trypsin

Figure 4B. Alignment of amino acid residues for horridum toxin, human pancreatic kallikrein, human plasma kallikrein, and trypsin. Residues which are shaded are identical to horridum toxin. Alignments were made to maximum homology.

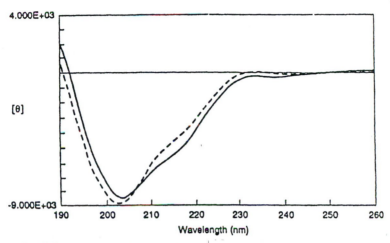

Figure 5. CD spectra of gila toxin (-) and horridum toxin (- - -) in the range of 190 to 260 nm.

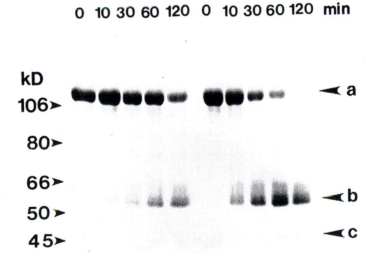

Figure 6. Degradation of high molecular weight kininogen with plasma kallikrein (left) and with gila toxin (right) examined by SDS-PAGE.

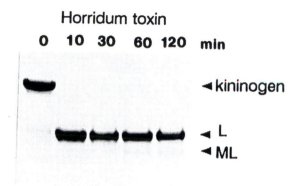

Figure 7. Degradation of high molecular weight kininogen by horridum toxin.

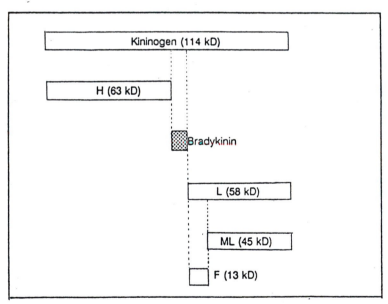

Figure 8. Diagram showing the hydrolysis products of kininogen when incubated with horridum toxin.

TABLE 1

Proteolytic Activity of Horridum Toxin and Gila Toxin Toward Synthetic Substrates and Effects of Inhibitors

Substrate	Activity of horridum toxin (units/mg)	Activity of gila toxin (units/mg)
N-Benzoylarginine ethyl ester (BAEE)	240	245
N-Tosylarginine methyl ester (TAME)	200	209
N-Acetyl-L-tyrosine ethyl ester (ATEE)	0	0
N-Benzoyl-Phe-Val-Arg-pNA (thrombin-like)	54	52
N-Benzoyl-Ile-Glu-Gly-Arg-pNA (factor X A)	0	0
N-Benzoyl-Val-Leu-Lys-pNA (human plasmin)	0	0
Val-Leu-Arg-pNA (urinary kallikrein)	0	0
Ile-Phe-Lys-pNA (plasmin-like)	0	0
Inhibited by DFP	Yes	Yes
Inhibited by EDTA	No	No

Unit is the amount of enzyme which hydrolyzes 1 μmol of substrate per minute.

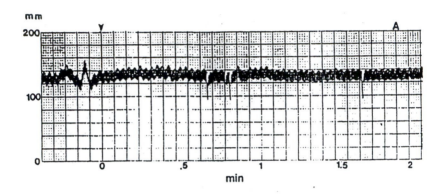

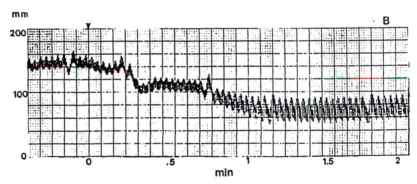

Figure 9. The effect of gila toxin on rat blood pressure. A, blood pressure after injection with normal saline. B, blood pressure after intravenous injection of gila toxin (0.5 μg/g; 20% of LD_{50}). Arrow indicates the point of sample injection.

concentration-dependent, indicating that the B_{40} peptide is indeed bradykinin (Figure 10).

Neither gila toxin nor horridum produced a fibrin clot when incubated with fibrinogen (data not shown). Fibrinogen has been used as a substrate by many investigators to study the action of venom proteases because these proteases often affect the blood coagulation pathway. The nature of fibrinogen hydrolysis gives some insight into the substrate specificity of horridum toxin. Fibrinogen is composed of three polypeptide chains: $A\alpha$, $B\beta$, and γ. It is therefore of interest to determine which of the chains are hydrolyzed by horridum toxin. Horridum toxin and gila toxin hydrolyzed the $A\alpha$ chain of fibrinogen faster than the $B\beta$ chain (Figure 11 A, C). The γ chain was resistant to hydrolysis for 12 h, although hydrolysis did occur after 18 h or incubation. Atroxase hydrolyzed both the $A\alpha$ and $B\beta$ chains, but not the γ chain (Figure 11B). The manner of hydrolysis by horridum toxin differed from that of atroxase, but was similar to that of gila toxin (Figure 11 A,B,C), which hydrolyzed all three chains.

Although gila toxin hydrolyzed kininogen, it did not hydrolyze the substrate for kallikrein, Val-Leu-Arg-pNA. Plasmin-like activity was also absent, as neither substrate tested, $\underline{N}$-benzoyl-Val-Leu-Lys-pNA and $\underline{N}$-benzoyl-Ile-Phe-Lys-pNA, was hydrolyzed.

Comparison to Other Kallikreins

Kallikreins exist as a tissue-type and as a plasma-type protein. A notable difference between tissue-type kallikreins and plasma-type kallikreins is the presence of residues 91-100 (Figure 4 A, B), MSLLENHTRQ, in tissue-type kallikreins and not in plasma-type kallikreins. This segment is frequently called the "kallikrein loop." Horridum toxin has segment 91-100, although the sequence is different from that of tissue-type kallikrein. Other similarities between horridum toxin and tissue-type kallikrein include the conservation of the cystine residues, the proline residues, and some large domains, which are characteristic of serine proteases. These sequences include the LTAAHC region at position 40-45, GWG at 155-157, GGKD at 209-212, and GDSGGPL at 216-222. In fact, the C-terminus has a high degree of homology with trypsin (a serine protease). Nevertheless, the Cys^{174}-Val^{175} motif seems to be a feature of tissue kallikreins and is absent from plasma kallikreins. The Cys^{29} in other tissue kallikreins is substituted by Trp in horridum toxins, whereas the Trp^{55} in human kallikrein is substituted by Cys in the toxins. However, the enzymatic activity is conserved, and there appears to be room for those residues in the structure. The absence of some characterisitic features of plasma kallikreins, such as the GLPLQ region at 48-52, Cys^{138}, and Lys^{225}-His^{226}, and Trp^{231}-Arg^{232}, also further substantiates the classification of horridum toxin as a tissue kallikrein. His^{44}, Asp^{109}, and Ser^{218} play important roles in catalysis by serine proteases. All these residues are present in horridum toxin as well, further corroboration that horridum toxin is also a serine protease. Taken together, the evidence indicates that horridum toxin and gila toxin are different types of kallikreins: structurally similar to tissue type, but having the substrate specificity of plasma kallikrein.

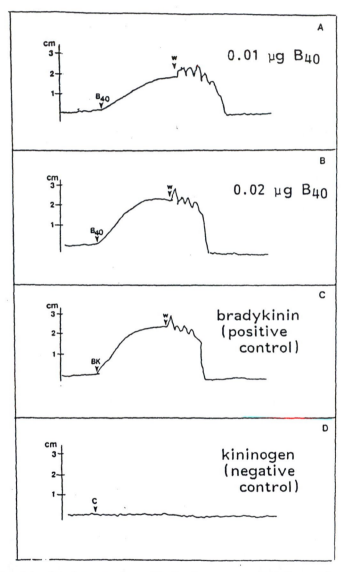

Figure 10. Stimulated contraction of isolated rat uterus smooth muscle. A, effect of 0.01 μg/g of B_{40}. B_{40} is the purified product bradykinin released from HMW kininogen by the action of gila toxin. W is the point of washing of muscle with buffer. B, effect of 0.02 μg of B_{40}. C, effect of 0.02 μg of bradykinin (positive control). D, effect of filtrate (potential spontaneously released product) from incubation of kininogen alone (negative control). The X-axis represents time.

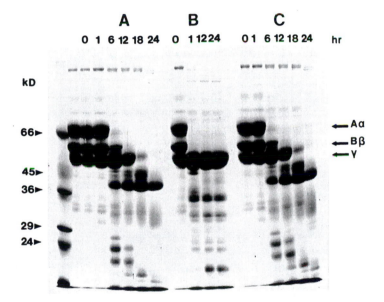

Figure 11. SDS-PAGE analysis of fibrinogen (2%) hydrolysis products after incubation for the indicated time intervals with (A) gila toxin, (B) atroxase, (C) horridum toxin.

Cleavage of Angiotensin I and II by Gila Toxin

Incubation of angiotensin I with gila toxin resulted in the degradation of angiotensin I, a hypertensive peptide originating from angiotensinogen. At zero time, only angiotensin I was visible (as peak a, Figure 12). As incubation continued, digestion of angiotensin I is demonstrated by the appearance of a new peak, peak b (Figure 12 Bb, Cb, Db). The amino acid sequence of peak b was determined and found to be Val-Tyr-Ile-His-Pro-Phe-His-Leu, demonstrating that the arginylvaline bond of angiotensin I was cleaved by gila toxin. Gila toxin also hydrolyzed angiotensin II and released a dipeptide from the NH_2-terminal end (data not shown). The cleavage of angiotensin I may be a contributing factor for the prolonged hypotensive action of gila toxin.

Toxicology

Both gila toxin and horridum toxins are lethal and have an LD_{50} of 2.9 and 2.6 $\mu g/g$ in mice, respectively. Fractionation of the toxins from crude venom did not significantly increase toxicity.

The action of horridum toxin on biological systems is significantly different from gila toxin. Horridum toxin causes exophthalmia, whereas gila toxin does not (Figure 13). The presence of two proteins with similar sequences but different biological activities is not unexpected in *Heloderma* venom. Other peptides, such as extendin-3 and extendin-4, differ only in two amino acids but have different interactions with their receptors.

Discussion

Lizard envenomation causes dizziness, hypotension, tachycardia, tissue damage, and ecchymosis. Anaphylaxis, severe periorbital swelling, lymphangitis, and lymphoadenites were also observed in lizard poisoning. Since a crude venom contains many varieties of proteins, it is hard to define which venom component induces which biological effect. So far, two toxins were isolated from lizard venom in our laboratory. Gila toxin is nonhemorrhagic but lethal to mice. Horridum toxin is hemorrhagic, lethal, and causes exophthalmia. Venom lethality is apparently caused by gila toxin or horridum toxin. One surprising fact is that both gila toxin and horridum toxin are proteolytic enzymes especially like kallikrein. Kallikrein is a relatively natural constituent of all animals. The amino acid sequence of both gila toxin and horridum toxin has high homology with kallikrein from mammarian source. Then a question is why the lizard kallikrein is so toxic while the mammarian enzyme is nontoxic. This question has to be solved eventually in the future by further investigation.

Another question is why horridum toxin exhibited exophthalmia while gila toxin did not, although the two toxins have some similarity in their amino acid sequences. Again, there is no answer at the moment.

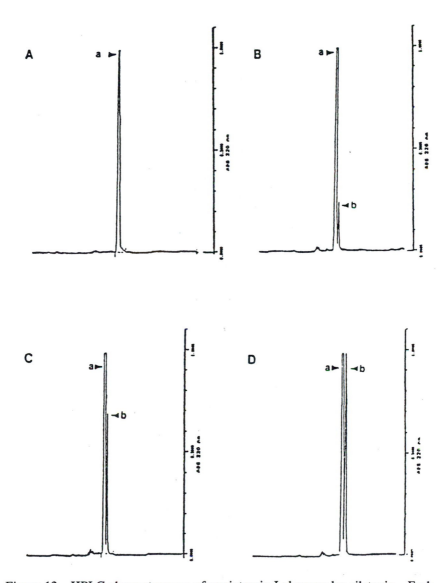

Figure 12. HPLC chromatograms of angiotensin I cleavage by gilatoxin. Each chromatogram represents angiotensin I (a) and degradation products (b) after 0-, 1-, 2-, and 6-h incubation times (A-D, respectively). Chromatography was performed by using a linear gradient of 0-50% acetonitrile in water containing 0.1% trifluoroacetic acid for 50 min at a flow rate of 1 ml/min.

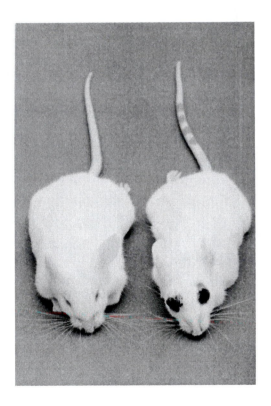

Figure 13. The exophthalmic effect induced by intravenous injection of horridum toxin. (A) control (saline) and (B) horridum toxin.

Literature Cited

1. Tu, A. T. In *Handbook of Natural Toxins*; Tu, A. T., Ed.; Marcel Dekker: New York, 1991, Vol. 5; pp. 755-773.
2. Eng, J.; Kleinman, W. A.; Singh, L.; Singh, G.; Raufman, J. P. *J. Biol. Chem.* **1992**, *267*, 7402-7405.
3. Raufman, J. P.; Jensen, R. T.; Sutliff, V. E.; Pisano, J. J.; Gardner, J. D. *Am. J. Physiol.* **1982**, *242*, G471-G474.
4. Vandermeers, A.; Vandermeers-Piret, M. C.; Vigneron, L.; Rathe, J.; Stievenart, M.; Winand, J.; Christophe, J. *Eur. J. Biochem.* **1991**, *196*, 537-544.
5. Vandermeers, A.; Gourlet, P.; Vandermeers-Piret, M. C.; Cauvin, A.; DeNeif, P.; Rathe, J.; Sroboda, M.; Robberecht, P.; Christophe, J. *Eur. J. Biochem.* **1987**, *164*, 321-377.
6. Parker, D. S.; Raufman, J. P.; O'Donohue, T. L.; Bledsoe, M.; Yoshida, H.; Pisano, J. J. *J. Biol. Chem.* **1984**, *259*, 11751-11755.
7. Utaisincharoen, P.; Mackessy, S. P.; Miller, R. A.; Tu, A. T. *J. Biol. Chem.* **1993**, *268*, 21975-21983.
8. Hendon, R. A.; Tu, A. T. *Biochemistry* **1981**, *20*, 3517-3522.
9. Nikai, T.; Imai, K.; Sugihara, H.; Tu, A. T. *Arch. Biochem. Biophys.* **1988**, *264*, 270-280.
10. Komori, Y.; Nikai, T.; Sugihara, H. *Biochem. Biophys. Res. Comm.* **1988**, *154*, 613-619.
11. Komori, Y.; Nikai, T.; Sugihara, H. *Biochim. Biophys. Acta* **1988**, *967*, 92-102.
12. Datta, G.; Tu, A. T. *J. Peptide Res.* **1997**, *50*, 443-450.

NERVE AGENTS AND DOPING COMPOUNDS

Chapter 20

Overview of Sarin Terrorist Attacks in Japan

Anthony T. Tu

Department of Biochemistry and Molecular Biology, Colorado State University, Fort Collins, CO 80523–1870

Multiple sarin attacks in Tokyo subway trains by Aum Shinrikyo on March 20, 1995, resulted in 12 deaths and nearly 6000 injuries sending shock waves throughout the world. Actually, a less publicized case of sarin attacks by Aum Shinrikyo in Matsumoto produced great shock, fear, and uproar in Japan the year before on June 27, 1994. The Matsumoto attack resulted in several deaths and 600 injuries. The attacks in Matsumoto and the Tokyo subways were deliberate acts of terrorism planned by the new religious sect Aum Shinrikyo.

Nerve gas is a major component of chemical weapons arsenals in many countries. The use of sarin nerve gas in Matsumoto marked the first use of chemical weapons in a non-combat situation by a terrorist group. The Matsumoto attack took place in an apartment complex and its surrounding homes while the occupants were sleeping.

Sarin was being produced by the Aum Shinrikyo sect in Kamikuishiki, Japan. A sarin gas leakage occurred in July 1994; however, the Japanese police could not pin down the source, or identity, of the poisonous gas. In November of the same year there was another leakage, and this time the police identified methylphosphonic acid as a degradation product from the soil nearby. This finding could hold Aum Shinrikyo responsible for the manufacturing of sarin nerve gas. However, until they used the gas in a terrorist attack, no crime was committed under former Japanese law.

Aum Shinrikyo was not only dealing with sarin; the assassination of a person with VX gas using a plastic tube attached to a syringe took place before the Tokyo subway incidents. The Japanese police identified VX degradation products from the victim's serum approximately one year after the assassination took place. Using degradation products as a method for detection is a powerful tool.

304

In this article, an overall look at sarin production, the terrorist attacks of Aum Shinrikyo, and the treatment of the injured will be addressed.

Background

Aum Shinrikyo

In February 1984, Shoko Asahara started the Aum Shinsen Club with only 15 members. Developing at an astonishing rate, this club swelled in membership to 1300 members only three years after its initiation. The club was so large that it opened a branch office in Osaka, Japan. In July 1987, the name was changed to "Aum Shinrikyo," which means "Aum Supreme Truth Sect." The development of Aum Shinrikyo accelerated unhindered, and in 1989 the group obtained a legal title as a religious sect from the Tokyo Metropolitan Government. Obtaining a legal title had a tremendous effect on the subsequent development of the sect. In 1994 the membership swelled to 10,000 in Japan, 30,000 in Russia, and several thousand members spread throughout the rest of the world. Various facilities were established in Japan, and offices overseas included such cities as New York, Frankfurt, and Moscow.

Ultimate Objective

The ultimate objective behind the sarin production was to overthrow the present Japanese government and establish the new government of Aum Shinrikyo. They chose sarin as the major chemical in their arsenal to carry out this objective. In order to gain governmental control, Aum Shinrikyo planned the following:

a. Spray sarin from the air over the parliament building using a large helicopter. For this purpose a Russian helicopter was purchased.

b. A special force of several hundred loyal troops would attack parliament and other important Japanese government officials and kill all the parliament members and high-ranking government officials. In order to carry out this objective, they actively recruited members from the Self-Defence Forces (SDF) and manufactured AK-47 automatic rifles in large quantities at the 10th Satyan building in Kamikuishiki near Mt. Fuji.

Terrorist Attacks

Matsumoto Sarin Poisoning Incident

The manufacturing of sarin started under the supervision of Mr. Tsuchiya, who within 3 months had produced 5 kg and after 6 months developed 20 kg. Mr. Tsuchiya was trained as a physical organic chemist and obtained an M.S. degree from Tsukuba University. Aum Shinrikyo decided to test the sarin's effectiveness by attempting the assassination of judges in Matsumoto who were not friendly to the group's cause.

On the morning of July 27, 1994, an unknown poisonous gas was released outside of an apartment complex in Matsumoto. The outside temperature was warm and therefore many residents of the apartment complex slept with their windows open. The poisonous gas entered the rooms through the open windows, resulting in 7 deaths and 600 injured.

At the time of the poisoning no one knew the exact nature of the substance. However, shocking news was released by the Japanese police on July 3, 1994, after they uncovered evidence pointing toward sarin as the unknown assailant. A water sample in a small pond from the garden of a nearby house was tested to isolate sarin. To many Japanese, sarin was unheard of prior to this attack, and the news caused a shock wave of questions throughout the whole country. An investigation of the source began immediately.

Detection of Sarin From Soil

Japanese police launched a massive search for the origin of the sarin, but with little success. The questioning of almost everyone who purchased organophosphate compounds began, including many scientists at various universities. The search yielded few answers.

After Aum Shinrikyo set a sarin production target of 70 tons, large-scale manufacturing began. Due to a worker mishandling the materials during production, sarin precursors leaked from the production site to the surrounding soil. On July 9th and 15th the police received many complaints from village people near the 7th Satyan or nearby buildings; however, Aum Shinrikyo flatly denied responsibility for the noxious gas (Figure 1).

Japanese police learned that sarin could be detected from the soil through the detection of its degradation products. On November 16, 1994, poisonous gas leaked again from the 7th Satyan, and this time the police actively went after those degradation products which could prove the manufacturing of sarin. Using gas chromatography - mass spectrometer (GS-MS), they found methylphosphonic acid, and this was enough to firmly establish the scientific evidence of sarin production. This evidence came four months before the Tokyo subway incident.

After the Tokyo subway sarin incident, many staff members of the Aum Shinrikyo cult were arrested. It then became clear to the authorities that the leaks in 1994 were indeed from the cult's facilities. An Aum Shinrikyo spokesman claimed that the U. S. Army released the gas in order to destroy the sect.

Preparation for Tokyo Subway Attack

The decision to attack the Tokyo subway system with sarin was made early in the morning March 18, 1995. Sarin was produced from its immediate precursor difluoromethylphosphonate, $CH_3(PO)F_2$, at the Jivaka Prefab near the 7th Satyan (Figure 2). For the reaction to produce sarin the workers used three necked flasks filled with hexane, diethylaniline, and difluoromethylphosphonate. The reaction temperature was carefully controlled, and isopropyl alcohol was added drop by drop by a worker named Seiji Tashita.

Figure 1. A farm house nearby the Seventh Satyan where sarin was manufactured by Aum Shinrikyo. Farmers complained that they could not breathe because of poisonous gas leak from the Seventh Satyan.

Figure 2. The author standing in front of the manufacture sites of sarin in Kamikuishiki.

  Right: The 7th Satyan.
  Middle: Kushiti Galva Prefab. The sarin made here was used in Matsumoto City.
  Left: Jivaka Prefab. The sarin produced here was used in Tokyo Subway.

Within a 24-hour period between 6 and 7 liters of highly impure sarin (30%) were produced. Knowing that the sarin they had produced was crude made the cult anxious; the anxiety was exacerbated by a fear of a police raid on the plant. Deciding to strike Tokyo in order to distract the police away from the plant, the cult transferred sarin to 11 nylon-polyethylene bags 50x70 cm. To practice puncturing the bags, they ruptured water-filled bags with the tip of an umbrella.

In the attacks on Matsumoto and the Tokyo subways, the sarin had a slightly different composition. For the Matsumoto incident, sarin was made from phosphorus trichloride (PCl_3) and the product was purified. The gas used in the Tokyo subway was made directly from difluoromethylphosphonate and contained a small amount of diisopropylmethylphosphonate, a common byproduct of sarin production (Figure 3).

Asahara ordered the selection of several members to carry out the mission. The people selected were Yasuo Hayashi, Kenichi Hirose, Mabito Yokoyama, Toru Toyada, and Ikuo Hayashi. At 3 p.m. in the afternoon of March 18, 1995, they assembled at the office of Hideo Murai. This meeting was held two days prior to the Tokyo subway incident. The meeting consisted of planning which trains to attack, the routes of arrival and departure. Those trains closest to the Tokyo Metropolitan Police Headquarters were selected for the attack; in particular the Hibiya, Marunouchi, and Chiyoda Lines. The simultaneous release of sarin on all the trains was to take place at the chosen time of 8 a.m. due to the "rush hour" commute that would create the highest amount of chaos.

The Strike of the Subways

For each action group there was one person selected to deliver the bags to the subway train station, and one person was selected to plant the bags and rupture them with an umbrella. Sarin was placed on five different subway trains on March 20, 1995. The times of sarin release were as follows:

1. Hibiya Line from Kitasenjyu to Megro - slightly after 7:46 a.m.
2. Hibiya Line from Naka Megro to East Zoo Park - 7:59 a.m.
3. Marunouchi Line from Ikebukuro to Ogikubo - 7:59 a.m.
4. Chiyoda Line from Abiko to Yoyogi - slightly after 7:48 a.m.
5. Marunouchi Line from Ogikubo to Ikebukuro - 8:01 a.m.

The release of sarin caused panic, confusion, and agony to the passengers. The casualty figures of the respective trains were as follows:

1. Hibiya Line to Megro - 7 deaths, 2475 injured
2. Hibiya Line to East Zoo Park - 1 death, 532 injured
3. Marunouchi Line to Ogikubo - 1 death, 358 injured
4. Chiyoda Line - 2 deaths, 231 injured
5. Marunouchi Line to Ikebukuro - no deaths, 200 injured

Many important Japanese government buildings are located in the area where the attack took place (Kasumigaseki). The trains were filled with commuters going toward Kasumigaseki to work, and obviously the cult's targets were people going toward the Police Station and other government buildings.

Sarin is a deadly nerve agent, but these attacks resulted in only 13 deaths (although about 6000 were injured). The low mortality is due to the impurity of the sarin the cult used. According to a Japanese police analysis, the sarin used in the Tokyo subway contained only 35% sarin. This was a fortunate factor within this unfortunate incident. The Tokyo subway system has a ventilation system where the air inside the subway was constantly replaced with fresh air. This helped lower the mortality of the victims. After the attack, the contaminated train was detoxified using 5% sodium hydroxide solution by Ground Self Defense Forces of Japan.

Chemical Composition Detected From Tokyo Subway Sarin

Three hours after the terrorist attacks on the Tokyo subway the Japanese police announced that the poison was sarin. The Japanese police initially used GC-MS to find that the Tokyo subway sarin contained sarin (35%), diisopropylphosphono-fluoridate, triisopropylphosphoric acid, diisopropylphosphonic acid, and diethylaniline. The last compound is a catalyst or stabilizing agent to accelerate sarin formation by neutralizing the hydrogen chloride formed in the early stage of sarin production.

Raids on Aum Shinrikyo Facilities

On March 22, 1995, the Japanese police raided Aum Shinrikyo's compound in Kamikuishiki and other cult facilities throughout Japan. A large quantity of various chemical compounds was confiscated (Table 1). Most of the chemicals captured in the raids were those used in the manufacture of sarin, VX, mustard gas, and other biological weapons. No intact sarin was found. Apparently Aum Shinrikyo destroyed all of its sarin stockpile to eliminate evidence. However, police identified the degradation products from the vessels and nearby soils. This evidence is as incriminating as sarin itself.

Another Sarin Attack

The most dramatic sarin terrorist attack was on the Tokyo subway (March 20, 1995), but a previous one was in Matsumoto on June 27, 1994. But Aum Shinrikyo used sarin even before these events. In March 1994, a cult member tried to kill Mr. Taisaku Ikeda, a religious leader of Sokagakkai considered a rival, by spraying sarin from equipment mounted on a wagon similar to the one used in Matsumoto. But the spray machine malfunctioned and their own member, Mr. Niimi, an operator of the machine, was poisoned. Immediately they used an oxime (PAM), and Niimi recovered. From this incident they learned that PAM was an effective antidote for sarin. They also learned the defect of their spray machine, and subsequently developed a better one and successfully used it in Matsumoto four months later.

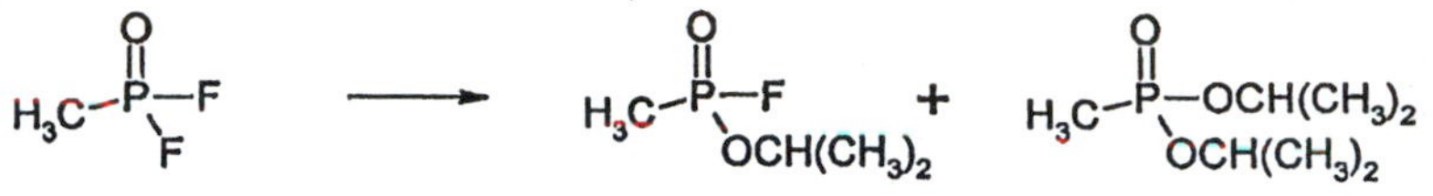

Sarin   Diisopropyl methylphosphonate

Figure 3. Production of sarin and its byproducts.

Table 1. Chemicals Confiscated from Aum Shinrikyo's Compounds in 1995

Compounds	Quantity	Possible use
Acetone	—	Solvent
Acetonitrile	206 half-liter bottles	Solvent
Acetylcholinesterase	—	Interaction with sarin
Activated charcoal	—	Protection against sarin, for purification
Aluminum chloride	—	—
Ammonium chloride	—	—
Atropine	—	Drug for sarin
Calcium chloride	1.2 tons	Drying agent
Carbon monoxide	10 2-m cylinders	—
Chloroform	2 tons	Solvent
Diethylaniline	—	Solvent
Dimethylamino ethanol	—	Preparation for choline
Dimethylmethylphosphonate	—	Preparation for saline
Ether	100 20-l drum cans	Solvent
Ethylalcohol	2 tons	Solvent
Glycerol	60 tons	Preparation for explosive
Hydrogen fluoride	—	Preparation for sarin
Isopropylalcohol	several drum cans	Preparation for sarin
Magnesium chloride	—	—
Magnesium nitrate	50 bags	Preparation for explosive
Methanol	large quantity	Solvent
Methyliodide	—	Preparation for sarin
Nitric acid	1.5 tons	Preparation for explosive
Peptone	200 cans of 30×30×50 cm	Media for bacteria
Phenylacetonitrile	—	Preparation for amphetamine
Phosphorus trichloride	50 tons	Starting compounds for sarin
Potassium iodide	—	Preparation for sarin
2-PAM (2-pyridine-aldoximethiodide)	600 ampules	Drug for sarin
PB (pyridostigmine)	—	Prophylactic drug for sarin
Sodium carbonate	—	Neutralization agent for sarin
Sodium cyanide	80 kg	Preparation for tabun
Sodium fluoride	10 tons	Preparation for sarin
Sodium hydroxide	—	Neutralization agent for sarin
Sodium hypochlorite	—	Oxidation agent for sarin
Sulfur	—	Preparation for VX
Sulfuric acid	320 bottles	For chemical reactions

Sarin Production by Aum Shinrikyo

Prior to the Tokyo subway attack, Aum Shinrikyo made about 50 kg of sarin. More detailed production data are shown here:

November 1993 - 20 kg
December 1993 - 3 kg
February-March 1994 - 30 kg

The first one was a trial preparation, and the last preparation was used for the terrorist attack in Matsumoto on June 27, 1994. Mass production of sarin with 70 tons as the target production quantity was started after September 1994.

Sarin was produced in five steps starting from phosphorus trichloride.

After the seizure of the 7th Satyan, the Japanese police detected trimethylphosphate, dimethylmethylphosphonate, monoisopropylmethylphosphonic acid, and methylphosphonic acid. Trimethylphosphate $[PO(OCH_3)_3]$ is the oxidation product of trimethoxyphosphorus. Monoisopropylmethylphosphonic acid is the degradation product of sarin. Methylphosphonic acid is broken down from dichloromethylphosphonate, difluoromethylphosphonate, and monoisopropylmethylphosphonic acid (1).

Treatment

Sarin attaches to acetylcholinesterase (AChE) rendering the enzyme nonfunctional (2, 3, 4). The drug of the oxime type compound will release the enzyme by combining with sarin (Figure 4). 2-PAM is frequently used for the treatment of sarin and other nerve gases.

Another treatment is based on the administration of atropine. When one is poisoned by sarin, AChE does not function. This means that acetyl-choline attached to the acetylcholine receptor is not hydrolyzed. This causes excitement of the nerve and the muscle. Atropine attaches to the receptor by replacing acetylcholine (Figure 5). Atropine's structure is different from that of acetylcholine, but the distance from C=O group to N atom is 7°A for both compounds. This is why atropine can attach to the acetylcholine receptor.

In Japan, 2-PAM is mainly kept in rural hospitals because the drug is used for organophosphate insecticide poisoning that occurs in the countryside. Therefore, at the time of the Tokyo subway sarin attack, many hospitals in Tokyo City did not possess the drug needed for the treatment. An emergency request was made to the pharmaceutical companies to ship both 2-PAM and atropine to Tokyo's hospitals. Even with these shipments, the absolute quantity of the drugs for all sarin terrorist attack victims was not enough. For this reason, only critically ill victims received drug treatment.

The drug, 2-PAM, would effectively raise the blood cholinesterase level in 2-3 days (5, 6). Without 2-PAM, it took two weeks to recover to the normal blood acetyl cholinesterase level (7). When 2-PAM was combined with atropine, the recovery was even more satisfactory.

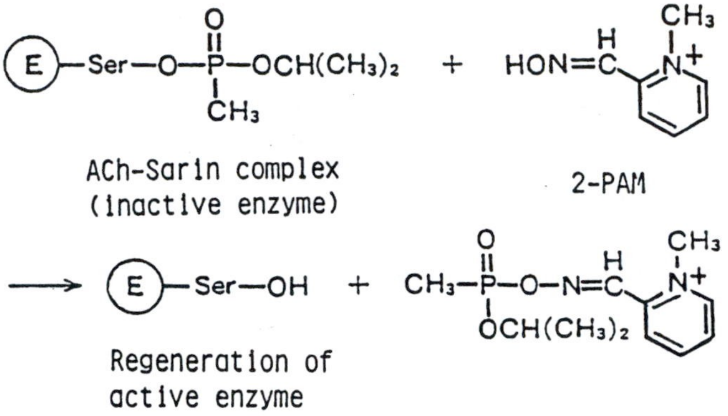

312

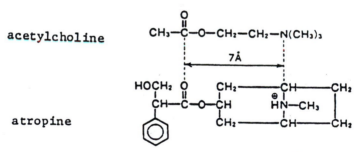

ACh-Sarin complex
(inactive enzyme)

2-PAM

Regeneration of
active enzyme

Figure 4. Detoxification mechanism of sarin by a drug, PAM.

acetylcholine

$$CH_3-C-O-CH_2-CH_2-N(CH_3)_3$$

7Å

atropine

Figure 5. Similarity of atropine and acetylcholine.

Many sarin victims showed shrinking of the pupil (miosis), as this effect is commonly observed for patients of organophosphate poisoning. The eyes are very sensitive to sarin, and many victims suffered temporary loss of vision (*8, 9*). Among many hospitals in Tokyo, St. Luke's International Hospital treated the largest number of patients. On the day of March 20, 1998, 640 patients were admitted to the Emergency Department alone (Figures 6, 7). Signs and symptoms of 111 patients moderately to severely injured from sarin were miasis, eye pain, blurred vision, dim vision, conjunctival injection, tearing, dyspnea, coughing, chest oppression, wheezing, tachypnea, nausea, vomiting, vomiting, diarrhea, headache, weakness, fasciculations, numbness of extremities, decrease of consciousness level, vertigo and dizziness, convulsion, running nose, nose bleeding, sneezing, agitation (*9, 10, 11*) (Figure 8).

Assassination Using VX

VX is one of the most toxic nerve agents known and has great penetrating action through the skin. The lethality is 0.1 mg-min/m^3 for VX compared to about 100 mg-min/m^3 for sarin. VX has probably never been used in combat; however, it was used by Aum Shinrikyo. VX was produced by Aum Shinrikyo during September to December of 1994 under the supervision of Tsuchiya. In total, 340 grams of VX were produced.

In December 1994, Mr. Chyujin Hamaguchi, a 28-year-old man, was murdered on a street in Osaka, Japan, because he soured on the cult. He was attacked by two men who sprinkled the nerve agent on his neck. The VX was dropped from a plastic tube attached to a syringe. He chased them for about 100 yards before collapsing, dying 10 days later without ever coming out of a deep coma. An autopsy indicated that the plasma cholinesterase level was low, and contraction of the pupil was observed in one eye, but not in the other. At that time nobody knew who killed him or for what purpose. It was suspected that the victim was poisoned somehow with an organophosphate pesticide. However, the cause of death was classified after cult members arrested for the Tokyo subway attack confessed to the killing. Fortunately, the victim's blood was saved and analysis was made using GC-MS by the Osaka Police Forensic Science Laboratory, Osaka, Japan. They found O-methyl-ethylphosphonate CH_3-(PO) (OC_2H_5) (OH), and another metabolite, $CH_3SCH_2N[CH(CH_3)_2]_2$ (*12, 13*). This murder is the only successful assassination with VX ever documented.

Although Mr. Hamaguchi's murder is the most documented case involving VX, the cult members also used VX in another assassination attempt. They tried to kill a lawyer who opposed them. They mixed VX with hair oil and squirted the mixture into a keyhole hoping the lawyer would then touch the nerve agent on his contaminated key. The Japanese police later found VX degradation products in the keyhole.

When VX is degraded in soil, several compounds are obtained, and they are CH_3-(PO) (OC_2H_5) (OH) and methylphosphonic acid $[CH_3(PO) (OH)_2]$. The formation of methylphosphonic acid is not limited to VX, but sarin also produces methylphosphonic acid when metabolized in the human body.

314

Figure 6. A large number of ambulances assembled to rescue sarin victims in Tokyo after terrorist attack in subway.

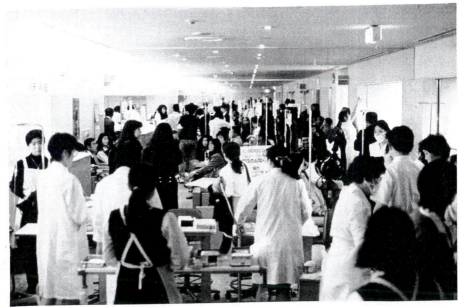

Figure 7. A large influx of patients made treatment very difficult.

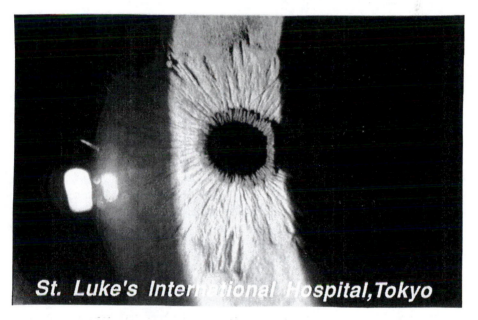

Figure 8. Shrinkage of pupil after exposed to sarin.

Aum Shinrikyo's Biological Weapons

Aum Shinrikyo also released other toxic agents. One incidence was the release of anthrax in Tokyo in 1994 (2). Anthrax is the most feared biological agent in the world, but no significant poisoning took place except many people complaining of a bad smell.

Botulinum (C. botulinum) was released on March 15, 1995, in a subway station in Tokyo. This was five days prior to the sarin attack in the subway. Botulinum was released by spraying, using two spray machines. At that time no one realized that Aum Shinrikyo had ever used such a dreadful bacteria, although the Japanese police discovered spraying machines left in the subway station.

In both the anthrax and botulinum cases, there was little damage to the public. This is worth the attention of toxicologists, pathologists, and microbiologists. Although these two types of bacteria are considered to be deadly, there was little damage. Both bacteria produce toxins or spores only when the organisms are grown in anaerobic conditions or in stressed nutrient conditions.

Trials of Cult Members

The trials of cult members are currently underway in Japan. Low-ranking members of the Aum Shinrikyo cult were tried first, and many were sentenced but without jail time. Some were jailed but for only a few years. Trials for serious crimes committed by the members who were directly associated with the terrorism are currently underway--jail time is expected. The sentencing will be proportional to the seriousness of their crimes. One of the senior members, Ikuo Hayashi, a physician who also participated in planting sarin in the Tokyo subway, was sentenced to life in prison. The authorities felt Dr. Hayashi repented for the crime he committed, and because he cooperated with the police to crack down on the crimes committed by another cult member, the death sentence was not requested.

Revival of Aum Shinrikyo

In Japan, since 1952 there has been a law called "Prevention of Destructive Activity." Because of the horrifying acts that the sect performed, the Japanese government tried to apply this law to disband the Aum Shinrikyo permanently and in 1996 requested that the Security Review Committee outlaw the organization. However, the Committee felt that all the members involved in the terrorist activities had been rounded up and the Aum Shinrikyo was no longer a threat to the public security. As long as they were only involved in religious activity, the sect could retain the legitimate status of a religious organization.

Since then, the Aum Shinrikyo has been continuing religious activity and expanding the organization network nationwide in Japan.

Acknowledgments

The author expresses thanks for information supplied by Dr. Y. Seto, Ms. N. Tsunoda, Dr. H. Tsuchihashi, Dr. T. Okumura, and Dr. K. Maekawa. The author also expresses his gratitude for a medal given to him by the Research Institute of Police Science for his assistance to the Japanese police. The author admires Japanese scientists' competence in dealing with the sarin incidents even though they did not have prior knowledge about sarin.

Literature Cited

1. Seto, Y. In *Current Topics in Forensic Science*, Proc. 14th Meeting of the International Association of Forensic Sciences; Takatori, T.; Takasu, A., Eds.; Tokyo, Japan, 1977, Vol. 4, pp. 35-38.
2. Tu, A. T. *J. Mass Spectrom. Soc. Jpn.*, **1996**, *44*, 293-320.
3. Tu, A. T. *Proc. Third Congress of Toxicology in Developing Countries*, Cairo, Egypt, 1996, Vol. 1, pp. 81-108.
4. Tu, A. T. *Proc. 6th CBW Protection Symposium*, Stockholm, Sweden, 1998, Vol. 1, pp. 13-18.
5. Seto, Y.; Tsunoda, N.; Ohta, H.; Nagano, T. In *Current Topics in Forensic Science*, Proc. 14th Meeting of the International Association of Forensic Science; Takatori, T.; Takasu, A., Eds.; Tokyo, Japan, 1997, Vol. 4, pp. 35-38.
6. Minami, M. *Program of Japanese Clinical Toxicology*, Annual Meeting, Kagawa Medical University, Kagawa, Japan; **1996**, pp. 43-50.
7. Maekawa, K. *Jpn. J. Tox.*, **1997**, *10*, 39-41.
8. Okumura, T.; Suzuki, K.; Fukuda, A.; Kohama, A.; Takasu, N.; Ishimatsu, S.; Hinohara, S. *Acad. Emergency Med.*, **1998**, *5*, 613-624.
9. Okumura, T.; Takasu, N.; Ishimatsu, S.; Mianoki, S.; Mitsuhashi, A.; Kunada, K.; Tanaka, K.; Hinohara, S. *Disaster Med./Concept.*, **1996**, *28*, 129-134.
10. Inoue, N.; Makita, Y. *Jpn. J. Exp. Med.*, **1994**, *71*, 144-148.
11. Inoue, N. *Neurol. Med.*, **1995**, *43*, 308-309.
12. Katagi, M.; Nishikawa, M.; Tatsano, M.; Tsushihashi, H. *J. Chromatogr.* B, **1997**, *689*, 327-333.
13. Tsuchihashi, H.; Katagi, M.; Nishikawa, M.; Tatsumo, M. *J. Anal. Toxicol.*, **1998**, *22*, 383-388.

Chapter 21

Toxicological Analysis of Victims' Blood and Crime Scene Evidence Samples in the Sarin Gas Attack Caused by the Aum Shinrikyo Cult

Y. Seto, N. Tsunoda, M. Kataoka, K. Tsuge, and T. Nagano

National Research Institute of Police Science, 6–3–1, Kashiwanoha, Kashiwa, Chiba 277–0882, Japan

The sarin gas attacks committed by the AUM SHINRIKYO group in Matsumoto city and in the Tokyo subway system resulted in 19 deaths and numerous injuries. The National Research Institute of Police Science has been engaged in forensic investigations into these incidents. Both plasma butyrylcholinesterase (BuChE) and red blood cell (RBC) acetylcholinesterase (AChE) activities were measured in victim's blood samples. Chemical analyses of the victim's blood, water, soil and wipe samples were performed by organic solvent extraction, followed by gas chromatography-mass spectrometry with or without *tert.*-butyldimethylsilylation, and, as a result, sarin and related compounds, including hydrolysis products, derived from sarin were positively identified.

Lethal nerve gas attacks in Matsumoto city in the summer of 1994 and in the Tokyo subway system in the spring of 1995 caused a great deal of shock with regard to the illegal use of the chemical warfare agent "Sarin" against a defenseless public (*1*). The Chemistry Section of National Research Institute of Police Science has been engaged in forensic investigations related to both sarin terrorisms caused by Japanese Cult, AUM SHINRIKYO, as well as other Cult-related incidents. In this paper, outlines of the sarin incidents and toxicological analysis performed throughout our forensic investigation are presented.

Outlines of Sarin Incidents caused by AUM SHINRIKYO.

In Japan, the terrorist acts involving chemical warfare agents occurred over a period about one year, in 1994 to 1995. In addition to the Matsumoto and Tokyo Subway Sarin Gas Attacks, several other terrorist activities against individuals occurred. From the interrogation of the suspects arrested, it appeared that these murders and attacks were committed using VX agent, sarin, hydrogen cyanide, and forensic examinations support these conclusions.

In 1993, the cult leader "Asahara" directed cult members to initiate the mass production of sarin. In November 1993, the Cult succeeded in the synthesis of sarin, and released sarin gas in Matsumoto city. Their sarin production facility was again activated in Autumn, 1994. The cult also succeeded in the synthesis of VX-agent. However, after the Matsumoto Sarin Gas Attack, police investigators suspected that the incident was committed by the cult. Aware of criminal investigation activities underway, the cult stopped operating the chemical plant in early 1995, and attempted to conceal evidence of the mass production of sarin.

Matsumoto Sarin Incident. The first sarin gas attack occurred in a quiet residential area of Matsumoto City, at midnight June 27, 1994. At 11:30 p.m. Matsumoto Police station received an urgent report from an ambulance team of the City Fire Defense Bureau that some patients were transported to a hospital. The cause of injury was not clear. The Nagano Prefecture Police Headquarters immediately started to rescue the injured and conduct investigations. Five deceased residents were discovered at the site and two victims who had been injured and hospitalized subsequently died. About 270 people were treated in the hospital. Carcasses of dogs, sparrows, a dove and a large number of caterpillars were found in the area under some trees. In addition, discolored grass was observed in the garden, and dead fish and crayfish were found in a nearby pond. Nearly all casualties were discovered in a sector-shaped residential area within a radius of 150 meters from the center near the pond. Persons near open window or in air-conditioned rooms were severely affected. Some victims observed a slow-moving fog with a pungent and irritable smell. Typical symptoms of the victims were darkened vision, ocular pain, nausea, miosis and severely decreased serum cholinesterase (ChE) activities. Autopsy findings showed intense postmortem lividity, miosis, pulmonary edema, increased bronchial secretion and congestion of the parenchymatous organs.

Tokyo Subway Sarin Incident. Sarin mass terrorism occurred inside trains on three Eidan-subway lines during the morning rush-hour peak on March 20, 1995. Passengers and station personnel were overcome by fumes in the train cars, as well as within Kasumigaseki and Tsukiji stations. The Tokyo Metropolitan Police Department immediately began to rescue the injured, to conduct traffic control and to investigate the incident. Evidence samples, such as containers, newspapers and other remains were collected, and the Forensic Science Laboratory identified sarin by gas chromatography-mass spectrometry (GC/MS) in a short time. This conclusion was possible because of our previous experience in the Matsumoto Incident. The cult decided to use sarin in trains on the three subway lines, all of which stop at Kasumigaseki station near the Metropolitan Police Department. Twelve passengers and station personnel were killed, and approximately 5000 people were injured. Typical symptoms were darkened vision, ocular pain, nausea, miosis, hyperemia and nosebleed. Autopsy findings were nearly identical to those in the Matsumoto Sarin Incident.

The sarin manufacture case. Sarin was manufactured in the Cult facilities in Kamikuishiki, Yamanashi Prefecture near Mt. Fuji. On 1994 July, inhabitants in the vicinity of a cult facility noticed some foul smells. Later, a sarin hydrolysis product was detected from a soil sample taken near the site. Furthermore, police seized evidence which showed that cult dummy companies had purchased a large quantity of the raw materials for manufacturing sarin. On March 22, 1995, a simultaneous raid of the cult facilities was launched by 2500 police personnel in connection with the confinement of the notary public manager to death. The search uncovered a large amount of chemicals such as phosphorus trichloride, as well as a

320

chemical plant in No. 7 Satyam building. Nearly all the cult perpetrators were arrested within two months. In the chemical plant building, various types of equipment required for the manufacture of chemicals were found. Through the investigation of the plant facilities, the police concluded that a very strong suspicion existed for the construction and operation of the plant used to produce sarin.

Laboratory techniques performed in forensic investigation.

The special nature of nerve gases restricted police efforts for an early solution. First, nerve gases are objects of The Law on the Ban of Chemical Weapons and the Regulations of Specific Substances, which was enacted in May 1995 in Japan, but only after the attacks occurred. Second, nerve gases are highly toxic, and lethal in trace amounts. Third, they are volatile. Handling of samples requires caution, and it is difficult to detect nerve gases because of rapid evaporation. Fourth, nerve gases are chemically labile and easily hydrolyzed (2) to alkylmethylphosphonate. In order to verify nerve gas exposure, it is necessary to identify the degradation products, rather than the nerve gas instead (3). However, sarin is toxic at very low levels and covalently binds to hydrolytic enzymes (4), therefore it is difficult to detect isopropylmethylphosphonate (IMPA) in the blood of a victim. The detection of free IMPA (5) or sarin (6), liberated from a ChE adduct, may be useful for the verification of sarin poisoning. However, a decrease in blood ChE activity remains the index of sarin exposure.

We performed toxicological analysis of crime scene evidence samples. First, we carried out ChE assays on victim's blood samples (7). We then carried out chemical analysis of victim's blood, water, soil and wipe samples, by organic solvent extraction and GC/MS. The analytical system used in our forensic investigation makes it possible to detect trace levels of toxic substances, and also provides chemical proof for usage, possession and production of chemical warfare agents.

Cholinesterase assay. Serum BuChE activities have been assayed for clinical diagnosis of liver function, and for evaluation of the metabolism of muscle relaxants. In blood, there is another type of enzyme, RBC AChE (9). This enzyme is genetically identical with neuronal AChE, and a better marker for monitoring the extent of exposure by anticholinesterase agents (10). A decrease in blood ChE activity represents a better index of exposure to nerve gases. In the human body, sarin irreversibly inhibits ChE activity. The nucleophilic oxygen of the active site serine residue attacks the phosphorus atom, resulting in covalent bonding. Antidotes such as N-methyl-2-pyridinealdoxime reactivate ChE activity. However, phosphonylated enzyme undergoes aging, from the dealkylation of the alkoxyl radical, and is resistant to reactivation with a half life of several hours.

Blood samples were collected from 19 fatal casualties at autopsy, and their ChE activities were measured within a few days. Blood samples drawn from normal adults were used as controls. ChE activities were measured by a modified Ellman's method (11, 12). Blood is fractionated into plasma and RBC by centrifugation. One hundred μl of the fraction is diluted with distilled water (50-fold for plasma, 100-fold for RBC) and added to 3 ml of 0.5 mM 5,5'-dithiobis(2-nitrobenzoic acid) solution in 5 mM sodium phosphate (pH 7.7). The reaction was started by the addition of 40 μl of 156 mM butyrylthiocholine for plasma and 10 μl of 156 mM acetylthiocholine for RBC. After a 15 min incubation at 25°C, the enzyme reaction was terminated by addition of 50 μl of 12 mM eserine, and the absorbance was measured at 405 nm. ChE activity was calculated from the ε value for 5-thio-2-nitrobenzoate (13500). One unit (U) represents ChE activity sufficient to hydrolyze 1 μmol of substrate per min. For a blank, the eserine solution was added prior to the enzyme reaction.

Chemical analysis. In water, nerve gases are easily hydrolyzed, producing characteristic methyl phosphorus bonding compounds which are metabolically stable, water soluble and never found in nature. These alkylmethylphosphonates are unique for the original nerve gases. They are ultimately hydrolyzed to methylphosphonic acid (MPA). Therefore, detecting the hydrolysis products can provide indirect proof for the existence of nerve gas. Samples wiped with cotton were extracted with dichloromethane and then with water. After the water fraction is adjusted to a neutral pH, both the dichloromethane and water fractions are combined, and again shaken. Soil samples were extracted with water and then with dichloromethane. Blood samples were deproteinized with perchloric acid, and the resulting supernatant were adjusted to a neutral pH, and extracted with dichloromethane. Nerve gases and their intermediates and byproducts are extracted into the organic solvent fraction, and after concentration under mild conditions, subjected to GC/MS. The hydrolysis products of nerve gases are extracted into the aqueous fraction, and after derivatization, subjected to GC/MS.

Compounds in the organic solvent fraction can be simultaneously separated using apolar DB-5 capillary column and by multi-step temperature program starting from an oven temperature of 45°C. Two types of GC detections are typically used. MS with electron impact ionization (EI) and isobutane chemical ionization (CI) provides both chemical and structural information. Atomic emission detection is highly selective for monitoring phosphorus-containing compounds. Molecules in the cavity are degraded to their elements and the atomic emission can be detected under selected wavelength. Figure 1 shows total ion chromatogram of 300-900 ppm of sarin, dimethylmethylphosphonate (DMMP), trimethylphosphate (TMPO) and diisopropylmethylphosphonate (DIMP). The later three are typical precursors or byproducts of sarin synthesis. EI-mass spectra offered representative fragmentation patterns and CI-mass spectra gave quasi-molecular ions as base peaks (Figure 2).

Hydrolysis products of nerve gases in the aqueous fraction are obtained by evaporation of the water, and *N*-methyl-*tert*.-butyldimethylsilyltrifluoroacetamide and acetonitrile are added, heated and injected into the GC/MS (*13*). Figure 3 shows a total ion chromatogram of *tert*.-butyldimethylsilylated hydrolysis products (30-45 ppm) of nerve gases, ethylmethylphosphonate (EMPA) from VX agent, IMPA from sarin, pinacolylmethylphosphonate (PMPA) from soman, and MPA. The mass spectra offered representative fragmentation pattern on EI-MS and quasi-molecular ions on CI-MS (Figure 4).

Results of forensic investigation.

Matsumoto Sarin incident. Sarin-inhibited ChE shows no reactivation (*4*) and the regeneration of blood ChE activities requires more than one month (*14*). Therefore, the measured ChE values should reflect the activity levels just after sarin poisoning. Both RBC AChE and plasma BuChE activities (Figure 5, ●) were drastically decreased for all 7 fatal casualties ($p < 0.05$), compared to the control blood levels (RBC AChE: 4.35 (average) ± 1.27 (standard deviation) U/ml, plasma BuChE: 3.46 ± 0.86 U/ml). The RBC AChE activities were decreased significantly more than the counterpart plasma BuChE activities, which is consistent with both *in vivo* and *in vitro* experiments (*14*).

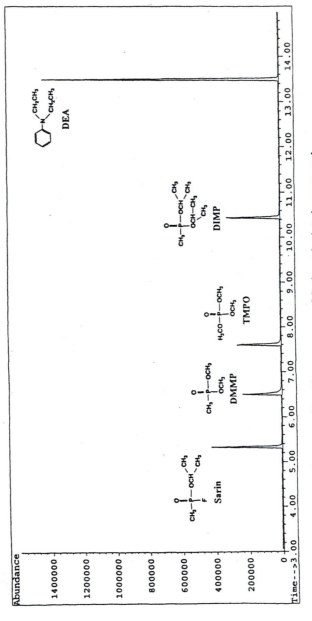

Figure 1. Gas Chromatogram of Sarin and related compounds.

Sarin was detected in one sample of the pond water, and also in the nasal mucosa of one victim. This represents the first case where sarin is detected in a sample from a victim. We have no authentic "sarin" as a standard. However, observed retention index (817) is very similar to values quoted in the literature (15). The EI-mass spectrum was also identical to previously published data (16). The CI-mass spectrum gave quasimolecular ions of m/z 141. Furthermore, one peak was observed under atomic emission detection with a phosphorus emission line near the position of the retention index of 817. IMPA and MPA were detected in numerous samples from victim's blood, as well as wipe samples. In addition, from the discolored blades of grass, a high level of fluoride and chloride was identified by capillary electrophoresis.

Allegedly, the perpetrators sprayed the lethal gas from evaporator-type spray containers. They fabricated the spraying device by modifying a refrigerator-car, which contained a heating-pot and fan. In the Cult chemical plant, sarin was produced by adding isopropyl alcohol to a mixture of methylphosphonyl difluoride and methylphosphonyl dichloride. Because the reaction conditions were in error, the yield of sarin was low, and by-products were produced. After storage for nearly 3 months, this solution was released by evaporation with forced heating over a 10 minute period. High levels of sarin hydrolysis products were detected from even injured victims, and high levels of DIMP were also detected. This suggests that the sarin solution was composed of not only sarin itself but also the other sarin synthesis precursors and byproducts.

Tokyo Subway Sarin Incident. Victims who immediately escaped from the crime scenes were given first aid. If patients receive a prescription of oxime antidotes within a few hours, during which the aging of inhibited ChE does not occur (17), the blood ChE activity levels of such patients can be remarkably restored. As shown in Figure 5 (O), both RBC AChE and plasma BuChE activities were extremely lowered ($p < 0.05$) for the 6 fatalities, 5 of whom died immediately. For the other 5 victims, both activities were not necessarily decreased, and 4 of these victims died in the hospital after 21th March. Compared to the activity levels in the Matsumoto Incident, the degree of decrease was not significant. This may be due to the fact that the victims received prompt and adequate treatment soon after poisoning. In the case of acute pesticide poisoning, plasma BuChE activities are generally decreased more significantly than RBC AChE (18). In contrast, in the Sarin incidents, RBC AChE activities were decreased to a greater extent than plasma BuChE.

Sarin-containing vinyl bags were used for mass terrorism. The perpetrators boarded subway trains with these plastic bags, and released the gas by puncturing them with the metal tips of umbrellas. Sarin, n-hexane and N,N-diethylaniline (DEA) were identified as major components from a sarin bag which was obtained at the scene. IMPA was detected in blood samples from only two victims. Allegedly, in the Cult chemical laboratory, sarin was produced by adding isopropyl alcohol to 1.4 kg of methylphosphonyl difluoride, using n-hexane as solvent and DEA as the acid neutralyzer. The resultant solution, about 7 liters, was divided into 11 bags. From forensic investigation, the concentration of sarin was determined to be about 30%, and this value is supported by the testimony of suspects.

Sarin manufacture case. From the police investigation of the chemical plant, synthetic route of sarin mass manufacture was disclosed. The protocol was comprised of 5 stages. In the 1st stage, phosphorus trichloride was reacted with methanol to produce trimethylphosphite. In the 2nd stage, trimethylphosphite was converted to DMMP via a thermal rearrangement reaction. In the 3rd stage, DMMP was reacted with phosphorus pentachloride with heating, to produce

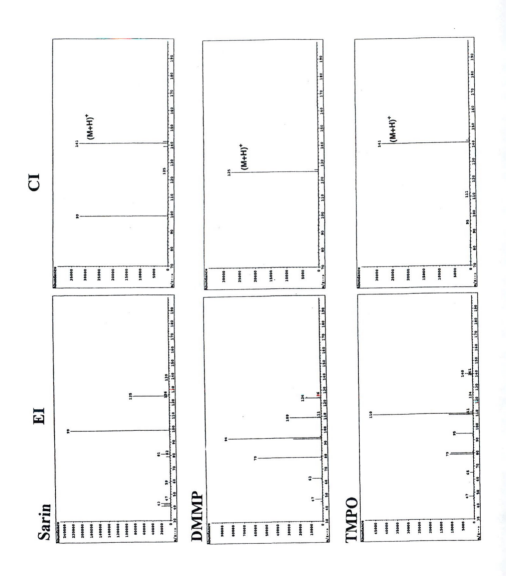

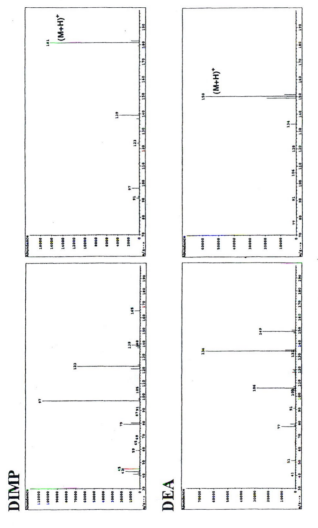

Figure 2. Mass spectra of Sarin and related compounds.

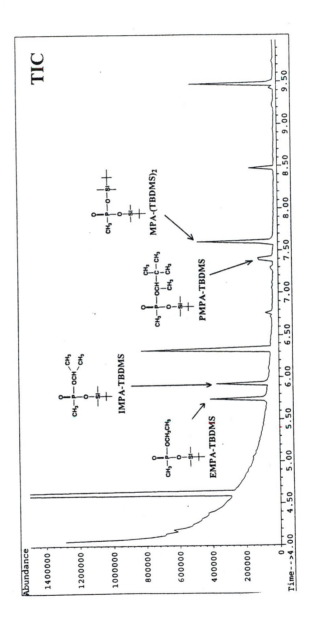

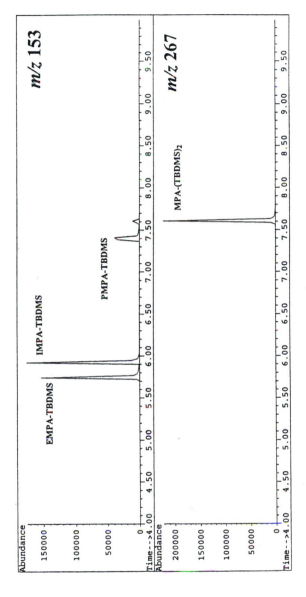

Figure 3. Gas Chromatogram of *tert.*-butyldimethylsilyl derivatives of nerve gas hydrolysis products.

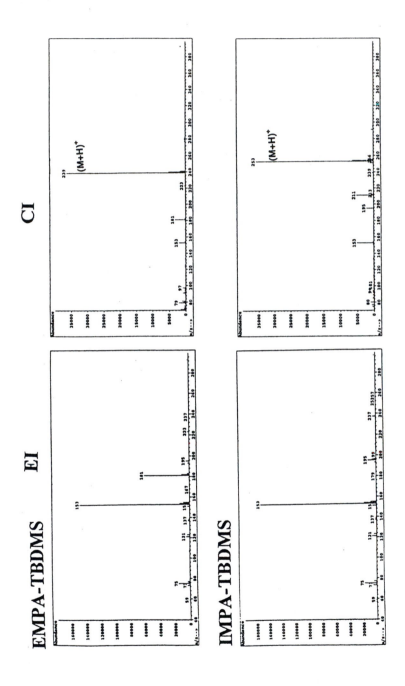

329

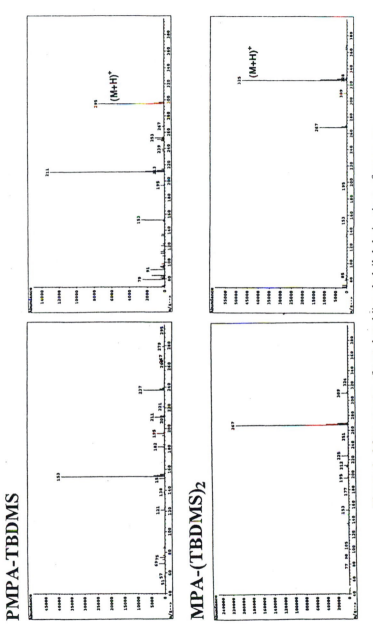

Figure 4. Mass spectra of *tert.*-butyldimethylsilyl derivatives of nerve gas
hydrolysis products.

330

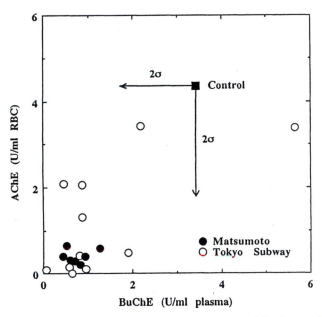

Figure 5. Blood AChE and BuChE activity levels of fatal casualties in sarin gas attacks.

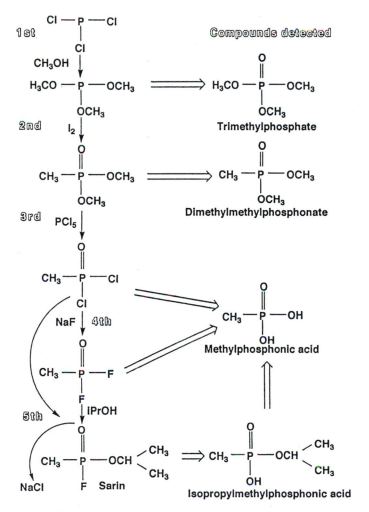

Figure 6. Chemical verification of sarin manufacture in AUM SHINRIKYO chemical plant.

methylphosphonyl dichloride. In the 4th stage, methylphosphonyl dichloride was reacted with sodium fluoride to produce methylphosphonyl difluoride. In the final stage, methylphosphonyl difluoride and methylphosphonyl dichloride were mixed with isopropyl alcohol to produce sarin. From the wiped samples taken from the 1st stage equipment, TMPO, n-hexane and DEA were detected. From the 2nd stage equipment, TMPO, DMMP, iodine and DEA were detected. From the 3rd stage equipment, MPA, DEA, phosphorus oxychloride and sodium chloride were detected. From the 4th stage equipment, MPA, DEA, sodium chloride and sodium fluoride were detected. From the final stage equipment, IMPA, MPA, DEA, DMMP and sodium chloride were detected. As shown in Figure 6, from the chemical analysis of evidence samples taken from the scene of manufacturing plant, stable substances derived from the precursors and the products and by-products corresponding to synthetic routes, have been identified, which provides both evidence and verification of sarin synthesis by the Cult.

References

1. National Police Agency. *White Paper on Police 1995 and 1996*, Government of Japan, Tokyo.
2. Stewart, C.E.; Sullivan, Jr, J.B. In *Hazardous Materials Toxicology, Clinical Principles of Environmental Health*; Sullivan, Jr, J.B.; Krieger, C.R. Eds.; Williams & Wilkins; Baltimore, 1992, pp.986-1014.
3. Kingery, A.F.; Allen, H.E. *Toxicol. Environ. Chem.*, **1995**, *47*, 155.
4. Tayler, P.; Radic, Z. *Ann. Rev. Pharmacol. Toxicol.*, **1994**, *34*, 281.
5. Nagao, M.; Takatori, T.; Matsuda, Y.; Nakajima, M.; Iwase, H.; Iwadate, K. *Toxicol. Appl. Pharmacol.*, **1997**, *144*, 198.
6. Polhuijs, M.; Langenberg, J.P.; Benschop, H.P. *Toxicol. Appl. Pharmacol.*, **1997**, *146*, 156.
7. Seto, Y; Tsunoda, N.; Ohta, H.; Nagano, T. In *Current Topics in Forensic Science: Proceedings of 14th Meeting on the International Association of Forensic Sciences, August 26-30, 1996, Tokyo, Japan*; Takatori, T.; Takasu, A. eds.; Shunderson Communications: Ottawa, Canada, 1997, Vol. 2; pp 178-179.
8. Seto, Y. In *Current Topics in Forensic Science: Proceedings of 14th Meeting on the International Association fo Forensic Sciences, August 26-30, 1996, Tokyo, Japan*; Takatori, T.; Takasu, A. eds.; Shunderson Communications: Ottawa, Canada, 1997, Vol. 4; pp 35-38.
9. Tayler, P. *J. Biol. Chem.*, **1991**, *266*, 4025.
10. Zavon, M.R. *Int. Arch. Occup. Environ. Hlth*, **1976**, *37*, 65.
11. T. Shinohara, T.; Seto, Y. *Rept. Natl. Res. Inst. Police Sci.*, **1985**, *38*, 178.
12. Seto, Y.; Shinohara, T. *Arch. Toxicol.*, **1988**, *62*, 37.
13. Kataoka, M.; Tsunoda, N.; Ohta, H.; Tsuge, K.; Takesako, H.; Seto, Y. *J. Chromatogr. A*, **1998**, *824*, 211.
14. Grob, D.; Harvey, J.C. *J. Clin. Invest.*, **1958**, *37*, 350.
15. Hancock, J.R.; Peters, G.R. *J. Chromatogr.*, **1991**, *538*, 249.
16. Sass, S.; Fisher, T.L. *Org. Mass Spectrom.*, **1979**, *14*, 257.
17. Sidell, F.R.; Groff, W.A. *Toxicol. Appl. Pharmacol.*, **1974**, *27*, 241.
18. A. Silver. *The Biology of Cholinesterases*; North-Holland Publishing Co., Amsterdam, 1974.

Chapter 22

An Attack with Sarin Nerve Gas on the Tokyo Subway System and Its Effects on Victims

Y. Ogawa[1,4], Y. Yamamura[2], H. Ando[3], M. Kadokura[1], T. Agata[1],
M. Fukumoto[1], T. Satake[1], K. Machida[1], O. Sakai[1], Y. Miyata[3],
H. Nonaka[3], K. Nakajima[1], S. Hamaya[1], S. Miyazaki[1], M. Ohida[1],
T. Yoshioka[1], S. Takagi[1], and H. Shimizu[1]

[1]Jikei University School of Medicine, Minato-ku, Tokyo 105–8461, Japan
[2]Professor Emeritus, St. Marianna University School of Medicine,
Miyamae-ku, Kawasaki 216–0015, Japan
[3]Criminal Investigation Laboratory, Metropolitan Police Department,
Chiyoda-ku, Tokyo 100–8929, Japan

On the morning of March 20, 1995, the Tokyo subway system was attacked with nerve gas. Liquid, in plastic bags, left on the subway cars was analyzed and sarin, hexane, and N,N-diethylaniline were detected as the main components. The health effects of victims were studied through questionnaire and hospital records. Plasma cholinesterase levels were used as exposure indicators. Muscarine-like symptoms appeared generally and could be used as early warning signs. Nicotine-like symptoms and effects on central nervous system appeared in more severely exposed cases suggesting that they can be used as severity indicators. Muscarine-like effects to the eye and respiratory system must be induced by the direct contact of sarin gas to mucous membranes. On the other hand, nicotine-like effects are caused by a systemic exposure to sarin.

On the morning of March 20, 1995, the Tokyo subway system was attacked with nerve gas. According to an announcement from the Tokyo Metropolitan Fire Department, twelve people were killed and 5498 were injured (1). Checking the hospital patients suffering from the attack, we found that shrinkage of the pupil was a common symptom, and that plasma cholinesterase levels of many patients were under the normal range. These symptoms indicated that some kind of organophosphorus substance must be the culprit, and it strongly suggested the use of nerve gas. This in-

[4]Current address: National Institute of Industrial Health, 6–21–1 Nagao, Tama-ku, Kawasaki 214–8585, Japan.

cident reminded us of the Matsumoto's case, which was thought of as sarin attack executed by a religious cult Aum Shinrikyo in June 27, 1994.

The liquid left in the plastic bags on the subway cars was analyzed, and sarin was detected as a major component. There are a few reported incidents of accidental exposure to sarin (2-4) and a single report on experimental exposure (5), but there had been no study of mass exposure to sarin until the attack occurred at Matsumoto (6,7). Reports of severe cases (8-11), relation between pupil size and acetylcholinesterase (AChE) level (12,13), and a survey of subjective symptoms in patients (14,15) affected by the Tokyo attack have already been reported to journals, but there has not yet been a formal evaluation of the relation between the occurrence of symptoms and dose of exposure to sarin. In this study, health effect of victims at acute phase was studied and its relation to the exposure dose was analyzed.

Analyzing Sarin

Attackers left plastic bags filled with liquid in the subway cars. Volume of liquid in each bag was about 600 ml and was separated into two layers. The top layer was a transparent liquid of about 200 ml, and the bottom layer was a dark brown liquid measuring about 400 ml. A group of Metropolitan Police lead by one of our authors analyzed these materials using GC/MS analysis, NMR, and other methods.

GC/MS Analyses.   The total ion chromatogram of the material left in the subway cars was obtained using temperature programmed methods and capillary column chromatography (Figure 1). From retention time analysis and mass spectrum data of standard material of each substance we identified diisopropyl-fluoro-phosphonate (DFP) at a retention time of 6 minutes, diisopropyl-methyl-phosphonate (DIMP) at a retention time of 7.2 minutes, a small peak of triisopropyl-methyl-phosphonate (TIP) at a retention time of 8.2 minutes, and a huge peak of N,N-diethylaniline (DEA) at a retention time of 8.8 minutes.

A peak at 4 minutes retention was suggested to be sarin from the peaks of its fragmented ions in EI-GC/MS (electrical ionization mode) spectrum, from which we could detect demethylated substance of sarin at m/z 125 and methyl-fluoro-phosphonate dihalide at m/z 99, we could not detect the ion of sarin itself which must be at m/z 140. From CI-GC/MS (chemical ionization mode) we could detect molecular-related ion (M+1) of sarin at m/z 141.

NMR Studies.   Using ^{31}P-NMR spectrum to detect phosphorus atom signals we received a spectrum as shown in Figure 2. Signals at −10.42 ppm and 29.32 ppm were identified from their standard materials as DFP and DIMP respectively. A signal at 29.62 ppm (d, J=1037 Hz) was within the range of the chemical shift of sarin (16), the

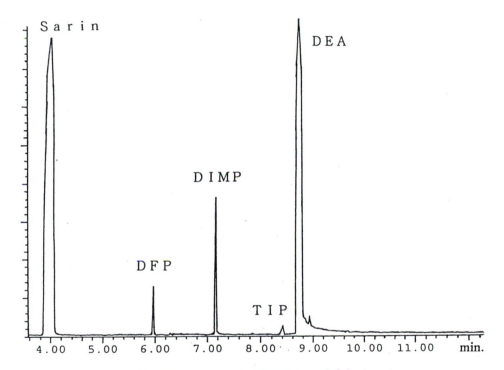

Figure 1. Total ion chromatogram of the substance left in the subway cars.

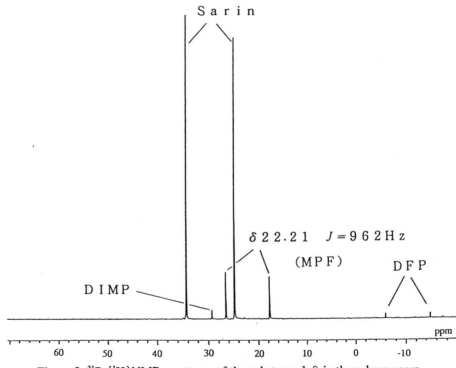

Figure 2. ^{31}P-{^{1}H}NMR spectrum of the substance left in the subway cars.

signal was at 28.95 ppm (d, J_{PF}=1036 Hz) (17); we concluded that this signal was from the phosphorus atom of sarin. These substances identified by NMR studies were comparable with those identified by the GC/MS analyses. Unfortunately we could not identify the substance from which we received the phosphorus signal at 22.21 ppm (d, J=96.2 Hz).

Analysis of Sodium Ethoxide Reaction Products. Sodium ethoxide was added to the material left in the subway cars and mixed. The water soluble portion and ether soluble portion were separated and both portions were treated to induce *tert*-butyldimethylsiryl (TBDMS) derivatives (18) and analyzed using EI-GC/MS.

We could not detect the peak of sarin but could detect the peak of methyl-ethyl-phosphonate isopropyl ester in the mass spectrum analysis of ether soluble portion. In the spectrum of the water soluble portion we detected TBDMS derivatives of methylphosphonate monoethyl ester, methylphosphonate (MPA), and isopropyl-methyl-phosphonate (IMPA).

The reactions which occurred between the materials and sodium ethoxide can be explained by the mechanism shown in Figure 3. So called Williamson's synthesis shows that sodium ethoxide reacts with halogens but not with hydroxy residue (19). Accordingly, sarin was converted to methyl-ethyl-phosphonate isopropyl ester, but MPA and IMPA were not affected and remained intact.

. The important aspect of this reaction was that methylphosphonate monoethyl ester was produced, which means that a content of the material in the bags was methylphosphonate-fluoride (MPF). Considering this result and the structure of MPF, the signal in the ^{31}P-NMR spectrum which we could not identify must have been from MPF.

Concentration of the Contents of the Material. The existence of sarin in the materials left on the subway cars was proved using GC/MS analyses, ^{31}P-NMR studies and other methods.

^{31}P-NMR spectra were measured in ^{1}H decoupling mode. Sarin and its related chemicals do not have hydrogen atom bound to phosphorous atom directly, and furthermore, the phosphorous atoms were coordinated by four atoms other than hydrogen. This means by stereochemistry that there are no influential hydrogen atoms near the phosphorus atoms. These facts suggests that there is little change in ^{31}P-NMR signals by nuclear Overhauser effect (NOE).

Using trimethylphosphonate as an internal standard for the phosphorous contents of the liquid left in the subway car, we analyzed it with ^{31}P-NMR. Combining these results and the results obtained from the ^{1}H-NMR spectrum, and gas chromatography, the concentrations (w/w) of phosphorous compounds sarin, MPF, DIMP, and DFP were found to be 35%, 10%, 1%, and 0.1% (trace amount) respectively, and those of hexane and DEA, which were used as solvents, or a reaction promoter, were 16% and 37% respectively.

Health Effects

The main substances used in the attack were DEA, sarin, and hexane. Clinically manifested symptoms suggested the effects of organophosphorus anticholinesterase compounds. The toxic effects of DEA are similar to those of aniline but less severe. The typical sign of exposure is cyanosis due to methemoglobinemia (20). The acute toxic effects of hexane are euphoria, dizziness, and numbness of limbs, but in those toxic cases the concentration of inhaled hexane is very high and its odor would be irritating (21); this was not the case in the subway attack. According to the collected facts on the attack, the main agent responsible for the toxic symptoms in the victims was decided to be sarin.

We collaborated with doctors from 7 hospitals to conduct a survey about the health effects of sarin on people suffering from the attack. This survey was intended to study their subjective symptoms as they occurred at acute phase, and its time course plus the relations between the incidence of those symptoms and the dose of the exposure.

Exposure Indicator. We needed a dose indicator of the exposure to sarin. Organophosphorus agents bind to AChE and inactivate them, which is an irreversible reaction with nerve gas. These reactions produce; 1) stimulation of muscarinic receptor responses at autonomic effector organs; 2) stimulation, followed by depression or paralysis, of nicotinic receptor at all autonomic ganglia and skeletal muscle; and 3) stimulation, with occasional subsequent depression, of cholinergic receptor sites in the central nervous system (CNS) (22). Due to this relation, reduced amount of AChE activity correlates with the intensity of manifested effects in the acute phase (5,23).

We used the activity of butyrylcholinesterase or benzoylcholinesterase in plasma, also known as plasma cholinesterase (ChE), as a dose indicator of exposure either than the activity of AChE (24). This enzyme is not a specific esterase for acetylcholine but it can be inactivated by organophosphorus agents (22). Accordingly, the change of this dose level in plasma does not directly correlate with the toxic symptoms by organophosphorus agent, but rather it works as a dose indicator of the exposure; further more, the activity of this enzyme in plasma has been commonly measured in clinical situations when organophosphorus pesticide poisoning was suspected (25).

Methods of ChE activity measurement were different among hospitals. To construct a common indicator we used the ratio of ChE level to lower limit value of normal range in percent (%ChE) instead of ChE level itself which was defined as follows.

$$\%ChE = \frac{100\times(ChE\ level)}{(lower\ limit\ value\ of\ normal\ range\ of\ ChE\ level)}$$

%ChE was compared with the reduction rate of ChE level (rChE) , which is defined as follows.

$$rChE = \frac{\{(level\ after\ two\ years) - (level\ at\ the\ day\ of\ the\ attack)\}}{(level\ after\ two\ years)}$$

rChE correlates well with the dose of exposure to organophosphorus agents and is considered as a reliable indicator (23,26). Figure 4 shows the relation between %ChE and rChE among 32 victims. It shows a good linear relationship between the two variables especially under 100%.

Subjects. As summarized in Table I, 1089 people suffered from the attack and had medical treatment at 7 hospitals. All of these people were listed and at the end of May 1995 questionnaires were mailed to them. Among those people 681 subjects responded to the questionnaire and agreed to participate in our survey. The response rate was 62.5%. Participants consisted of 369 men and 312 women, with an average age of 40.2 (14-72) and 28.2 (16-70) respectively. Parenthesis shows the range. We checked the hospital records further and selected those people who had their ChE level measured on March 20 and before treatment, because PAM (pyridine-2-aldoxime methyl chloride; pralidoxime) recovers ChE level drastically within a few hours (27). Finally 454 subjects remained, 259 men and 195 women, with an average age of 39.8 (16-70) and 28.2 (16-70), respectively.

Figure 5 shows the time course of the cases treated by PAM. The low ChE level cases, treated with-in a certain hour following the exposure (our cases were all within

Table I. Study Subjects

	Male	Female	Total	
Mailed			1089	
Respondent	369	312	681	(62.5%)
	(40.2)	(28.2)		
Subjects with ChE	259	195	454	(41.6%)
	(39.8)	(28.2)		

( ) under the number: Average age

( ) outer right column: Rate to mailed number in percentage

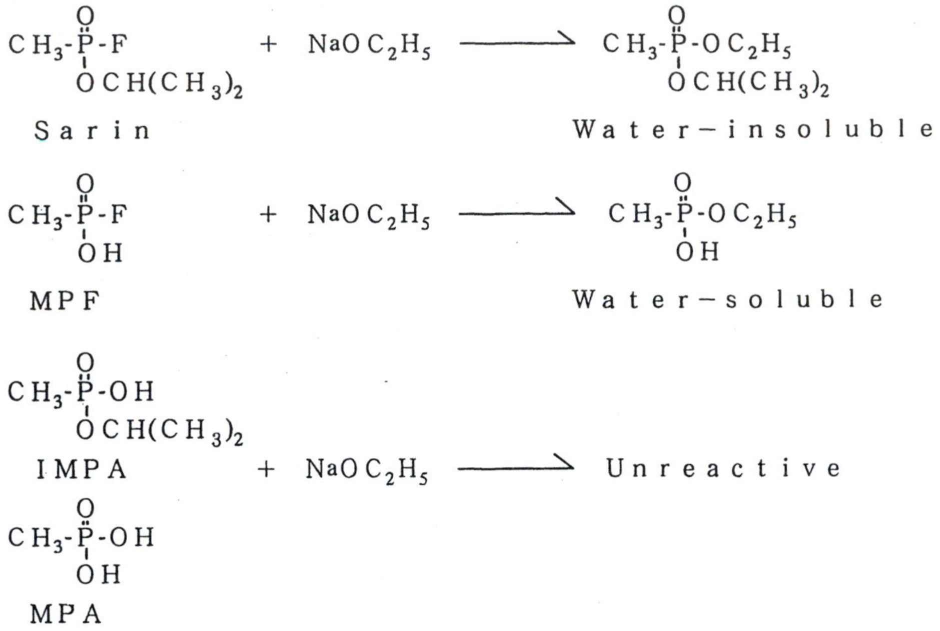

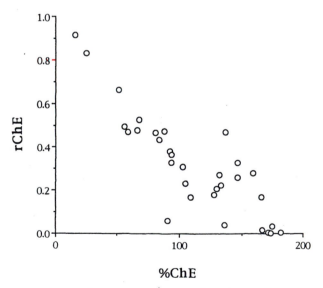

Figure 3. Results of sodium alkoxide reactions with the substance left in the subway cars.

Figure 4. Relation between %ChE and rChE among 32 victims.

341

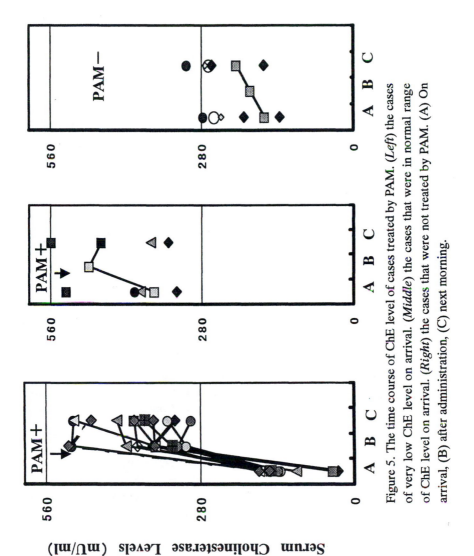

Figure 5. The time course of ChE level of cases treated by PAM. (*Left*) the cases of very low ChE level on arrival. (*Middle*) the cases that were in normal range of ChE level on arrival. (*Right*) the cases that were not treated by PAM. (A) On arrival, (B) after administration, (C) next morning.

4 hours), quickly recovered to the normal range. However, recovery of untreated cases and that of normal level cases were not prominent.

Figure 6 shows the histogram of the %ChE of our study group. Proportion of the people whose %ChE were under100% was 28.2%. The 454 subjects were classified into two groups. High %ChE group (H), %ChE of whose members were over or equal 100%, and low %ChE group (L), %ChE of whose members were below 100%.

Questionnaire. The questionnaire we used can be divided into two parts. The first part was asking about the exposure situations the participants experienced: times when they were aware of the involvement, where (in the subway car, on the platform, or at other places) they were involved, and when and how they accessed doctors for medical care. The second part asked about the symptoms which the participants experienced.

The symptoms constituted three groups: 1) muscarine-like effects which show over secretion of glands, bronchoconstriction, contraction of sphincter muscle of iris and ciliary muscle, increased motility of gastrointestinal tract, and urinary bladder dysfunction: sneeze, rhinorrhea, cough, nasal speech, sore throat, dyspnea, dimness of vision, constricted visual field, eye pain, increased lacrimation, blurred vision, nausea, vomiting, headache, increased sweating, increased salivation, anorexia, abdominal pain, diarrhea or tenesmus, frequent and involuntary micturition; 2) nicotine-like effects which shows impaired muscle movements: dyspnea, double vision, slurred speech, disturbed mouth movement, muscular twitching or cramps, gait disturbance, difficulty in standing, dysphagia, generalized weakness, muscle pain; and 3) effects on central nervous system (CNS): dizziness, headache, slurred speech, gait disturbance, difficulty in standing, generalized weakness, anorexia, insomnia. Other selected symptoms included eye irritation, dys-osmia, numbness of extremities, nasal bleeding and subfever. The total symptoms we asked about were 37. Among them, 23 questions were from the questionnaire used at the survey on the incident of Matsumoto (7), the other questions were added by us referring to the study of Grob (2). The occurrence of these subjective symptoms were asked in four categories; symptoms which occurred during the whole course, symptoms which appeared earliest, symptoms which disappeared earliest, and symptoms which still remain.

Statistical Analysis. Comparison of group means was first tested by ANOVA and then comparison between two groups was tested by Bonferroni correction. For comparisons of prevalence of symptoms among subjects belong to group L and group H, statistical significance was determined by Fisher's exact p-value. SAS statistical package was used for calculation and analysis.

Results. The distribution of the places where subjects were involved were shown

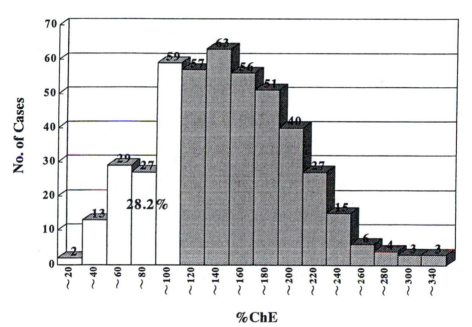

%ChE

Figure 6. Histogram of %ChE from our study group.

on Table II. Among 681 people 31.7% were involved in the subway cars and 43.3% were involved on the subway platform. Five hundred and ninety four people visited doctors to get medical care on the same day. Median visiting time of 594 people was 11:00 (8:00 - 22:00), which is approximately 3 hours after the incident. The methods used to reach clinics or hospitals were: by themselves 63.6%, by ambulance 14.3%, and with some one's help 13.6% (Table III).

The proportion of cases belonging to group L was unevenly distributed among the groups of four places, the proportion in the subway car was the highest. Mean %ChE of the group involved in the subway car was significantly low compared to the group involved on the platform. Mean symptom count, which was defined as a number of positive symptoms during the whole course, of the group involved in the subway car was significantly high compared with those of the group involved on the platform, the group involved in the station and the group involved at other places (Table IV).

The frequency of symptoms which occurred during the whole course in percentage were; dimness of vision 64.8, headache 52.3, dyspnea 45.5, constricted visual field 42.3, rhinorrhea 36.4, cough 35.4, eye irritation 33.9, blurred vision 30.5, nausea 27.3, and eye pain 25.6. Frequency of symptoms which appeared most early in percentage were; dimness of vision 35.8, dyspnea 28.9, cough 23.5, headache 17.5, constricted visual field 14.8, rhinorrhea 13.4, eye irritation 12.0, blurred vision 9.5, sore throat 8.8, and nausea 8.1. Frequency of symptoms which disappeared most

Table II. The distribution of places where subjects were involved
(Results from 681 subjects)

Places	%
In the subway car	31.7
On the platform	43.3
In the station	15.2
Other places	9.7

Table III. Ways to reach medical facilities
(Results from 681 subjects)

Ways	%
By ambulance	14.3
With someone's help	13.6
By himself (had symptoms)	53.8
By himself (had no symptoms)	9.8
Others	8.2

Table IV. Prevalence of Group L Subjects, Mean %ChE, and Mean Symptom Count
(SC) by Places
(Results from 454 subjects)

	n	% of group L[a]	%ChE		SC	
(A) In the subway car	152	36.1	122.9[1]	(55.2)	7.93[2]	(5.05)
(B) On the platform	195	26.1	140.8	(56.0)	6.34	(4.16)
(C) In the station	64	20.3	142.5	(55.2)	5.56	(3.25)
(D) Other places	42	21.4	145.8	(51.4)	5.38	(4.03)
Missing	1					

Number in ( ): SD

[a]: p<0.05 by Chi-square test: not evenly distributed

[1]: p<0.05 between (A) and (B)

[2]: p<0.05 between (A) and (B), (A) and (C) , and (A) and (D)

early in percentage were; cough 20.6, dyspnea 19.2, rhinorrhea 18.1, headache 12.9, nausea 11.6, dimness of vision 10.6, sore throat 6.8, eye irritation 6.5, increased lacrimation 6.2, and constricted visual field 6.0. Frequency of symptoms which still remained after two months in percentage were; headache 7.6, blurred vision 6.5, generalized weakness 5.3, eye pain 4.6, eye irritation 4.4, insomnia 3.7, rhinorrhea 3.4, dimness of vision 2.6, constricted visual field 2.2, and dyspnea 2.1 (Table V).

Comparing the symptoms which occurred during the whole course between group L and group H, occurrences of following symptoms were significantly high in group L. The ratio of the prevalence of a symptom among group L to that among group H are shown in the parenthesis. Disturbed mouth movement (∞), dys-osmia (10.1), muscular twitching or cramps (5.6), difficulty in standing (4.9), vomiting (4.9), slurred speech (4.2), insomnia (3.6), dysphagia (3.5), gait disturbance (3.1), numbness of extremities (2.1), nausea (2.0), increased lacrimation (1.8), subfever (1.8), dizziness (1.8), blurred vision (1.7), generalized weakness (1.5), constricted visual field (1.4), and dyspnea (1.3) (Table VI).

Comparing those symptoms which remain between group L and group H, occurrences of following symptoms were significantly high in group L. Shown in the parentheses is the ratio of the prevalence of a symptom among group L to that among group H. Double vision (∞), insomnia (4.6), blurred vision (2.3), and headache (1.8) (Table VI).

Discussion. Median sampling time of the blood was 3 hr after exposure. The

Table V. Prevalence rates of subjective symptoms in percentage
(Results from 681 subjects)

Symptoms	Whole Course	First appeared	First disappeared	Still remains
Dimness of vision	64.8	35.8	10.6	2.6
Headache	52.3	17.5	12.9	7.6
Dyspnea	45.5	28.9	19.2	2.1
Constricted visual field	42.3	14.8	6.0	2.2
Rhinorrhea	36.4	13.4	18.1	3.4
Cough	35.4	23.5	20.6	2.5
Eye irritation	33.9	12.0	6.5	4.4
Blurred vision	30.5	9.5	4.8	6.5
Nausea	27.3	8.1	11.6	0.4
Eye pain	25.6	6.5	2.2	4.6
Generalized weakness	21.0	4.1	2.5	5.3
Increased lacrimation	19.8	5.3	6.2	0.4
Sore throat	19.8	8.8	6.8	1.5
Dizziness	15.1	3.8	4.7	1.0
Gait disturbance	12.6	3.1	5.1	0.3
Insomnia	12.5	0.6	0.1	3.7
Anorexia	12.3	0.4	0.9	0.7
Subfever	11.9	1.0	1.3	1.0
Double vision	10.9	2.6	1.0	1.0
Difficult in standing	10.3	2.5	2.9	0.1
Numbness of extremities	10.3	1.8	3.4	0.9

Symptoms	Whole Course	First appeared	First disappeared	Still remains
Vomiting	8.4	2.1	2.3	0.3
Increased sweating	5.7	0.7	1.5	0.7
Sneeze	5.3	2.3	2.3	1.0
Slurred speech	4.6	0.4	0.6	0.1
Diarrhea or tenesumus	4.0	0.0	0.4	0.7
Dysphagia	3.4	0.3	0.1	0.3
Nasal speech	3.2	0.9	0.7	0.1
Disturbed mouth movement	2.6	0.6	0.4	0.1
Muscular twitching or cramps	2.6	0.4	1.0	0.0
Muscle pain	2.2	0.4	0.0	0.9
Abdominal pain	1.9	0.4	0.7	0.3
Cutaneous sign	1.6	0.0	0.1	0.6
Frequent or involuntary micturition	1.6	0.0	0.4	0.0
Dysosmia	1.3	0.3	0.0	0.4
Nasal bleeding	1.0	0.0	0.6	0.0
Increased salivation	1.0	0.1	0.4	0.0

Table VI. Prevalence Rates of Subjective Symptoms and The Ratios of those of Group L to those of Group H
(Results from 454 subjects)

Symptoms	Whole Course			Still remains		
	Incidence	Ratio	Confidence interval (95%)	Incidence	Ratio	Confidence interval (95%)
Dimness of vision	70.2	1.12		3.7	1.06	
Headache	54.6	1.12		9.4	1.83*	1.03 - 3.24
Dyspnea	48.4	1.37**	1.13 - 1.65	2.2	1.09	
Constricted visual field	46.4	1.40**	1.15 - 1.70	2.4	0.56	
Rhinorrhea	41.8	1.09		3.3	0.39	
Cough	37.6	1.17		1.7	0.36	
Eye irritation	35.4	1.18		4.8	0.95	
Blurred vision	34.8	1.73**	1.35 - 2.21	7.2	2.39*	1.24 - 4.59
Nausea	29.7	2.09**	1.60 - 2.75	0.2	0.00	
Eye pain	27.0	1.36		5.0	0.89	
Generalized weakness	23.1	1.56*	1.11 - 2.19	7.0	0.99	
Increased lacrimation	22.6	1.89**	1.36 - 2.64	0.4	2.54	
Sore throat	20.9	1.01		0.6	0.00	
Dizziness	15.1	1.84**	1.19 - 2.84	1.1	0.63	
Gait disturbance	14.3	3.16**	2.02 - 4.93	0.2	0.00	
Insomnia	13.4	3.66**	2.29 - 5.85	3.7	4.66**	1.76 - 12.3
Anorexia	12.3	1.41		0.4	2.54	
Subfever	14.5	1.87**	1.20 - 2.92	1.3	2.54	

*: p<0.05, **: p<0.01

Symptoms	Whole Course			Still remains		
	Incidence	Ratio	Confidence interval (95%)	Incidence	Ratio	Confidence interval (95%)
Double vision	12.7	3.13**	1.94 - 5.04	0.6	∞*	-
Difficult in standing	11.6	4.95**	2.91 - 8.41	0.0		
Numbness of extremities	11.4	2.18**	1.31 - 3.61	0.8	2.54	
Vomiting	10.3	4.93**	2.79 - 8.70	0.2	0.00	
Increased sweating	6.3	1.55		0.4	2.54	
Sneeze	5.0	0.53		0.4	0.00	
Slurred speech	5.2	4.24**	1.90 - 9.45	0.2	∞	
Diarrhea or tenesumus	4.4	1.37		0.8	0.84	
Dysphagia	4.1	3.50**	1.44 - 8.50	0.4	0.00	
Nasal speech	3.0	1.91		0.2	0.00	
Disturbed mouth movement	2.4	∞**	-	0.2	∞	
Muscular twitching or cramps	3.5	5.60**	1.98 - 15.8	0.0		
Muscle pain	3.0	1.91		1.1	0.63	
Abdominal pain	2.4	0.25		0.2	∞	
Cutaneous sign	1.9	2.03		0.4	2.54	
Frequent or involuntary micturition	1.5	1.91		0.0		
Dysosmia	1.1	10.18**	1.14 - 90.2	0.4	0.00	
Nasal bleeding	1.3	1.27		0.0		
Increased salivation	1.3	2.54		0.0		

*: p<0.05, **: p<0.01

activity of ChE quickly reaches lowest level after acute exposure, and then the restoration starts. The starting time is 3 to 10 hours following exposure. The amount of the recovery of ChE level is only around 10% during first 24 hours (2). Our data of ChE level were therefore, not far from the lowest levels of ChE activities which occurred after the acute exposure.

Validness of using %ChE as an exposure indicator is supported by the result that %ChE decreased among the people exposed in the subway car compared to those people exposed on the platforms. Symptom count defined as a count of positive symptoms among the functional symptoms questioned is often used as an indicator for self-perceived illness (28,29). Symptoms asked in our questionnaire were all functional and our symptom count might be related to general illness. This was supported by the fact that highly exposed group had the highest symptom counts. Our study also showed that symptom count was more reliable as an exposure indicator compared with %ChE activity.

It was reported (30,31) that in the case of A720S train running Hibiya line, in which sarin was released when it reached Akihabara station one stop before Kodenma-chyo station, one of the passengers kicked out the plastic bag on to the platform at Kodenma-chyo station which attacked many people who get off at the station. Passengers of A750S train were the most severely affected, because it was the fourth train to have reached Kodenma-chyo station after the A720S train and had to stop. It stopped because the line had already stacked up by the former trains. Most of the passengers were waiting in the car hoping to start moving again, so they were exposed to sarin in the car. After that they were told to get off the car and wait on the platform, because of the need to clear the platform for the next car A738S which had been waiting between Akihabara and Kodenma-chyo. When the A750S had left and the A738S came in, the station attendant told all passengers to evacuate the car, and from the underground station. Heavily exposed victims were exposed in the car and on the platform around the position where the plastic bags filled with sarin were left in the A720S train. Most people who were exposed at the other part of the platform were suspected to have milder exposure. Table VII supports this estimation, which shows that 210 subjects who suffered at Kodenma-chyo station shows the slight increase in symptom count among victims who thought that they were exposed in the car. But %ChE level showed no prominent difference between the groups but, however, the value of group A (the car) was lower compared with group B (the platform) and C (the station). Group D (other places) showed the lowest level. These subjects were mainly exposed by helping the victims, including the station personnel.

Symptoms related to contraction of sphincter muscle of iris and ciliary muscle, bronchoconstriction, and secretion of airway first appeared and the latter two symptoms disappeared in early stage. Headache, symptoms relating to the eye, and generalized weakness prolonged until two months after the attack in some cases.

Table VII. Prevalence of Group L Subjects, Mean %ChE, and Mean Symptom Count
(SC) by Places among Victims at Kodennma-chyo Station
(Results from 210 subjects)

	n	% of group L	%ChE		SC	
(A) In the subway car	33	39.3	117.5	(48.0)	8.57[1]	(4.22)
(B) On the platform	195	34.4	125.7	(49.0)	6.87	(4.17)
(C) In the station	64	23.8	125.3	(37.5)	6.45	(3.35)
(D) Other places	42	38.4	106.2	(33.9)	7.07	(3.86)

Number in ( ): SD

[1]: p<0.05 between (A) and (B) , and (A) and (C)

Most of the symptoms which occurred generally (dimness of vision, headache, dyspnea, constricted visual field, rhinorrhea, cough, blurred vision) were muscarine-like effects. Although these symptoms did not show evident dose-effect relationships, these could be used as early warning signs. Nicotine-like effects which appeared in some people were generalized weakness, gait disturbance, double vision, difficulty in standing, muscular twitching or cramps, and disturbed mouth movement. People in group L have manifested these nicotine-like effects and CNS effects (insomnia) more frequently. These symptoms could be used as severity indicators. Sweating and salivation were rare complaints among the victims. These observations were incompatible with the early published papers (2,3,5) which reported that the both symptoms also appeared along with other muscarine-like effects from early stage.

Table VIII shows the comparison of symptoms occurred along the course of attack between our report and other reports. The result obtained from the cases of St. Luke's International Hospital (14), who were victims of the same Tokyo sarin attack, show that the prevalence of symptoms were quite similar except for eye irritation 52.5%, eye pain 52.3%, and insomnia 31.2% compared with our results 33.9%, 25.6%, and 12.5% respectively. The results obtained from Matsumoto's cases (6) show that rhinorrhea was most frequent (51.3%) and dimness of vision (34.2%), headache (34.8%), and constricted visual field (16.4%) were less frequent compared with our results. These suggest that miosis was more severe in Tokyo's cases. This may be explained by the fact that Tokyo's cases were exposed in an underground facility which is closed space equaling a higher exposure level compared with Matsumoto's cases who were exposed mainly at out door. Direct contact of sarin gas to eye mucus membrane must be more serious in Tokyo's cases compared with Matsumoto's cases, this might explain some part of the differences. The purity of

352

Table VIII. Comparing in prevalence rates of subjective symptoms among three studies

Symptoms (subjects)	Our study (681)	St. Luke's (474)	Matsumoto (155)
Dimness of vision	64.8	77.8	34.2
Headache	52.3	60.3	34.8
Dyspnea	45.5	44.3	31.5
Constricted visual field	42.3	45.4	16.4
Rhinorrhea	36.4	44.7	51.3
Cough	35.4	36.9	26.9
Eye irritation	33.9	52.5	21.0
Blurred vision	30.5	42.6	21.0
Nausea	27.3	34.0	9.2
Eye pain	25.6	52.3	23.6
Generalized weakness	21.0		11.8
Increased lacrimation	19.8		5.2
Sore throat	19.8		22.3
Dizziness	15.1		11.1
Gait disturbance	12.6		4.6
Insomnia	12.5	31.2	
Double vision	10.9		4.6
Numbness of extremities	10.3		7.8
Vomiting	8.4		5.2
Sneeze	5.3		17.1
Dysphagia	3.4		6.5
Nasal speech	3.2		9.2
Disturbed mouth movement	2.6		2.6
Muscular twitching or cramps	2.6		1.9

sarin used at Matsumoto was said to be higher, which was about 70% (from personal communication), than the sarin used at Tokyo but the high concentration worked only in a limited area where seven people were killed (7). The difference in occurrence of insomnia between our cases and St. Luke's International Hospital's cases was explained by the fact that St. Luke's International Hospital's was the largest hospital near the site and many severe cases must have been carried there. This explanation could be supported by the fact that 5 unconscious patients admitted and 2 deaths occurred in this hospital compared with no death cases in our hospitals and also the proportion of patient whose %ChE was under 100 was 38.8% which is higher than that of our subjects (32).

Conclusion

In the attack, sarin was used and poisonous symptoms related to muscarine-like effects generally appeared. These could be recognized as early warning signs. Symptoms relating to nicotine-like effects and effects on the central nervous system appeared in severely exposed cases suggesting that they can be used as severity indicators. Muscarine-like effects to the eye and respiratory system must be induced through the direct contact of sarin gas to the mucous membranes. On the other hand, nicotine-like effects occur due to systemic exposure to sarin.

This survey was conducted by a multidisciplinary team under the direction of Prof. H. Shimizu, Department of Public Health and Environmental Medicine, Jikei University School of Medicine. The position of co-authors belonging to Jikei group are as follows. O. Sakai, former president of Jikei University Hospital; Y. Ogawa, M. Kadokura, T. Agata, and M. Fukumoto, Department of Public Health and Environmental Medicine; T. Satake, Department of Anesthesiology; K. Machida, Department of Laboratory Medicine; K. Nakajima, Nakajima Clinic; S. Hamaya, Kyobashi Hospital; S. Miyazaki, Kobikichyo Clinic; M. Ohida, Kiba Hospital; T. Yoshioka, Tsukishima-Samaria Hospital; S. Takagi, Daiyabiru Clinic.

Literature Cited

1. Yoshida, S. *Trend in Fire Prevention* **1997**, *188*, 52.
2. Grob, D. *A. M. A. Arch. Intern. Med.* **1956**, *98*, 221.
3. Sidell, F. R. *Clin. Toxicol.* **1974**, *7*, 1.
4. Rengstorff, R. *Arch. Toxicol.* **1985**, *56*, 201.
5. Grob, D.; Harvey, J. C. *J. Clin. Invest.* **1958**, *37*, 350.
6. Morita, H.; Yanagisawa, N.; Nakajima, T.; Shimizu, M.; Hirabayashi, H.; Okudera, H.; Nohara, M.; Midorikawa, Y.; Mimura, S. *Lancet* **1995**, *346*, 290.

7. *The report on the survey of noxious gas poisoning in Matsumoto;* The medical committee for the toxic gas intoxication of Matsumoto local medical council, Ed.; Matsumoto, 1995.

8. Suzuki, T.; Morita, H.; Ono, K.; Maekawa, K.; Nagai, R.; Yazaki, Y. *Lancet* **1995**, *345*, 980.

9. Nozaki, H.; Akikawa, N.; Shinozawa, Y.; Hori, S.; Fujishima, S.; Takuma, K.; Sagoh, M. *Lancet* **1995**, *345*, 980.

10. Nozaki, H.; Aikawa, N. *Lancet* **1995**, *345*, 1446.

11. Yokoyama, K.; Ogura, Y.; Kishimoto, M.; Hinoshita, F.; Hara, F.; Yamada, A.; Mimura, N.; Seki, A.; Sakai, O. *JAMA* **1995**, *274*, 379.

12. Masuda, N.; Takatsu, M.; Morinari, H.; Ozawa, T. *Lancet* **1995**, *345*, 1446.

13. Nozaki, H.; Hori, S.; Shinozawa, Y.; Fujishima, S.; Takuma, K.; Kimura, H.; Suzuki, M.; Aikawa, N. *Intensiv. Care Med.* **1997**, *23*, 1005.

14. Ishimatsu, S.; Tamaki, S.; Nakano, M.; Matsui, I.; Hinohara, S. *Nihon-Iji-Shinpo* **1995**, *3720*, 32.

15. Maekawa, K. *Igaku-no-Ayumi* **1996**, *177*, 731.

16. *NMR 1: Recommended operating procedure of identification of treaty-related compounds by NMR spectrometry;* The Ministry of Foreign Affairs of Finland, Ed.; Recommended operating procedures for sampling and analysis in the verification of chemical disarmament; The Ministry of Foreign Affairs of Finland: Helsinki, 1994, pp 1-8.

17. Greenhalgh, R.; Weiberger, M. A. *Canad. J. Chem.* **1967**, *45*, 495.

18. Tsunoda, N. *J. Chromatogr.* **1993**, *637*, 167.

19. *Synthesis and reaction of organic compounds I;* The Chemical Society of Japan, Ed.; Maruzen: Tokyo, 1978, Vol. 14; pp 568-572.

20. Benya, T. J.; Cornish, H. H. In *Aniline*; Clayton, G. D.; Clayton, F. E., Eds.; Patty's industrial hygiene and toxicology 4 Ed.; John Wiley & Sons, Inc.: New York, 1994, Vol. II B; pp 982-984.

21. Cavender, F. In *Hexanes;* Clayton, G. D.; Clayton, F. E., Eds.; Patty's industrial hygiene and toxicology 4 Ed.; John Wiley & Sons, Inc.: New York, 1994, Vol. II B; pp 1233-1235.

22. Taylor, P. In *Anticholinesterase agents;* Hardman, J. G.; Limbird, L. E., Eds.; Goodman and Gilman's the Pharmacological Basis of Therapeutics 9 Ed.; McGraw-Hill Companies, Inc.: New York, 1996, pp 161-176.

23. Christenson, W. R.; Van Goethem, D. L.; Schroeder, R. S.; Wahle, B. S.; Dass, P. D.; Sangha, G. K.; Thyssen, J. H. *Toxicol. Let.* **1994**, *71*, 1390.

24. Whittaker, M. In *Cholinesterase;* Bergmeyer, H. U., Ed.; Methods of Enzymatic Analysis 3 Ed.; Verlag Chemie: Weinheim, 1984, Vol. IV(X); pp 63-74.

25. Yamauchi, M.; Toda, G. *Sougo-Rinsyo* **1993**, *42*, 834.

26. Burtan, R. C.; Haimes, S. C. In *Organophosphates;* Clayton, G. D.; Clayton, F. E.,

Eds.; Patty's industrial hygiene and toxicology 4 Ed.; John Wiley & Sons, Inc.: New York, 1994, pp 3190-3192.

27. Namba, T.; Nolte, C. T.; Jackrel, J.; Grob, D. *Ame. J. Med.* **1971**, *50*, 475.

28. Blaxter, M. *Lancet* **1987**, *8549*, 30.

29. Foppa, I.; Noack, H.; Minder, C. E. *J. Clin.l Epidemiol.* **1995**, *48*, 941.

30. *Sarin attack on the Tokyo metro;* Asahi-Shinbun: Tokyo, March 16, 1996.

31. Murakami, H. *Underground*; Kohdanshya: Tokyo, 1997.

32. Hinohara, S. *Nihon-Iji-Shinpo* **1995**, *3706*, 47.

Chapter 23

Preparedness against Nerve Agent Terrorism

Tetsu Okumura[1,4], Kouichiro Suzuki[2], Atsuhiro Fukuda[2], Shinichi Ishimatsu[3],
Shou Miyanoki[1], Keisuke Kumada[1], Nobukatsu Takasu[3], Chiiho Fujii[2],
Akitsugu Kohama[2], and Shigeaki Hinohara[3,4]

[1]Emergency Department, St. Luke's International Hospital, Tokyo, Japan
[2]Department of Acute Medicine, Kawasaki Medical School, Kurashiki, Japan
[3]St. Luke's International Hospital, Tokyo, Japan

On the morning of March 20, 1995 sarin was released in
the Tokyo Subway System. As a result, 12 people died
and 5500 more were sickened. There had never been
such a large-scale act of urban terrorism caused by nerve
gas. The most important measure in fighting nerve agent
terrorism is preparedness. Public organizations and
hospitals must have decontamination facilities. EMTs
and the medical staff in hospitals should have personal
protective equipment. Hospitals should stock antidotes
and public organizations need multi-ventilator systems.
Local communities must have protocols for chemical-
and-radiation contaminated victims, multi-channel and
multi-directional information networks regarding
hazardous materials, and must practice repeatedly and
regularly. The close follow-up of sarin victims is also
extremely very important.

As the world has learned, urban terrorism can take many forms: bombing,
biological agent attacks, chemical agent attacks, and nuclear weapon
attacks. On the morning of March 20, 1995 the Tokyo Subway System
was filled with a noxious substance later identified as a diluted form of the
nerve gas, sarin. A total of five subway commuter cars were affected
during the Monday morning rush hour. There were 12 fatalities and 5500
more were sickened. From a worldwide historical perspective, there has
never been such a large-scale disaster caused by a nerve gas in peacetime.
This disaster in Japan was a wake-up call for the rest of the world. The
most important measure against urban terrorism using nuclear, biological,
or chemical agents is preparedness. Terrorists always have an eye pealed
for the weak points of an urban community. In this attack, the lack of
preparedness against nerve agents was exposed. This article throws light
on what problems existed from the viewpoint of preparedness, and
discusses preparedness against nerve agent terrorism. We also believe that

[4]Current address: Department of Acute Medicine, Kawasaki Medical School, Kurashiki,
Japan.

most of the preparation tactics that can be used in an urban terrorism situation are useful for other chemical or nuclear disasters.

Background

The attack occurred at approximately 8:00 a.m. According to a traffic white paper, the capacity of subway trains between 8:00 am and 8:40 am reaches more than 200% (passengers cannot read opened newspapers, only magazines). Later police announcements noted the terrorists respectively carried diluted sarin solution in plastic bags into the subway trains and simultaneously stuck the sharpened tip of an umbrella into these bags. There were 12 fatalities and 5500 more persons were sickened.

St. Luke's International Hospital, a private 520-bed facility, received the greatest number of victims because it is located within 3 km of the affected subway stations. Fig. 1 shows the affected Tokyo Subway System and the location of St. Luke's Hospital.

When the first emergency call from the Tokyo Metropolitan Ambulance Control Center, which is under the administration of the Tokyo Metropolitan Fire Department came into our Emergency Center at 8:16 a.m., it was reported that there had been a gas explosion at a subway station. We, therefore, started to prepare for burns and carbon monoxide poisoning. At 8:28 a.m. the first patient complaining of eye pain and visual darkness arrived at the emergency department on foot from one of the subway stations. The first ambulance arrived at 8:43 a.m. In the first hour about 500 patients including three cardiopulmonary arrest on arrival patients were rushed to the Emergency Center. The directors of the hospital declared a disaster-oriented system, which included canceling of all routine operations and outpatient examinations at 9:20 a.m. More than 100 doctors and 300 nurses and volunteers were mobilized to care for the patients (1). Table 1 shows the time course of events with details. Even in the hospital passageways, the victims were receiving IV infusion.

At St. Luke's Hospital, triage was done mainly in the Emergency Center. We divided victims into three categories, mild: ambulatory victims who presented with only eye symptoms, moderate: non-ambulatory victims who presented with other systemic signs and symptoms, and severe: victims needing mechanical ventilation. Mild patients were closely observed with an IV line for several hours in the outpatient department and sent home. Moderate patients were admitted to wards. Severe patients were moved to our ICU. The number of mild cases with only eye symptoms, who were released after a half day observation, was 528. The number of moderate cases with other systemic signs and symptoms, who were admitted, was 107. The number of severe cases requiring intubations and artificial ventilation was 5 (2).

Table 2 shows the symptoms of the patients. The most prominent sign was miosis, followed by headache, visual darkness, eye pain, dyspnea, nausea, cough , and throat pain. The most significant change in chemical laboratory data was low plasmal cholinesterase values. The ability to manage samples reached its limitation. Usually we use a computer system, but the system broke down due to the enormous number of samples. Therefore results had to be reported manually. As for

358

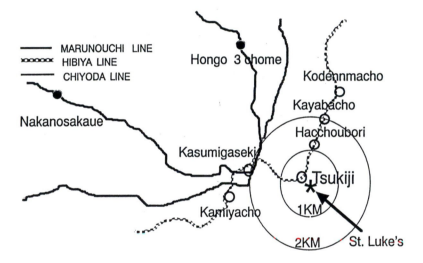

O Subway stations where many casualties arose

Figure 1. Affected Tokyo Subway System

Table 1: Time Course with Information

Time	Events	Information source	Information
7:55	The attack occurred simultaneously in a number of locations.		
8:16		The Tokyo Metropolitan Fire Department	Explosions occurred at subway stations.
8:25	First victim came to ER on foot.	Victims	There was no explosion.
8:40	First ambulance came.		Many people collapsed in the subway station.
8:43	First cardiopulmonary arrest patient came by a private vehicle.		
	More than 500 victims rushed to the ER.		
9:20	All routine operations and outpatient examinations were canceled.	The Tokyo Metropolitan Fire Department	Cause material is acetonitrile.
9:40	Cholinesterase level returned: very low.		
	Pralidoxime was started to be given for severe patients.	President of Shinshu University Hospital	Sarin intoxication is suspected.
10:30		Doctor from the Self Defense Forces Hospital.	Sarin intoxication is suspected.
11:00	First press conference was given.	TV news:	Cause material is sarin.
12:00	Doctor conference was held to standardize the triage and treatment.	police announcement (There was no direct information from police.)	
14:00	Mildly affected patients were sent home.		
17:00	Transportation of the admitted patients was rerejected by the Tokyo Metropolitan Fire Department.	The Tokyo Metropolitan Fire Department	Ambulances are not available for the victims' transport to other hospitals until midnight.
18:00	Germany, France, and England offered the dispatch of rescue teams.		
20:00	Final reconfirmation of admitted patients' information (name, address, severity, etc)was done.		
22:00	The list of the patients was announced.		

SOURCE: Adapted with permission from Reference 1. Copyright 1998.

Table 2: Patient Signs and Symptoms

Signs and Symptoms	% (number of patients)

Ophthalmological Si/Sx:

Miosis:	90.5% (568 / 627)
Eye Pain:	37.5% (235 / 627)
Visual Darkness:	37.6% (236 / 627)
Blurred Vision:	17.9% (112 / 627)
Conjunctival Injection :	6.7% (42 / 627)
Tearing:	4.3% (27 / 627)

Respiratory Si/Sx:

Dyspnea:	29.2% (183 /627)
Cough:	18.8% (118 / 627)
Chest Oppression:	12.0% (75 /627)
Wheezing:	1.0% (6 / 627)

Gastrointestinal Si/Sx:

Nausea:	26.8% (168 / 627)
Vomit:	14.7% (92 / 627)

Neurological Si/Sx:

Headache:	50.4% (316 / 627)
Easy Fatigability:	15.2% (95 / 627)
Fasciculation:	7.0% (44 / 627)
Extremity's Numbness:	2.9% (18 / 627)
Vertigo & Dizziness:	2.7% (17 / 627)
Consciousness Disturbance:	2.4% (15 / 627)

ENT Si/Sx:

Throat Pain:	18.3% (115 / 627)
Runny Nose:	15.2% (95 / 627)

Psychological Si/Sx:

Agitated State:	5.7% (36 / 627)

other chemical laboratory data, 11 % of the patients had higher than normal CPK values. Blood cell counts indicated leukocytosis. Arterial blood gas assays revealed many cases of respiratory alkalosis with the exception of the cardiopulmonary arrest on arrival patient. Of the 57 cases who had an ECG taken, four cases showed mild QT elongation without the necessity of therapy. Four cases had bradycardia, which responded well to atropine sulfate (2).

Fig.2 shows the distribution of plasma cholinesterase levels. The severity of the condition of patients correlated closely with the plasma cholinesterase levels.

Ninety-six percent of the patients received atropine sulfate for the treatment of miosis. We were careful not to overdose and it was used until the first signs of thirst and tachycardia appeared. After nerve gas became the suspected cause we decontaminated the admitted patients by having them change clothes and shower. Administration of pralidoxime (PAM) was also started within three hours of initial chemical exposure in 95 % of the admitted cases. PAM was especially effective for fasciculation. For convulsion, we used diazepam in eight cases, although its effect was short. For injected conjunctiva, topical steroidal eyedrops were used. Two cases experienced acute stress disorder (ASD). For these severe ASD patients, a mild antidepressant was used (2) and respiratory stabilization was mandatory to the well being of the patients. The duration of a hypoxic state has the most influence on the patient's prognosis. That means that not only hospital care, but also prehospital care is very important.

Within two to four days after the Sarin Attack, 95% of the hospitalized patients had recovered and were subsequently discharged with satisfactory relief of their complaints.

Decontamination

There was no field decontamination when the Attack occurred. The Tokyo Metropolitan Fire Department established Haz-Mat Units in 1990. At the time of the attack, 10 units were deployed in several fire stations in Tokyo. These were all dispatched to the affected subway stations. They were mainly engaged in analysis of the cause material. The cause material was analyzed as acetonitrile. Later this analysis was proved to be wrong. Japanese Self Defense Forces carried out decontamination of the affected subway trains and stations, but they did not decontaminate victims.

We were unable to satisfactorily decontaminate victims in the hospital , especially mild cases. In part, this was because it took time to determine the cause of the victims' illness and, in part, because we did not have enough space for the victims to shower and change clothes. However, we should be able to recognize some chemical agents and carry out decontamination earlier when we see mass casualties complaining of the same signs and symptoms. In Japan there are very few hospitals with a room specifically for decontamination. There was no room set aside for decontamination in St. Luke's Hospital. Therefore, secondary exposure of the medical staff arose from sarin vapor on the victims' clothing . Many

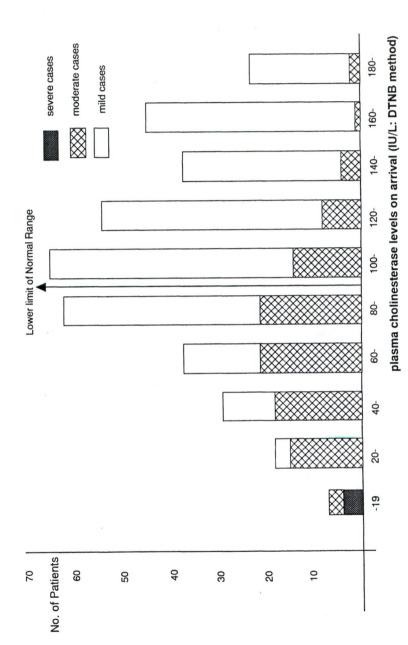

Figure 2. Distribution of Plasma Cholinesterase Levels

cases of secondary exposure were especially noted in poorly ventilated area of the hospital (3).

Decontamination; i.e., removal of the cause material, is a very essential concept in chemical disasters. Nerve agents are no exception. Decontamination should be thought a matter to be considered at any stage and site. It should be done either in fields or outside of the hospitals, since if mass casualties arise, some victims without decontamination will come directly to the hospital from the disaster site.

Therefore public organizations, such as the armed forces or fire departments, must have some mobile facilities where victims can shower and change clothes. These facilities could be utilized not only for dealing with nerve agents, but also for other chemical disasters and nuclear disasters. Prevention of hospital contamination is also very important. Therefore all hospitals should have decontamination facilities. The decontamination area of a hospital can be located either inside or outside of the hospital. An indoor decontamination area would be ideal, but decontamination outside is simpler and less expensive than equipping an indoor decontamination area (4). A parking lot would be one good candidate. The waste water from showering should be kept separated and should not be allowed to run over the pavement into sewer drains.

Protective Equipment

The Tokyo Haz-Mat Units used after the Sarin Attack had protective masks, suits, boots, and gloves, but ordinary EMTs had no protection. Without decontamination and protection, 135 of 1,364 EMTs involved in rescue efforts experienced secondary exposure.

At St. Luke's Hospital, the medical staff wore usual medical masks and surgical latex gloves after it was recognized the cause material was a nerve agent. But the efficacy of their protection was questionable. Better ventilation of the accommodation room was more effective against the secondary exposure. Vapor sarin from victims' clothes could not be avoided.

Protection from nerve agents is also important. Two types of respiratory protection are available: cartridge respirators and supplied air respirators. Cartridge respirators are especially useful for decontamination outside hospitals. They are inexpensive, portable and easy to use and store. However, they must be fit tested, someone must decide which type of cartridge to use based on the materials causing injury to victims, and they cannot be used for long time. Air-powered respirators require no fit testing but need an air supply and hoses.

Chemical resistant suits are necessary for decontamination in both prehospital and hospital settings. After this attack, protective clothing has been included in ordinary ambulances, but few hospital in Japan keep any of this clothing. Ordinary latex gloves offer little chemical protection. Cox has recommended nitrile and Viton as materials for the gloves of medical staff who care for chemically contaminated victims. He has also noted, however, that some classes of chemicals rapidly penetrate nitrile. In such cases gloves made of Viton, which is quite bulky though, are indicated (4). Every hospital should prepare such personal protective equipment (PPE).

Stockpiling of Drugs and Instruments

Antidote storage is important for mass casualties of poisonings. Initially, we had stored 100 ampoules of PAM and 1030 ampoules of atropine sulfate. These were sufficient for the initial treatment, but our pharmaceutical department had to make an additional order to wholesale dealers at the early stage. In total, we used 700 ampoules of PAM and 2800 ampoules of atropine sulfate (1).

In Japan, organophosphoral poisoning is most common in rural areas. Organophosphates are used for agricultural chemicals. If they are ingested by farmers attempting suicide, these persons will be taken to rural hospitals. In the Tokyo Metropolitan area , where organophosphoral poisoning is rare, few hospitals have sufficient stockpiles of antidotes, especially PAM. PAM was transported from a maker in Osaka to Tokyo by air and super express trains. Fortunately, St. Luke's Hospital happened to have a stockpile of these antidotes because organophosphoral poisoning patients had been admitted just before the attack.

Stockpiling of antidotes is troublesome for hospitals, because they must be regularly exchanged for new supplies before their expiration date. This is not cost-beneficial. WHO has divided antidotes into three (A, B, and C) according to the extent of emergency in its guidelines for the Poison Information Centers. Category A includes antidotes which should be given within thirty minutes, category B those which should be given within two hours, and category C those which should be given within six hours. In preparing for possible chemical disasters, every regional mainstay hospital should store antidotes according to WHO's classification and the geographical situation. Every hospital should stock highly emergent antidotes (WHO classification A), for example, CN-kits.

Fortunately, the cause material in the Tokyo Attack was "diluted" form of sarin. Therefore, the ratio of victims requiring intratracheal intubation among all the victims was low. But, if sarin had been used in pure form, the circumstances for the victims' treatment would have been greatly changed. In such a situation, an enormous number of intratracheal tubes, bags, and ventilators would have been required. At the early stage, manual bagging can be done in such a situation. A multiple ventilator system would be useful. Such a system has already come on market. Public organizations, such as the armed forces or fire departments, must also have such a system.

Information Network

Initially, we thought that the patients' illness including miosis was cased by some kind of organic phosphorus, which is often included in agricultural chemicals. However, we were puzzled as to why it had happened in the subway.

On the day of the Sarin Attack, medical information mainly came to St. Luke's Hospital from three sources (1). Information came first from the president of Shinshu University Hospital, who had experience with treatment of the Matsumoto Sarin Incident victims, via telephone and facsimile. Simultaneously, it came from a doctor sent from the Self Defense Forces Hospital. The third source was TV news. It took three hours for the police to announce officially that the material was sarin, but the police did not inform us directly.

We also tried to contact the Japan Poison Information Center regarding treatment, but the telephone circuit had already became overloaded due to the enormous number of calls from hospitals. There are only two Poison Information Centers to handle inqiries from all over the Japan.

Information on the cause material and its treatment should be well organized and distributed promptly. The information network ought to be mutidirectional and multi-channel (1). In the public trial for those participating in the Tokyo Subway Sarin Attack it was revealed that the laboratory of the Tokyo Metropolitan Police Department discovered the cause material to be sarin at 9:55 a.m., but an official announcement concerning the cause material was not made on TV until 11:00 a.m. This delay shows the level of the comprehension regarding information on chemical disasters in Japan.

Protocols, Training and Exercises

Disaster planning exists in Japan, but this is mainly for natural disasters. In the Tokyo Subway Sarin Attack, the concerned organizations acted independently without central coordination. Therefore, no substantial information networking was established among organizations (5).

Every community and hospital must include preparations for chemical disasters and nuclear ones in their disaster planning. Cox advocates that it is best to have one protocol to handle all these situations. Because situations involving radiation or chemical contamination patients will be rare, two different protocols could result in confusion (4).

Protocols without training and practice are worthless. Training in decontamination and PPE procedures must be done regularly and repeatedly. In local communities, every concerned organization, including the police, the armed forces, the fire department, Haz-Mat teams, and hospitals, must meet and communicate with each other in disaster drills.

The importance of further follow-ups of the victims

Learning from experience with past cases is important preparedness for the future. In other words, this is also preparedness for future cases. A lot is unknown about the long term effects of nerve agents. In animals, some literature has described such long-term effects, but there is little information about the effects in humans. No definite conclusions have been made concerning them. And whether initial treatment can change these effects or not is not known.

We sent questionnaires to the 660 victims treated at St. Luke's on the day of the attack one years after the attack. Of these, 303 victims replied, and 46% still had some symptoms. Regarding physical symptoms, 18.5% of the victims still complained of eye problems, 11.5% of easy fatiguability, and 8.6% of headache. As for psychological symptoms, 12.9% of them complained of fear of subways, 11.6% indicated the fears concerning their escape from the attack (6). A medical check-up was carried out one year after the attack on 133 victims who wanted such a check-up. Since we could find no abnormalities in any cases, can these symptoms be explained completely by post traumatic stress disorder (PTSD) ? We felt the necessity to investigate subtle subclinical neurological changes.

Therefore, we carried out a study of subclinical neuropsyco-behavioral effects with the Department of Public Health of Tokyo University. Table 3 shows the items. The results were published in journals (7, 8, 9).

In these studies, it was suggested that a delayed effect on the vestibulocerebellar system was induced by acute sarin poisoning, that females might be more sensitive than males (7), and that asymptomatic sequelae to sarin exposure, rather than PTSD, persist in the higher and visual nervous systems beyond the turnover period of ChE. Sarin may have neurotoxic actions in addition to the inhibitory action on brain ChE (8). Chronic effects on pychomotor performance are supposed to be caused directly by acute sarin poisoning. On the other hand, psychiatric symptoms and fatigue appeared to result from PTSD induced by exposure to sarin (9). These studies are number-limited and are preliminary study, but they indicate the necessity for close follow-ups of victims.

Fig. 3 shows the residual symptoms two years after the Tokyo Sarin Attack, and two most major symptoms in each category. We sent questionnaires to the victims. Eighty-eight percent still had some symptoms. Fortunately, there were no apparent physical abnormalities except for the victims' eyes. We should continue to study long-term effects of sarin clinically and subclinically. To know the actual conditions of the victims is the first step in making clear the pathogenicity of nerve agents. We believe that new and more effective treatment for victims of nerve agents must be promoted.

Table 3: Examined Items

WHO-NCTB: Neuro-Core Test Batteries

NES: Neurobehavioral Evaluation System

Evoked Potentials

   VEP: Visual Evoked Potential

   BAEP: Brainstem Auditory Evoked Potential

ERP: Event-Related Potential

   P300

RR Interval Variability in ECG

PB: Postural Balance

367

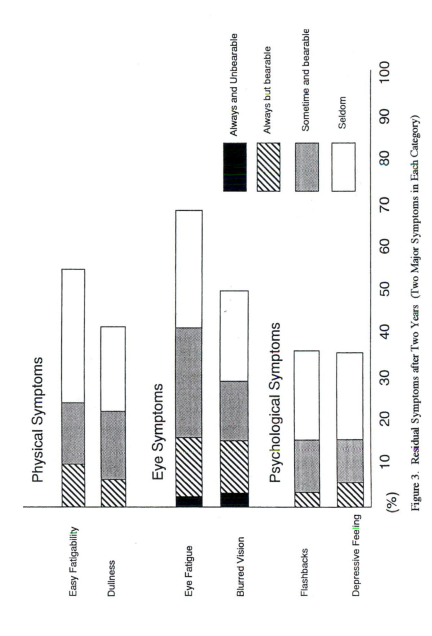

Figure 3. Residual Symptoms after Two Years (Two Major Symptoms in Each Category)

References:

1. Okumura, T.; Suzuki, K.; Fukuda, A. et al. The Tokyo Sarin Attack: Disaster Management, Part 2: Hospital Response. *Acad Emerg Med.* **1998**, *5*, 618-24.
2. Okumura, T.; Takasu, N.; Ishimatsu, N. et al. Report on 640 Victims on The Tokyo Subway Sarin Attack *Annal. Emerg Med.* **1996**, *28*, 129-35.
3. Sanoyama T.; Monden M.; Ohbu S. et al. The investigation of Secondary Exposure at St. Luke's International Hospital. *Nihon-Iji-Shinpo.* **1995**, *3727*, 17-9.
4. Cox, R.D. Decontamination and Management of Hazardous Materials Exposure Victims in the Emergency Department. *Annal. Emerg. Med.* **1994**, *23*, 761-770.
5. Okumura, T.; Suzuki, K.; Fukuda, A. et al. The Tokyo Sarin Attack: Disaster Management, Part 1: Community Emergency Response. *Acad Emerg Med.* **1998**, *5*, 613-7.
6. Ishimatsu, S.; Tanaka, K.; Okumura, T. et al. Result of the Follow-up Study of the Tokyo Subway Sarin Atack (1 Year after the Attack) [Abstract]. *Kyuukyu-Igakkai-shi.* **1996**, *7*, 567.
7. Yokoyama, K.; Araki, A.; Okumura, T. et al. Preliminary Study on Delayed Vestibulo-cerebellar Dysfunction in Tokyo Subway Sarin Poisoning in Relation to Gender Difference: Frequency Analysis Postural Sway. *J. Occup. Environ. Med.* **1998**, *40*,17-21.
8. Murata K.; Araki S.; Okumura, T. et al. Asymptomatic Sequelae to Acute Sarin Poisoning in the Central and Autonomic Nervous System 6 Months after the Tokyo Subway Attack. *J. Neurol.* **1997**, *244*, 601-606.
9. Yokoyama, K.; Araki, S.; Okumura, T. et al. Chronic Neurobehavioral Effects of Tokyo Subway Sarin Poisoning in Relation to Posttraumatic Stress Disorder. *Arch. Environ. Health* **1998**, *53*, 249-56.

Chapter 24

Determination of Metabolites of Nerve Agent O-Ethyl-S-2-Diisopropylaminoethyl Methylphosphonothioate (VX)

H. Tsuchihashi, M. Katagi, M. Tatsuno, M. Nishikawa, and A. Miki

Forensic Science Laboratory of Osaka Prefectural Police H.Q., 1–3–18 Hommachi, Chuo-ku, Osaka 541–0053, Japan

A serum sample from the victim of the Osaka VX incident was analyzed according to our developed technique. GC-MS in full-scan EI and CI modes were used, and for more reliable confirmation, GC-tandem mass spectrometry was also employed. In the serum sample, not only ethyl methylphosphonic acid (EMPA) but also 2-(diisopropylaminoethyl) methyl sulfide (DAEMS) were detected. Additionally, in the animal experiment, administration of synthesized 2-(diisopropylamino) ethanethiol (DAET) to rats resulted in the production of DAEMS. These results documented that the techniques by GC-MS and GC-MS-MS were applicable to biological samples such as serum. These also provided the first reported unequivocal identification of the specific metabolites of VX in victim's serum, and furthermore, clarified a part of the metabolic pathway of VX in the human body.

O-ethyl S-2-diisopropylaminoethyl methylphosphonothioate (VX) strongly and readily binds to acetylcholinesterase and thereby inhibits this vital enzyme's normal biological activity in the cholinergic nervous system. It can be mass-produced by means of fairly simple chemical techniques and equipment, and the precursor compounds are inexpensive and readily available. Thus, it has been feared as "a poor man's atomic bomb", along with isopropyl methylphosphonofluoridate (sarin, GB) and pinacolyl methylphosphonofluoridate (soman, GD). Especially, VX, whose lethality against humans is estimated to be 0.1 mg·min/m^3 through the lungs, is over 10,000 times more poisonous than HCN gas, several hundred times as poisonous as sarin, and is thus thought to be among the most toxic nerve agents(1-4).

These nerve agents have been designated as chemical warfare agents (CWAs) and their use, possession and manufacture are banned by an international convention (1, 2). Nonetheless, they have been developed, manufactured and stockpiled, and have often been used in international conflicts (3, 5). More recently, CWAs have been used as tools for terrorism; Sarin has been used to commit indiscriminate murders in Matsumoto City in 1994, the Tokyo subway in 1995, and VX has been used to commit murder in Osaka in 1994. The incidents have led to mass destruction and disruption of the social order, and their strong killing and wounding properties have caused great shock and unrest globally.

With these incidents as a turning point, countermeasures to the eventual terrorism by biological and chemical warfare agents has been worked out mainly in Europe and U.S.A. Especially in the U.S.A., the Chemical/Biological Incident Response Force (CBIRF) has been organized to save victims and rapidly identify the causative compounds (6). However, there has not been sufficient preparation for such misuse of warfare agents in Japan.

The three nerve agents (VX, sarin and soman) are readily hydrolyzed to the corresponding alkyl methylphosphonic acids: VX to ethyl methylphosphonic acid (EMPA), sarin to isopropyl methylphosphonic acid (IPMPA) and soman to pinacolyl methylphosphonic acid (PMPA). The alkyl methylphosphonic acids are ultimately very slowly hydrolyzed to methylphosphonic acid (Figure 1)(4, 7, 8). Therefore, the detection of methylphosphonic acids has been usually performed for proof of the use of nerve agents.

The determination of methylphosphonic acids has been mainly studied by using high-performance liquid chromatography (HPLC) (9), ion chromatography (IC) (10-13), capillary electrophoresis (14), high-performance liquid chromatography/mass spectrometry (15, 16), and capillary electrophoresis/mass spectrometry (17) with no derivatization, and by HPLC, GC and GC-MS with derivatization (7, 8, 18-26). Most of the determinations were made from authentic standard samples and environmental samples such as soil and surface water.

On the other hand, for unequivocal proof of exposure to these nerve agents, it is necessary to detect the original nerve agents and their metabolites from biological samples such as blood, urine, and saliva collected from victims. In the human body, they are also thought to be enzymatically and/or spontaneously hydrolyzed to the corresponding methylphosphonic acids according to the pathways outlined in Figure 1. Up to now, there are no published papers on their metabolism in the human body. Additionally, there is hardly any available analytical method that allows us to determine the compounds in biological samples (13, 25-27).

Quite recently, we have analyzed the serum sample collected from a victim of the VX incident in December, 1994, in Osaka. In the analysis, several specific compounds could be detected which would be very important for the proof of exposure to VX (27).

Additionally, to confirm the result from the analysis of the serum sample, a precursor compound of a detected metabolite was synthesized and administered to rats. The rats' sera were collected and analyzed according to our developing methods.

In this paper, the resulting detailed data will be shown, leading not only to unequivocal proof of the use of VX, but also to clarification of a part of the metabolism of VX in the human body.

Case History of VX Incident in Osaka

A summary of the VX incident in Osaka is as follows. A office worker (28-year-old male) was allegedly attacked by two members of a cult-group. They sprinkled VX on his neck with a disposable syringe. He suddenly cried out and fell down on the street.

After 6 min, emergency medical technicians arrived at the site, and carried him to an emergency hospital under basic cardio-pulmonary resuscitation. He received strenuous therapy, but 10 days later he died.

He was the only VX murder victim ever documented in the world. Because of his symptoms such as miosis and a low level of cholinesterase activity, at first the poisoning with organophosphorus pesticides was suspected; it was also thought that he possibly might have been killed by a nerve agent. But, health personnel at the time thought of only sarin, and it was thought that with sarin the agent would have extended to a wider area. Thus, the cause of the death was not clarified by a postmortem examination, and the incident was treated as an unnatural death.

According to the confession of suspected persons about 6 months later, his death turned out to be due to the attack with VX . Other details such as the concentration, purity and amount of the sprinkled VX were still not clear.

Thus, we were formally requested to analyze his serum that had been collected from him approximately 1 hour after his exposure to VX, and had been kept frozen at -20℃ until the analyses were performed. The analyses were performed according to the method described in the **Experimental** section.

Experimental

Materials. EMPA was purchased from Aldrich (Milwaukee, WI, USA). The standard stock solution of EMPA was prepared in distilled water (1mg/ml), and adjusted to the appropriate concentration with distilled water or human serum immediately prior to use. An internal standard (I.S.), diphenylmethane (DPM), was purchased from Wako (Osaka, Japan), and the I.S. solution was prepared in acetonitrile (the concentration being 100 μ g/ml). The derivatization reagent used, N-methyl-N-(*tertiary*-butyldimethylsilyl) trifluoroacetamide (MTBSTFA) with 1% *tertiary*.-butyl-dimethylsilyl chloride (*t*-BDMSC), was purchased from GL Sciences (Tokyo, Japan).

2-(diisopropylamino)ethanethiol (DAET), 2-(Diisopropylaminoethyl)methyl sulfide (DAEMS) and an I.S., 2-(diisopropyl-aminoethyl)methoxide (DAEMO), were synthesized in our laboratory according to the procedures described later in this paper. The standard stock solution of DAEMS was prepared in distilled water (100 μ g/ml), and adjusted to the appropriate concentration with human serum immediately prior to use. The I.S. solution was prepared in dichloromethane (the concentration being 30 μ g/ml).

Potassium ethyl xanthate, sodium thiomethoxide and sodium methoxide were purchased from Wako, and 2-(diisopropylamino)ethyl chloride hydrochloride (DAEC・HCl) were purchased from Aldrich. They were used for synthesis of DAET, DAEMS and DAEMO.

Polyoxyethylene (20) sorbitan mono-oleate (Tween 80) was purchased from Wako. Other chemicals used were of analytical grade.

Synthesis of 2-(Diisopropylamino)ethanethiol (DAET) and 2-(Diisopropylamino-ethyl)methyl sulfide (DAEMS). DAET was synthesized according to the following procedure. A 50-ml round bottom recovery flask containing 1.3 g of potassium ethyl xanthate and 10 ml of absolute ethanol was fitted with an efficient water-cooled condenser and a dropping funnel that was charged with 1 g of free DAEC prepared from DAEC·HCl. DAEC was added dropwise to the stirred ethanol solution. After the addition was complete, the mixture was refluxed for 5 hours and ethanol was distilled. The residue was cooled and 10 ml of distilled water was added. The solution was extracted twice with 15 ml of dichloromethane. The extracts were combined, dried with anhydrous sodium sulfate and evaporated. The residue was transferred to a 50-ml round bottom recovery flask, and 5 ml of 20% aqueous potassium hydroxide and 15 ml of ethanol were added to the flask. Then, argon gas was introduced to fill the space of the flask, and the solution mixture was stirred at ambient for 2 hours. After neutralization of the reaction mixture with conc. HCl, the reaction mixture was extracted twice with 30 ml of dichloromethane. The extract was combined, dried with anhydrous sodium sulfate and evaporated. The resultant mixture was placed in a distillation flask and distilled under reduced pressure, and the fraction distilling at 105-108 ℃ (12 mmHg) was collected.

DAEMS was synthesized according to the following procedure. A 50-ml round bottom recovery flask containing 0.5 g of sodium thiomethoxide dissolved in 5 ml of absolute methanol was fitted with an efficient water-cooled condenser and a dropping funnel which was charged with 1 g of free DAEC prepared from DAEC·HCl. DAEC was added dropwise over a period of 10 min to the stirred methanol solution maintained at 10-15 ℃. After the addition was complete, the mixture was boiled under reflux for 2 hours. The reaction mixture was cooled and 5 ml of distilled water was added. The solution was extracted with 10 ml of dichloromethane. The extract was dried with anhydrous sodium sulfate and evaporated. The resultant mixture was placed in a distillation flask and distilled under reduced pressure, and the fraction distilling at 108-111 ℃ (27mmHg) was collected.

The I.S., 2-(diisopropylaminoethyl)methoxide (DAEMO), could be obtained by a similar procedure, using the appropriate amount of sodium methoxide in place of sodium thiomethoxide.

Unambiguous structural assignments and confirmation of the synthesized DAET, DAEMS and DAEMO were made on the basis of chemical shifts and mass spectra employing ^{1}H-NMR (Varian, Palo Alto, CA, U.S.A.) and GC-MS.

Gas Chromatography-Mass Spectrometry (GC-MS) and Gas Chromatography Tandem Mass Spectrometry (GC-MS-MS).
Analyses of the Victim's Serum. GC-MS was carried out on a JEOL JMS-SX102AQQ hybrid mass spectrometer (JEOL, Tokyo, Japan) interfaced to a Shimadzu GC-17A gas chromatograph (Shimadzu, Kyoto, Japan). A fused-silica capillary column Hewlett Packard Ultra 2 (column dimension and film thickness being 25m×0.32mm I.D. and 0.52 μ m) was used for separation. Injections were manually done in the splitless mode at 270℃. The column oven temperature was raised from 70℃ to 300℃ at 10℃/min for the analysis of EMPA for the analyses of volatile metabolites, the oven

temperature was maintained at 70℃ for 2 min and then raised at 10℃/min to 300℃. The temperature of the transferline between the gas chromatograph and the mass spectrometer was set at 250℃. High purity helium, at a linear velocity of 45cm/sec, was used as the carrier gas. The EI operating parameters were as follows: resolution, 1000; source temperature, 200℃; electron energy, 70eV; ionizing current, 300 μ A; ion multiplier, 1.0kV; and accelerating voltage, 10kV. The isobutane CI operating conditions were as follows: resolution, 1000; source pressure, 3 $\times 10^{-4}$ Pa; source temperature, 200℃; electron energy, 200eV; ionizing current, 300 μ A; ion multiplier, 1.2kV; and accelerating voltage, 10kV. Data were collected from m/z 50-800 at a scan rate of 0.5 sec/scan.

GC-CI-MS-MS was carried out on a JMS-SX102AQQ hybrid mass spectrometer (JEOL) interfaced to a Shimadzu GC-17A gas chromatograph (Shimadzu). The chromatographic conditions were the same as those for the GC-MS. Argon, at a pressure of 2.2 $\times$ 10^{-6} Torr, was used as the collision gas. The other MS-MS operating parameters were as follows: resolution, 1000; source pressure, 3 $\times 10^{-4}$ Pa; source temperature, 200℃; electron energy, 200eV; accelerating voltage, 10 kV; ionizing current, 300 μ A; collision energy, 50 eV; and ion multiplier, 1.2 kV. The product ion spectra were recorded from m/z 50 to 300 at a scan rate of 1 sec/scan using the protonated molecular ions as the precursor ions (m/z 176 and 239 in the analysis of DAEMS and the t-BDMS derivative of EMPA, respectively).

Analyses of Rats' Serum in Animal Experiment GC-MS was conducted on a Shimadzu GC-MS QP1100EX (Shimadzu, Kyoto, Japan). A fused-silica capillary column J&W DB-1 (column dimension and film thickness being 30m×0.32mm I.D. and 0.25 μ m) was used for separation. The column oven temperature was maintained at 50℃ for 2 min and then raised at 20℃/min to 260℃. Other GC conditions was same as those for analyses of the victim's serum The EI operating parameters were as follows: source temperature, 250℃; electron energy, 70eV and ion multiplier gain, 1.3. Data were collected from m/z 50-500 at a scan rate of 0.5 sec/scan.

Sample Preparation. Serum samples were prepared as follows (26)(Figure 2): A 1 ml volume of serum sample was extracted twice with 1 ml of dichloromethane (centrifugation facilitates the separation of layers). Both the organic and aqueous layers were separated. The organic layer was used for the determination of VX and its volatile metabolites, and the aqueous layer for the determination of water-soluble compounds such as EMPA.

The organic layer was dried with anhydrous sodium sulfate, transferred to a 13 × 100 mm screw-capped pyrex tube and evaporated carefully under a gentle stream of nitrogen at ambient temperature. The residue was dissolved into 100 μ l of dichloromethane and 5 μ l of DAEMO solution (I.S.) was added to it. A 1- μ l aliquot of the extract was injected into the GC-MS and GC-MS-MS.

The aqueous layer was deproteinized by adding 1 ml of acetonitrile, and the supernatant was separated by centrifugation. The supernatant was acidified with 1 ml of oxalate buffer (pH 1.68) and an additional 0.6 g of sodium chloride was added for

VX
Sarin
Soman
EMPA
IPMPA
PMPA
MPA
H2O

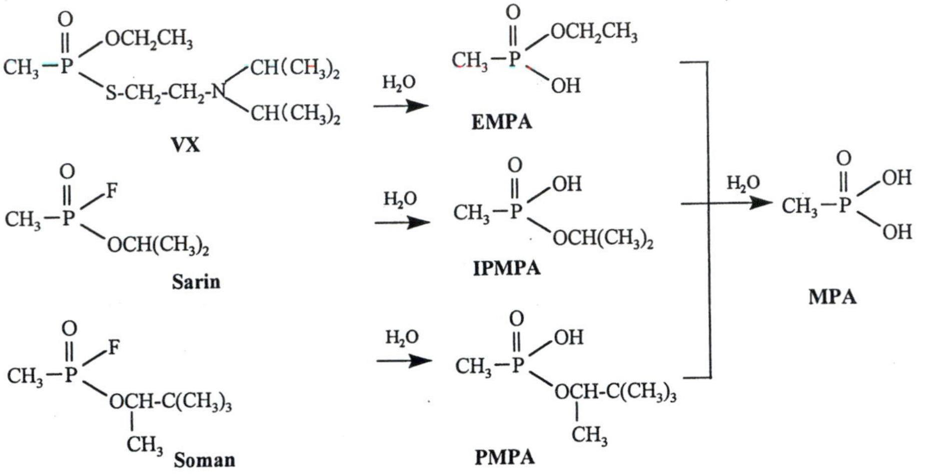

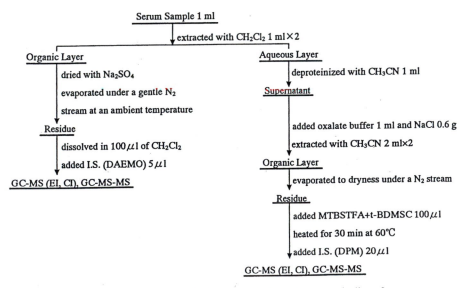

Figure 1. Generalized hydrolysis pathways for organophosphonate nerve agents.

Serum Sample 1 ml
extracted with CH2Cl2 1 ml×2
Organic Layer
dried with Na2SO4
evaporated under a gentle N2
stream at an ambient temperature
Residue
dissolved in 100 μl of CH2Cl2
added I.S. (DAEMO) 5 μl
GC-MS (EI, CI), GC-MS-MS
Aqueous Layer
deproteinized with CH3CN 1 ml
Supernatant
added oxalate buffer 1 ml and NaCl 0.6 g
extracted with CH3CN 2 ml×2
Organic Layer
evaporated to dryness under a N2 stream
Residue
added MTBSTFA+t-BDMSC 100 μl
heated for 30 min at 60°C
added I.S. (DPM) 20 μl
GC-MS (EI, CI), GC-MS-MS

Figure 2. Procedure for the determination of VX and its metabolites from serum sample.

salting-out (the addition of sodium chloride enables us to separate the aqueous layer and acetonitrile layer). The solution was extracted twice with 2 ml of acetonitrile (centrifugation facilitates the separation of layers). The organic layer was transferred to a 13×100 mm screw-capped pyrex tube and evaporated just to dryness under a stream of nitrogen at $60^{\circ}C$. The residue was derivatized by adding $100 \mu l$ of MTBSTFA with 1% t-BDMSC to the tube and then heating the tube at $60^{\circ}C$ for 30 min. At the end, 20 μl of DPM solution (I.S.) was added to the tube. A $1-\mu l$ aliquot of the reaction mixture was injected into the GC/MS and GC-MS-MS.

Animal Experiments. All adult male Sprague-Dawley (S-D) rats (255-280 g), obtained from Japan SLC (Hamamatsu, Japan), were 8 weeks of age, and were housed in an air-conditioned room at $24 \pm 2^{\circ}C$ for 1 week before use. DAET (100 mg) and Tween 80 were well mixed and diluted with distilled water. The mixture solution was injected in the rats' abdominal cavity in a dose of 20 mg/kg. In all experiments, control animals received the same amount of vehicle. At 10, 30, 60, 90, 120 and 180 min after injection, rats were killed by collection of blood from the heart under anesthesia with diethyl ether. The collected blood were immediately centrifuged at 3000 rpm, and were subjected to the analysis according to the methods previously described in the *Sample preparation* section (Figure 3).

Results and Discussion

Analyses of a Serum Sample Obtained from the Victim of VX Incident.
    GC-MS and GC-MS-MS Analyses of Separated Aqueous Layer. The separated aqueous layer was extracted with acetonitrile. After t-BDMS derivatization of the extract, the GC-MS analysis was conducted in both the full-scan EI and isobutane CI modes.
    The resultant total ion chromatogram (TIC) is depicted in Figure 4a. At a retention time of 8.4 min in the TIC, a fairly small peak appeared whose EI mass spectrum was characterized by a predominant fragment ion at m/z 153 and less intense fragment ions at m/z 181, 75 and 223 (Figure 4b). Careful examination of the spectrum revealed that the ions at m/z 75, 153, 181 and 223 originated from $[HOSi(CH_3)_2]^+$, $[CH_3PO(OH)OSi(CH_3)_2]^+$, $[CH_3PO(OC_2H_5)OSi(CH_3)_2]^+$ and $[CH_3PO(OC_2H_5)OSi-(CH_3)(CH_3)_3]^+$, respectively. Also, a CI mass spectrum obtained from the peak was dominated by the protonated molecular ion at m/z 239, suggesting that the molecular weight of the compound causing the peak was 238 (Figure 4c).
    For the confirmation of the result from the above mentioned GC-MS, the same sample was further analyzed by GC-CI-MS-MS, in which m/z 238 was selected as the precursor ion. Figure. 3d depicts the positive full-scan product ion spectrum produced from the peak described above. The spectrum which gives some characteristic product ions at m/z 73, 75, 153 and 195 corresponding to $[Si(CH_3)_3]^+$, $[HOSi(CH_3)_2]^+$, $[CH_3PO(OH)OSi(CH_3)_2]^+$ and $[CH_3PO(OC_2H_5)OSi(CH_3)_2CH_2]^+$, respectively. The obtained EI and CI mass spectra, product ion spectrum, and the retention time were quite identical with those of the t-BDMS derivative of the standard EMPA. According to the above data, EMPA was proven to be contained in the serum sample.

376

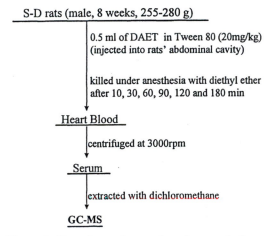

Figure 3. Experimental procedure for metabolism.

Additionally, to estimate the concentration of EMPA in the serum sample, a quantitative measurement by GC-MS in the selected ion monitoring (SIM) EI mode was carried out in the presence of DPM as the I.S. The calibration curve was constructed with spiked serum samples in which the concentration of the added EMPA was varied, and the predominant ions (m/z 153 for the t-BDMS derivative of EMPA and 168 for DPM) were chosen for the quantification.

The analysis showed good linearity throughout the concentration range from 0.1 to 5 μ g/ml for EMPA, and the concentration of EMPA in the serum sample estimated by using the calibration curve was 1.25 μ g/ml.

GC-MS and GC-MS-MS Analyses of Separated Dichloromethane Layer. The dichloromethane extract was directly analyzed by GC-MS in both the full-scan EI and isobuthane CI modes. The fragment ion at m/z 114 was known to be peculiar to the ester moiety in VX and its related compounds (28), and then, the ion at m/z114 was principally monitored. The resultant TIC is depicted in Figure 5a. Original VX was not detected in the sample. However, at the retention time of 7.1 min in the TIC, a fairly small peak was observed, whose EI mass spectrum was characterized by a predominant fragment ion at m/z 114 and less intense fragment ions at m/z 72, 75 and 128 (Figure 5b). Considering the data on the related compounds of VX previously described by P. A. D'Agostino et al. (28), careful examination of the spectrum disclosed that the ions at m/z 114, 72, 75 and 128 were due to $[(iPr)_2N=CH_2]^+$, $[(iPr)N=CH_2]^+$, $[CH_3SCH_2CH_2]^+$ and $[(iPr)(CH_3C=CH_2)NHC_2H_5]^+$, respectively. The CI mass spectrum was dominated by a protonated molecular ion at m/z 176, suggesting that the molecular weight of the desired compound was 175 (Figure 5c). Based on the above data, we assigned 2-(diisopropylaminoethyl) methyl sulfide (DAEMS) to this compound.

In addition, the sample was further analyzed by GC-CI-MS-MS, in which m/z 176 was selected as the precursor ion. The positive full-scan product ion spectrum produced from the desired compound is depicted in Figure. 5d. It shows a characteristic product ions at m/z 75 corresponding to $[CH_3SCH_2CH_2]^+$, representing the structure of DAEMS.

Confirmation and Quantitative Analysis of 2-(Diisopropylaminoethyl)methyl sulfide (DAEMS). For the confirmation of the assignment of DAEMS to the compound, DAEMS was synthesized in our laboratory according to the procedure described in the **Experimental** section and was subjected to the GC-MS and GC-MS-MS analyses. The data were collected under the same operating conditions as were done for the serum sample, and the resultant retention time and mass spectra were compared with those from the above assigned peak. The spectra and the retention time of the above assigned peak were almost identical with those of authentic DAEMS, thus confirming the assignment.

Additionally, to estimate the concentration of DAEMS in the serum sample, quantitative measurements by GC-MS in the SIM-EI mode were performed by using DAEMO as the I.S. The calibration curve was constructed with fortified serum samples in which the concentration of the added DAEMS was varied. Because the fragment ion

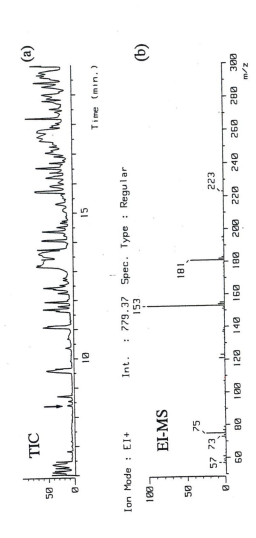

TIC

(a)

Time (min.)

50

0

10   15

Ion Mode : EI+   Int. : 779.37   Spec. Type : Regular

EI-MS

(b)

100

50

0

57 73

75

153

181

223

60   80   100   120   140   160   180   200   220   240   260   280   300

m/z

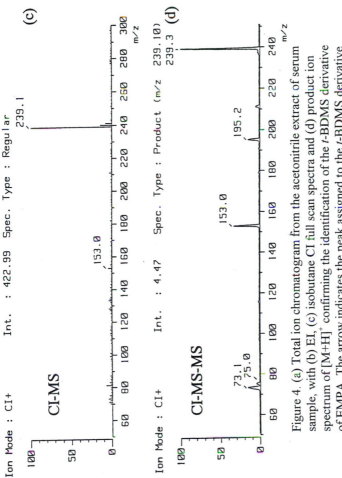

Figure 4. (a) Total ion chromatogram from the acetonitrile extract of serum sample, with (b) EI, (c) isobutane CI full scan spectra and (d) product ion spectrum of [M+H]⁺ confirming the identification of the *t*-BDMS derivative of EMPA. The arrow indicates the peak assigned to the *t*-BDMS derivative of EMPA.

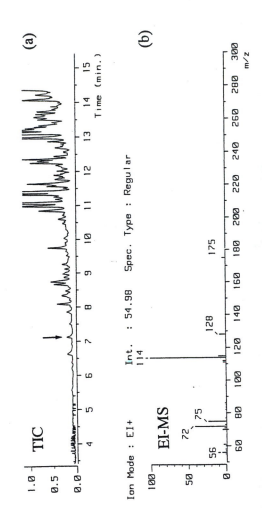

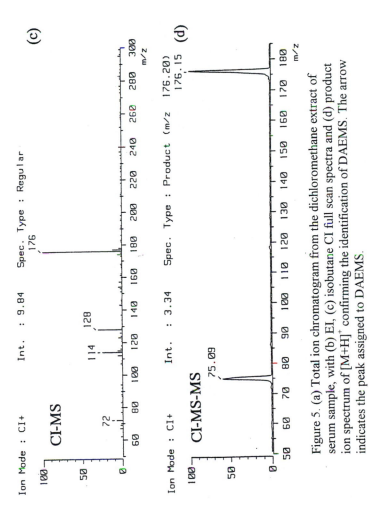

Figure 5. (a) Total ion chromatogram from the dichloromethane extract of serum sample, with (b) EI, (c) isobutane CI full scan spectra and (d) product ion spectrum of [M+H]+ confirming the identification of DAEMS. The arrow indicates the peak assigned to DAEMS.

at m/z 114 was present as the predominant ion in both the mass spectra of DAEMS and DAEMO, the ion at m/z 114 was selected for the quantification.

The analysis gave good linearity throughout the concentration range from 50 to 1000 ng/ml for DAEMS, and the concentration of DAEMS in the serum sample estimated by employing the calibration curve was 143 ng/ml.

Verification of Metabolic Pathway from DAET to DAEMS

In the human body, VX is thought to be enzymatically or/and spontaneously hydrolyzed to the EMPA and 2-(diisopropylamino)ethanethiol (DAET). In our GC-MS analyses, EMPA was detected, and DAEMS, the methylthioether of DAET, was also detected instead of DAET. It may be caused by the further metabolism of DAET to DAEMS. Then, to see whether DAET would be really metabolized into DAEMS in vivo, DAET was administered to rats, and metabolites in the serum were investigated according to the method descried in **Animal experiment** section. Additionally, quantitative measurements of DAEMS were also carried out for rats' serum by a similar method to that for victim's serum.

Every 10 to 180 min after injection, the expected DAEMS was detected, while the original DAET and other expected metabolites were not. Figure 6 shows the time course of DAEMS concentration in the rats' serum. The concentration of DAEMS reached a maximum at approximately 655 ng/ml within 10 min, fell rapidly with time, falling to the detection limit level at 180 min. This suggests that the metabolism of DAET to DAEMS would be extremely rapid and the resulting metabolite DAEMS would disappear from blood in a few hours. In the case of the victim's serum sample, the sample collection was done as soon as approximately an hour after his exposure to VX. The short duration between exposure and sampling probably facilitated the detection of DAEMS as well as EMPA in the serum sample.

In general, thiols are readily methylated and metabolized to the corresponding methylthioethers in the human body according to the biosynthetic reaction which was catalyzed by S-adenosyl-l-methionine-mediated thiol S-methyltransferase (29). Thus, it follows that VX would be also hydrolyzed to DAET, and the DAET was further metabolized immediately to DAEMS by methylation following the above mentioned enzyme system (Figure 7).

Conclusion

This is the first reported detection of DAEMS from a biological sample. We have successfully applied our previously developed method (26) that allows us to determine the hydrolysis products of nerve agents to the analyses of the serum sample obtained from a victim of the VX incident. In the analyses, DAEMS (concentration being 143 ng/ml) as well as EMPA (concentration being $1.25\ \mu$ g/ml) have been detected. Especially, DAEMS is considered to be the methyl-conjugated compound of the VX hydrolysis product (DAET) in human body.

To date, EMPA has been considered as an indicator of the use of VX. However, not only EMPA but also DAEMS has been detected from the serum sample in our analyses. The detection of both EMPA and DAEMS has provided more reliable proof of the use

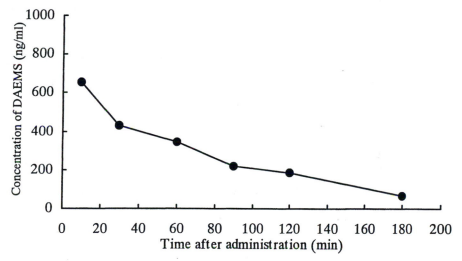

Figure 6. Time course of DAEMS concentration in the rats' serum.

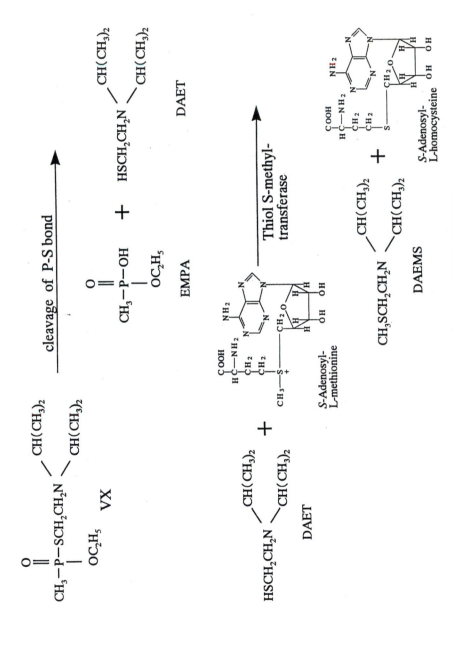

Figure 7. Metabolic pathway of VX to DAEMS.

of VX than that ever reported. Additionally, the animal experiment with DAET has demonstrated the main metabolic pathway of VX in human body.

However, the producing rate of DAEMS, the detectable period of DAET, etc. are not clear, and future studies on these subjects will be required to clarify the metabolism of VX.

Acknowledgments

The valuable suggestions and skillful technical assistance of Assistant Professor Dr. O. Muraoka and Mr. G. Tanabe (Kinki University) in synthesizing DAEMS and DAET is gratefully acknowledged.

Literature Cited

1. World Health Organization, *Health aspects of chemical and biological weapons, Report of a WHO Group consultants*, Geneva, 1970.
2. Compton, J. A. F., *Military Chemical and Biological Agents, Chemical and Toxicological Properties*, Telford Press, Caldwell, NJ, 1987.
3. Franke, S. ASA Newsletter **1994,** 42, 1.
4. Tu, A. T. *J. Mass Spectrom. Soc. Jpn,* **1996,** 44, 293.
5. Black, R. M.; Clarke, R. J.; Read, R. W.; Reid, M. T. J. *J. Chromatogr. A* **1994,** 662, 301.
6. Tu, A. T. *Kagaku* **1997,** 52(No.11),24.
7. Hirsjärvi, P.; Miettinen, J. K.; Paasivirta, J.; Kanolahti, E. (Editors), *Trace Analysis of Chemical Warfare Agents, An approach to the Environmental Monitoring of Nerve Agents*, Ministry of Foreign Affairs of Finland, Helsinki, 1981, pp. 27, 28, 37-39, 59-64, 72-79, 90-99.
8. Hirsjärvi, P.; Miettinen, J. K.; Paasivirta, J. (Editors), *Identification of Degradation Products of Potential Organophosphorus Warfare Agents. An Approach for the Standardization of Techniques and Reference Data*, Ministry of Foreign Affairs of Finland, Helsinki, 1980, pp.3-10,18-30 and appendices.
9. Ch. Kientz, E.; Verweij, A.; Boter, H. L.; Poppema, A.; Frei, R. W.; de jong, G. J.; Th. Brinkman, U. A. *J. Chromatogr.* **1989,** 467 385.
10. Bossle, P. C.; Reutter, D. J.; Sarver, E. W. *J. Chromatogr.* **1987,** 407, 399.
11. Kingery, A. F.; Allen, H. E. *Anal. Chem.:* **1994,** 66, 155.
12. Schiff, L. J.; Pleva, S. G.; Sarver, E. W.; in Mulik, J. D.; Sawicki, E. (Editors), *Ion Chromatografic Analysis of Environmental Pollutants*, Vol. 2, Ann Arbor Science Publishers, Ann Arbor, MI, 1979, pp. 329-344.
13. Katagi, M.; Nishikawa, M.; Tatsuno, M.; Tsuchihashi, H. *J. Chromatogr. B:* **1997,** 698, 81.
14. Oehrle, S. A.; Bossle, P. C. *J. Chromatogr. A* **1995,** 692, 247.
15. Wils, E. R. J.; Hulst, A. G. *J. Chromatogr.:* **1988,** 454 261.
16. Black, R. M.; Read, R. W. *J. Chromatogr. A* **1997,** 794, 233.
17. Kostiainen, R.; Bruins, A. P.; Häkkinen, V. M. A. *J. Chromatogr.* **1993,** 634, 113.
18. Tornes, J. Å.; Johnsen, B. A. *J. Chromatogr.* **1989,** 467, 129.
19. Daughton, C. G.; Cook, A. M.; Alexander, M. *Anal. Chem:* **1979,** 51, 1949.

20. Bossle, P. C.; Martin, J. J.; Sarver, E. W.; Sommer, H. Z. *J. Chromatogr.* **1983,** 267, 209.
21. Roach, M. C.; Ungar, L. W.; Zare, R. N.; Reimer, L. M.; Pompliano, D. L.; Frost, J.W. *Anal. Chem.* **1987,** 59, 1056.
22. Harvey, D. J.; Horning, M. G. *J. Chromatogr.* **1973,** 79, 65.
23. Bauer, G.; Vogt, W. *Anal. Chem.* **1981,** 53, 917.
24. Purdon, J. G.; Pagotto, J. G.; Miller, R. K. *J. Chromatogr.* **1989,** 475, 261.
25. Fredriksson, S.-Å.; Hammarström, L.-G.; Henriksson, L.; Lakso, H.-Å. *J. Mass Spectrom.* **1995,** 30, 1133.
26. Katagi, M.; Nishikawa, M.; Tatsuno, M.; Tsuchihashi, H. *J. Chromatogr. B* **1997,** 689, 327.
27. Tsuchihashi, H.; Katagi, M.; Nishikawa, M.; Tatsuno, M. *J. Anal. Toxicol.* **1998,** 22, 383.
28. D'Agostino, P. A.; Provost, L. R.; Visentini, J. *J. Chromatogr.* **1987,** 402, 221.
29. Weisiger, R. A.; Jakoby, W. B. *S-Methylation: Thiol S-methyltransferase*; in "Enzymatic basis of detoxication. Vol. II". Ed. by W. B. Jakoby, Academic Press, New York (1980) pp.131-40.

Chapter 25

Natural Compounds and Doping

R. Klaus Mueller[1], J. Grosse[1], A. Kniess[1], R. Lang[1], D. Schwenke[1],
D. Thieme[1], and R. Vock[2]

[1]Institute of Doping Analysis, Dresden, Germany
[2]Institute of Legal Medicine, Leipzig University, Leipzig, Germany

The majority of banned doping agents are synthetic
pharmaceuticals, but natural substances play also a double role:
- some are "weak points" of the analysis, being always
present in the body,
- other are used as replacements for banned agents with the hope
of similar effects.
Examples for the first category are the "classical" testosterone,
other hormones like erythropoietin EPO and human growth
hormone HGH, or morphine, strychnine, ephedrine.
Examples of the second category belong to the "grey zone doping"
phenomenon:
substances, somehow related to banned agents by their chemical
composition or by their activity, but not banned themselves.
Among them are creatine, carnitin, taurin, as well as "exotic"
substances like the ecdysones

The official doping definition of the International Olympic Committee (I.O.C.,
adopted by most nations and sports associations) is based on simple and clearly
stated motives – Doping contravenes the ethics of both sport and medical science.

I.O.C.-Doping Definition

It consists of the administration of prohibited classes of pharmacological agents
and/or the use of prohibited methods.
I. Prohibited classes of substances
    A Stimulants
    B Narcotics

C Anabolic agents
D Diuretics
E Peptide hormones

II. Prohibited methods: Blood doping,
                            manipulation of samples

III. Classes of drugs subject to certain restrictions:
        eg.   alcohol, cannabinoids, local anesthetics,
                corticosteroids, beta-blockers

The official Definition *defines* only classes of banned substances. The classes include synthetic as well as natural compounds, but those are by far not explicitly listed. Only examples are given for the banned groups, followed by the expression "...and related compounds" as e.g. for class I. C.

Article I Class C: Anabolic agents The Anabolic class includes anabolic androgenic steroids (AAS) and Beta-2 agonists.

1.   Anabolic androgenic steroids

clostebol   nandrolone
fluxymesterone   oxandrolone
metandienone   stanozolol
metenolone   testosterone*
                       and related substances.

* The presence of a testosterone (T) to epitestosterone (E) ration greater than six (6) to one (1) in the urine of a competitor constitutes an offence unless there is evidence that this ratio is due a physiological or pathological condition. e.g. low epitestosterone excretion, androgene production of tumor, enzyme deficiencies..

2.   Beta-2 agonists

clenbuterol   salmeterol
fenoterol   terbutaline
salbutamol   and related substances.
Because arguments arose in past years towards the legal meaning of the expression "related", this was additionally explained since 1994.

Natural compounds are to be found in the classes
        I. A   Stimulants   III.A   Ethanol
        I. B   Narcotics   III.B   Tetrahydrocannabinol THC
        I. C   Anabolic agents   III.C   Corticoids.
        I. E   Peptide Hormones

The officially defined doping agents are the target of doping control, which on the other hand is focused almost exclusively to the top athletes of olympic sport disciplines. There is almost no doubt, that doping plays a role also outside this sector - in disciplines like body building, in fitness clubs, among adolescents, perhaps in amateur sports. For this huge population potentially involved in doping practices, the misuse of officially banned agents bears only a part of the risks (of health damages, but not of detection and sanction). While the real role of doping customs cannot be known to the same extent in this area compared to the officially checked top-level sports, there are hints by anonymous surveys about a not negligible incidence.

The circumvention of the banned agents by the use of other substances (might their performance enhancing effect be real or illusionary) might be a temptation: their use would diminish the risk of detection and sanction for the controlled top-level athletes, while the expectation of a diminished risk to health might be attractive for both. So the official definition of banned doping agents does not necessarily cover the doping phenomenon completely: there is in our opinion a "grey zone" of doping attempts around it (Figure 1).

Even when accepting this "grey zone" as a challenge to discuss whether the single compounds should be tolerated or banned, there remain substances by the heading "supplementation" as another group of questionable position in the light of both the ability to enhance physical performance and of possible risks to health (mainly in overdosage). Substances like glucose, carnitin, creatin, taurin, amino acids, trace elements, vitamins, magnesium or other "minerals" are body constituents and constituents of nutrition. While they may require substitution after physical exhaustion, their use is certainly connected to performance enhancement, and overdoses may do harm to the health. But this was only mentioned for outlining the problem of "the role of natural substances to doping", - substances to be considered for supplementation/substitution can be neglected here with regard to their distance to "toxins" in the narrow sense.

So the first category of natural - potentially toxic - compounds to be considered in the context of doping are those included into the official doping definition.
Class I. A Stimulants contains natural as well as synthetic agents already as explicitly listed examples, and this would also apply to their "related compounds".

The alkaloids caffeine, ephedrine and cathine (together with analogues) offer a special problem due to their occurrence in widely distributed beverages or in pharmaceuticals as secondary ingredients. This led to limits of permitted concentrations in urine, whose surpassing is considered as a doping offence, while lower concentrations are tolerated. There is still some ongoing discussion, but this is no major problem. Synthetic stimulants like the amphetamine derivatives rather play a greater role, and the doping problem at all came at first to public attention by fatal cases after the misuse of such central analeptics decades ago, long before the anabolics and later the peptide hormones occurred.

Strychnine, the probably most toxic alkaloid among the banned natural stimulants, is obviously unimportant in the light of positive control samples.

Table I. Positive control samples of I.O.C. accredited Laboratories in 1995

A.	Stimulants	310
B.	Narcotics	34
C.	Anabolic Agents	986
D.	Beta Blockers	14
E.	Diuretics	59
F.	Masking Agents	3
G.	Peptides	9
H.	Others	224

Class I. B Narcotics is banned in contradiction to their performance decreasing action, because these agents permit the neglection of pain after overtraining or injuries, and provide therefore additional risks when applied for competitions. Mainly morphine is to be mentioned here, not so much due to its misuse incidence, but with regard to the need of another concentration limit: morphine is not only used as a prescribed (scheduled) drug, but is contained in foods (poppy cake and bakeries, even dairy products) and can appear in the urine in low concentrations even without intended (mis)use. Codeine (contained in numerous pharmaceutical preparations) is permitted.

Class I. C Anabolic agents (when introduced called Anabolic steroids) contains a number of synthetic anabolic androgenic analogues of testosterone and a second subgroup of (also synthetic) beta-2-agonists of differing chemical structure, which exert an anabolic action as a side effect. Altogether they have played a leading role in the statistics of detected doping cases since two decades, but the dominating problem is still the natural "mother substance" - the androgenic hormone testosterone itself. Permanent constituent of the male as well of the female body (although occurring in much lower concentrations in females), the proof of an external application presumes the differentiation between the "normal body content" and any artificial enhancement. Besides the differences between females and males, the physiological concentrations are varying considerably inter- and intra-individually, so that there are no "normal" concentration ranges suitable as means for such decisions between physiology and manipulation. This would be valid for blood, but even more it applies to urine - the only sample category routinely accessible for doping controls. The correlations with other physiological compounds of the so called "steroid profile" provide hints for external influences, but are not always sufficient to prove them.

Advantage was made of the fact, that the epimere epitestosterone - a normal accompanyon of testosterone biosynthesis - is completely inactive as an anabolic/androgenic; so it can be used for the normalisation of the testosterone urine concentration by determining the testosterone/epitestosterone (T/E) ratio. (Figure 2, 3).

Normally between 1 and 2, values above 6 are considered as suspicious for external application, with the consequence of additional analyses and research in each individual case to exclude any physiological reason for the elevated ratio and to undisputably prove a doping offense. Even the possibility of a second manipulation by application of epitestosterone together with the active testosterone was considered in the regulations: an upper limit of 150 ng/ml epitestosterone was set for the physiological variation.

With respect to two remaining problems:
      - rare cases of physiological elevations of T/E ratios beyond 6 or even 10 and
      - the possibility that somebody with normal ratios around 1 could carefully apply some testosterone until reaching the limit of 6
there has been proposed and tested a new principle for the detection of external application: the determination of the C12/C13 isotope ratio by a special arrangement of gas chromatography/oxidation/mass spectrometry. Products synthesized from other carbon sources give a different isotope ratio and indicate an external application, even if neither the testosterone concentration nor the T/E ratio are elevated. This principle is still under testing.

In any case, the actual regulations of I.O.C. demand subsequent checking of samples with elevated T/E ratios by

    -     comparison with previous results for the same individual
    -     drawing at least one more sample (unannounced),

to avoid wrong accusations if natural elevations occur at the individual athlete.

DOPING

Application of

prohibited substances	prohibited methods
A xenobiotics	B physiogical
(pharmaceuticals)	compounds

- examples of I.O.C. definition
  and related compounds

- "grey zone" of
  a   questionably related compounds
  b   permitted pharmaceuticals
      without medical reason
      and related   "SUBSTITUTION"

Figure 1. "Grey zone" of doping.

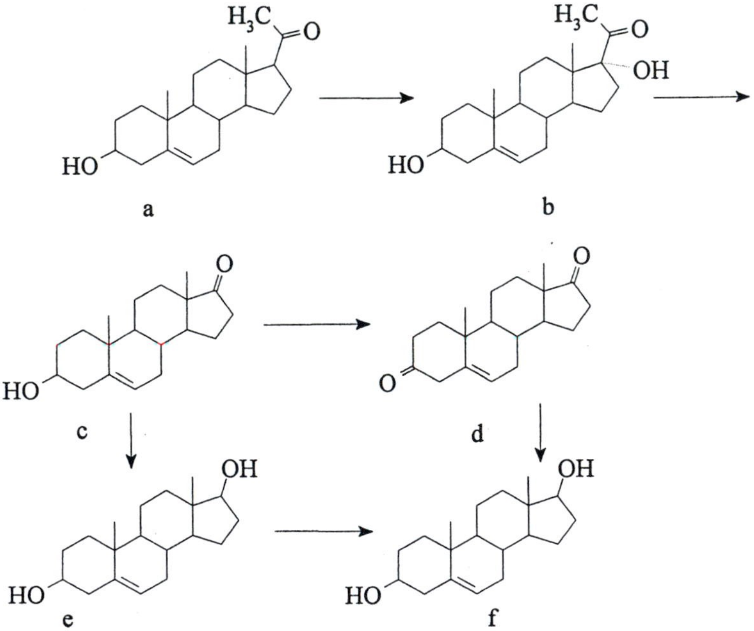

(a) Pregnenolon, 3ß-Hydroxy-5-pregnen-20-on
 (b) 17αHydroxypregnenolon, 3ß,17α-DiHydroxy-5-pregnen-20-on
 (c) Dehydroepiandrosteron, DHEA,3ß-Hydroxy-5-androsten-17-on
 (d) 4-Androsten-3,17.dion, (e) 3ß,17ß-Dihydroxy-5-androsten
 (f) Testosteron, 17ß-Hydroxy-4-androsten-3-on

Figure 2. Biosyntheses of testosteron

Quite recently, we observed a routine out-of-competition sample with a T/E > 6, when this compulsory checking clearly indicated a normal ratio before and after the simple evaluation of the questionable sample, and the Isotope Ratio MS underlined the interpretation as a doping offence (external source of the testosterone). The other anabolics are synthetic ones and shall not be discussed here. But quite recently, another problem was created by the occurrence of other "physiological" compounds, the precursors / accompanyons . of testosterone androstenedione, dihydroepiondrosterone DHEA, and dihydrotestosterone. Freely available in some countries (e.g. the U.S.A.), here also arises the need of distinguishing normal concentration ranges from such ones caused by external application.

Another way to circumvent the doping ban seems to be sought in the use of natural relatives of the anabolic steroids, e.g. ecdysones (insect hormones) or steroids from plants like protodioscin (Figure 4)

This is even more difficult with the peptide or glycoprotine hormones, mainly erythropoietin (EPO) and human growth hormone (HGH). Due to their ubiquitary presence in the body, their mere detection (possible e.g. with immunoassays) does not prove an external application or doping offence. There would be three alternatives for this prove:

	a	determination of normal concentration range, permitting to distinguish nonphysiological elevations
are	b	detection of slight differences in molecular composition (if there any) between the human and artificial compounds
	c	the combined evaluation of a number of correlated compounds (or factors, parameters) for the recognition of nonphysiological patterns.

The alternative a is strongly impeded by the natural scattering of normal levels - e.g. across several orders of magnitude for HGH. This is even more obvious in urine than in blood, because the variations of excretion superimpose to the "oscillations" in blood.

Slight differences of the structure (1) of the macromolecules (carbohydrate side chains) are the base for a possibility to separate (by electrophoresis so far) EPO produced by gene technology (bacteria) from human EPO. But this requires still too much sample, material and time, and it would immediately become irrelevant if the production of "human identical" EPO becomes possible.

The use of a pattern of parameters (2) around the misused hormone seems to be promising (insuline - like growth factor IGF1 and binding proteins like BP3 for

T/E Ratio's 1990-92
(All Athletes - Screen Results)

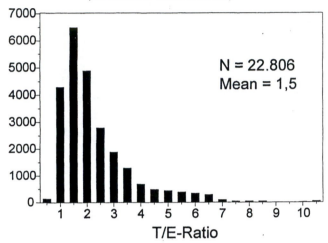

Figure 3. Testosterone/Epitestosterone Ratio

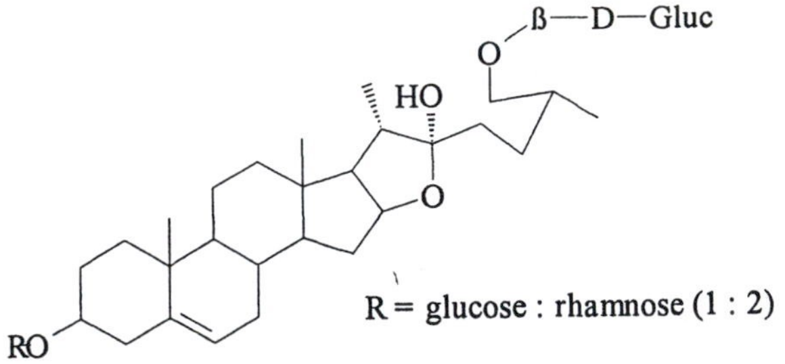

Figure 4. Protodioscin, main constituent of a remedy advertised via Internet as an anabolic.

HGH; Ferritin, Transferrin Receptor, hematocrit, hemoglobin, erythrocytes age distribution for EPO), but are still under scientific testing.

A compromise was introduced by some sport associations (e.g. cycling, skiing) by defining an upper limit for hematocrit or hemoglobin content of the blood: if these - finally the aim of EPO misuse - are strongly elevated, the athlete is prohibited to compete for medical reasons (risk of blood clogging, circulatory failure).

On the other hand, there are already on the market preparations mimicking EPO action: elevation of the oxygen transport capacity by the stimulation of the erythropoiesis, the formation of red blood cells.

Several papers recommend the use of velvet deer antlers preparations with this aim, and the officially distributed Russian pharmaceutical Hematogen (Pantocrin) obviously declares this target.

Current research is aimed to analytical methods permitting the prove of external administration of peptide hormones, because they represent still a "weak point" of doping control. The majority of doping agents – whether explicitly banned or belonging to the "grey zone", whether natural or synthetic – can be detected and identified with the instrumentation already in use.

References

1.   Wide, L.; Bengtsson, C., Berglund, B., Ekblom, B., Medicine & Science in Sports & Exercise. *DETECTION IN BLOOD AND URINE OF RECOMBINANT ERYTHROPOIETIN ADMINISTERED TO HEALTHY MEN* **1995**, *27*(11), 1569-1576.
2.   Birkeland, K.I.; Donike, M., Ljungqvist, A., Fagerhol, M., Jensen, J., Hemmersbach, P., Oftebro, H., Haug, E., International Journal of Sports Medicine. *Blood Sampling in Doping Control* **1997**, *18*, 8-12.

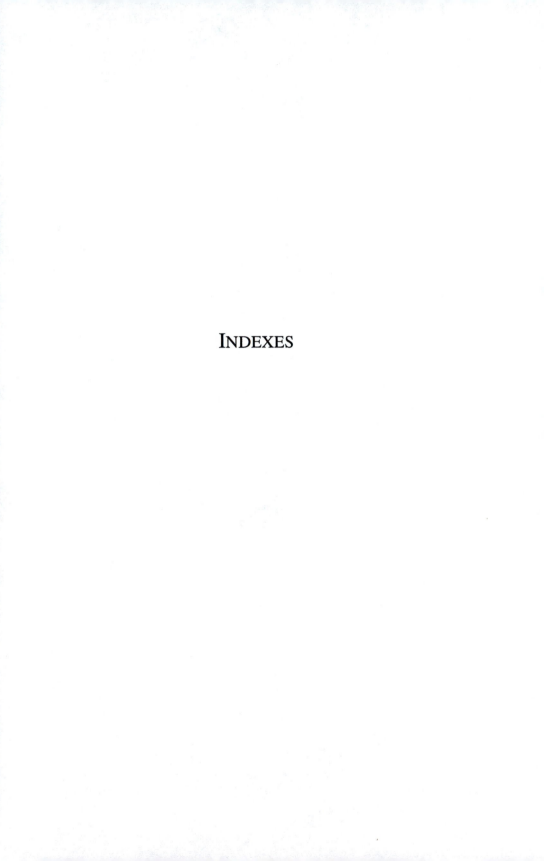

INDEXES

Author Index

Subject Index

401

402

404

effect of gila toxin on rat blood
  pressure, 294*f*
hypotensive agent, 285, 295
stimulated contraction of isolated rat
  uterus smooth muscle, 296*f*
See also Lizard venoms
Breast cancer
adhesion, invasion, and metastasis,
  276
angiogenesis, 277–278
binding sites for contortrostatin on
  breast cancer cells (MDA-MB-435),
  269
contortrostatin effects, 277
contortrostatin inhibiting growth and
  spontaneous metastasis of MDA-
  MB-435 in nude mouse model, 269,
  271*f*
daily injection of contortrostatin
  inhibiting breast cancer growth, 272*t*
human cells MDA-MB-435, 264
integrins and tumor metastasis, 276–
  277
testing effect of contortrostatin, 265
See also Disintegrins

C

Caffeine
problem due to occurrence, 389–390
See also Doping agents
Calcium
role in pardaxin-induced dopamine
  release, 25–27
See also Pardaxin
Calystegine alkaloids
assessing likely level of exposure in
  human diet, 132
assessing range of inhibition of
  mammalian digestive and lysomal
  glycosidases, 134
calystegine A3 causing vacuolation in
  Kupffer cells in liver, 134, 136*f*
chemistry review, 130–131

concentration giving 50% inhibition of
  rat digestive glycosidase activities,
  135*t*
content of commercially available
  potato products, 133*t*
content of edible fruits and vegetables,
  133*t*
discovery, 129–130
effects on mammalian liver lysosomal
  glycosidases, 135*t*
electron photomicrograph of mouse
  liver Kupffer cell, 136*f*
glycosidase-inhibiting alkaloids with
  potential pharmaceutical uses, 131
highest concentration in potato peels,
  135, 137
investigating theory of microbial
  break down or modification, 137
long-term cold storage altering ratios,
  132
lysomal storage disorders in cattle,
  131–132
potential for investigating toxicity
  cases in future, 137
questions about possible effects on
  humans, 134
structures, 130
study of over 70 potato varieties, 132
toxic effects of potent glycosidase
  inhibitors, 134
unusual aminoketal functionality, 131
wide range of therapeutic uses, 137–
  138
Cardiac effect, sudden death, 142
Cardiotoxins
biological properties, 222, 224
three-dimensional structures, 223*f*
See also Snake venom cardiotoxins
Cardols, phenolic lipids of pistachio
  hulls, 48–50
Cathine
problem due to occurrence, 389–390
See also Doping agents
Cattle
clinical effects of *Phalaris*

418

Highlights from ACS Books